EMERGING TRENDS OF PLANT PHYSIOLOGY FOR SUSTAINABLE CROP PRODUCTION

EMERGING TRENDS OF PLANT PHYSIOLOGY FOR SUSTAINABLE CROP PRODUCTION

Edited By
Zafar Abbas, PhD
Ajay Kumar Tiwari, PhD
Pradeep Kumar, PhD

APPLE ACADEMIC PRESS

Apple Academic Press Inc.   Apple Academic Press Inc.
3333 Mistwell Crescent   9 Spinnaker Way
Oakville, ON L6L 0A2 Canada   Waretown, NJ 08758 USA

© 2018 by Apple Academic Press, Inc.

First issued in paperback 2021

Exclusive worldwide distribution by CRC Press, a member of Taylor & Francis Group
No claim to original U.S. Government works

ISBN 13: 978-1-77-463076-1 (pbk)
ISBN 13: 978-1-77-188636-9 (hbk)

Library and Archives Canada Cataloguing in Publication

Emerging trends of plant physiology for sustainable crop production / edited by
Zafar Abbas, PhD, Ajay Kumar Tiwari, PhD, Pradeep Kumar, PhD.

Includes bibliographical references and index.
Issued in print and electronic formats.
ISBN 978-1-77188-636-9 (hardcover).--ISBN 978-1-315-10122-4 (PDF)
1. Plant physiology. 2. Sustainable agriculture. I. Abbas, Zafar (Senior associate professor), editor
II. Tiwari, Ajay Kumar, editor III. Kumar, Pradeep, 1982-, editor

QK711.2.E44 2018   571.2   C2018-900234-4   C2018-900235-2

Library of Congress Cataloging-in-Publication Data

Names: Abbas, Zafar, editor.
Title: Emerging trends of plant physiology for sustainable crop production / editors: Zafar Abbas, Ajay Kumar Tiwari, Pradeep Kumar.
Description: Waretown, NJ : Apple Academic Press, 2018. | Includes bibliographical references and index.
Identifiers: LCCN 2018000605 (print) | LCCN 2018002295 (ebook) | ISBN 9781315101224 (ebook) | ISBN 9781771886369 (hardcover : alk. paper)
Subjects: LCSH: Plant physiology. | Plant physiology--Research.
Classification: LCC QK714 (ebook) | LCC QK714 .E44 2018 (print) | DDC 571.2--dc23
LC record available at https://lccn.loc.gov/2018000605

Apple Academic Press also publishes its books in a variety of electronic formats. Some content that appears in print may not be available in electronic format. For information about Apple Academic Press products, visit our website at **www.appleacademicpress.com** and the CRC Press website at **www.crcpress.com**

CONTENTS

ABOUT THE EDITORS

Zafar Abbas, PhD
Senior Associate Professor and Chairman,
P. G. Department of Botany, G. F. College
(M. J. P. Rohilkhand University), Shahjahanpur,
Uttar Pradesh, India

Zafar Abbas, PhD, is a Senior Associate Professor and Chairman in the P. G. Department of Botany at G. F. College (M. J. P. Rohilkhand University) in Shahjahanpur, Uttar Pradesh, India. He has 40 years of research experience in plant and crop physiology, with a specialization in plant nutrition. He has attended several national and international seminars and conferences. At present, eight PhD students have completed their doctorate degrees under his supervision. Dr. Abbas is a life member and member of editorial boards of several Indian and foreign journals and societies, and he has authored a book and has published over 30 articles.

Ajay Kumar Tiwari, PhD
Scientific Officer, UP Council of Sugarcane Research,
Shahjahnapur, UP, India

Ajay K. Tiwari, PhD, is a Scientific Officer at the UP Council of Sugarcane Research, Shahjahanpur, UP, India. He has published 70 research articles, nine review articles in national and international journals, several book chapters in edited books, has also authored of several edited books published by Springer, Taylor and Francis and others. He has submitted more than 150 nucleotide sequences of plant pathogens to GenBank. He is a regular reviewer of many international journals as well as an editorial board member. He is the Managing Editor of the journal *Sugar Tech* and Chief Editor of the journal *Agrica*.

He has received several young researcher awards and was nominated for the Narshiman Award by the Indian Phytopathological Society, India. Dr. Tiwari is also the recipient of many international travel awards given by governmental agencies in India and others. He has visited Belgium, Brazil, China, Italy, Germany, and Thailand for conferences and workshops. Dr. Tiwari has been involved in the research on molecular characterization and management of agricultural plant pathogens for the last nine years. Currently he is working on molecular characterization of sugarcane phytoplasma and their secondary spread in nature. He is a regular member of several professional organizations, including the British Society of Plant Pathology, Indian Phytopathological Society, Sugarcane Technologists Association of India, International Society of Sugarcane Technologists, Society of Sugarcane Research and Promotion, and others. Dr. Tiwari earned his PhD from CCS University, Meerut, UP, India.

Pradeep Kumar, PhD
Assistant Professor, Department of Forestry,
North Eastern Regional Institute of Science and
Technology (Deemed University-MHRD),
Nirjuli (Itanagar),
Arunachal Pradesh, India

Pradeep Kumar, PhD, is an Assistant Professor in the Department of Forestry, North Eastern Regional Institute of Science and Technology (Deemed University-MHRD) (NERIST), Nirjuli (Itanagar), Arunachal Pradesh, India. Before joining NERIST, Dr. Kumar worked as a Research Professor in the Department of Biotechnology at Yeungnam University, South Korea. He was a postdoctorate researcher in the Department of Biotechnology Engineering at Ben Gurion University of the Negev, Beersheba, Israel, and awarded a PBC Outstanding PostDoc Fellowship for more than three years. His areas of research and expertise are wide, including plant pathology, bacterial genetics, insect-pest biocontrol, gene expression of cry genes, plant-microbe interaction, and molecular breeding. He has been honored with an international travel grant from Ben Gurion University of Negev, Israel, to attend an international conference. He is the recipient of a best paper presentation award as well as the Narasimhan Award from the Indian Phytopathological Society, India. He presented several oral and poster presentations at the various national and international conferences. He has published one book and has published more than 40 research articles, including original research papers in peer-reviewed journals and several book chapters with international publishers, including Springer, CABI, Bentham, and Apple Academic Press. He provides his service to many journals as an editorial board member, technical editor, and peer reviewer.

LIST OF CONTRIBUTORS

Zafar Abbas
PG Department of Botany G.F. College (M.J.P. Ruhilkhand University) Shahjahanpur – 242001, India

Sanchita Abhijita
Centre for Life Sciences, Central University of Jharkhand, Brambe, Ranchi – 835205, Jharkhand, India, Tel.: +919470416688, +919437295915, E-mail: pallavi.sharma@cuj.ac.in, anm@cuj.ac.in

Bhavin S. Bhatt
Shree Ramkrishna Institute of Computer Education and Applied Sciences, Surat, E-mail: bhavin18@gmail.com

Eric Bonneil
Institute for Research in Immunology and Cancer, Université de Montréal, PO Box 6128, Station Centre-Ville, Montréal, Québec, H3C 3J7, Canada

Bandana Bose
Department of Plant Physiology, Institute of Agriculture Sciences, Banaras Hindu University, Varanasi – 221005, India, E-mail: bbosebhu@gmail.com

Normand Brisson
Department of Biochemistry, Université de Montréal, PO Box 6128, Station Centre-Ville, Montréal, Québec, H3C 3J7, Canada

Robert Carpentier
Groupe de Recherche en Biologie Végétale, Université du Québec à Trois-Rivières, PO Box 500, Trois-Rivières, Québec, G9A 5H7, Canada

Fenisha D. Chahwala
School of Life Sciences, Central University of Gujarat, Sector 30, Gandhinagar, E-mail: fenisha_chahwala@yahoo.com

Sujit Das
Laboratory of Applied Stress Biology, Department of Botany, University of Gour Banga, Malda – 732103, West Bengal, India

Som Dutt
ICAR-Central Potato Research Institute, Shimla – 171001, HP, India, E-mail: birju16@gmail.com

Sharad K. Dwivedi
ICAR Research Complex for Eastern Region, Patna – 80 (0014). India, E-mail: sharad.dwivedi9736@gmail.com

Amit Kumar Gautam
Centre for Life Sciences, Central University of Jharkhand, Brambe, Ranchi – 835205, Jharkhand, India, Tel.: +919470416688, +919437295915, E-mail: pallavi.sharma@cuj.ac.in, anm@cuj.ac.in

Sridharan Govindachary
Groupe de Recherche en Biologie Végétale, Université du Québec à Trois-Rivières, PO Box 500, Trois-Rivières, Québec, G9A 5H7, Canada

A. Hasnat
Natural Products and Polymer Research Laboratory, Department of Chemistry, G.F. College (Affiliated to MJP Rohilkhand University), Shahjahanpur – 242001, U.P., India

Dildar Husain
Department of Botany, Maulana Azad Institute of Arts, Science and Humanities, Mahmudabad, Sitapur – 262003, U.P., India, E-mail; zafarabbas1255@yahoo.com

Ambuj Bhushan Jha
Crop Development Centre, Department of Plant Sciences, University of Saskatchewan, 51 Campus Drive, Saskatoon, SK S7N 5A8, Canada

David Joly
Groupe de Recherche en Biologie Végétale, Université du Québec à Trois-Rivières, PO Box 500, Trois-Rivières, Québec, G9A 5H7, Canada

Anoop Kumar
PG Department of Botany G.F. College (M.J.P. Ruhilkhand University) Shahjahanpur – 242001, India

Mahesh Kumar
Department of Plant Physiology, Institute of Agriculture Sciences, Banaras Hindu University, Varanasi – 221005, India, E-mail: bbosebhu@gmail.com

Prasann Kumar
Department of Plant Physiology, Institute of Agricultural Sciences, Banaras Hindu University, Varanasi, 221005, U.P., India, E-mail: prasann0659@gmail.com

Santosh Kumar
ICAR Research Complex for Eastern Region, Patna – 80 (0014). India, E-mail: sharad.dwivedi9736@gmail.com

Satish Kumar
Centre for Life Sciences, Central University of Jharkhand, Brambe, Ranchi – 835205, Jharkhand, India, Tel.: +919470416688, +919437295915, E-mail: pallavi.sharma@cuj.ac.in, anm@cuj.ac.in

Alexandre Maréchal
Department of Biochemistry, Université de Montréal, PO Box 6128, Station Centre-Ville, Montréal, Québec, H3C 3J7, Canada

Amarendra N. Misra
Centre for Life Sciences, Central University of Jharkhand, Brambe, Ranchi – 835205, Jharkhand, India, Tel.: +919470416688, +919437295915, E-mail: pallavi.sharma@cuj.ac.in, anm@cuj.ac.in

Sananda Mondal
Plant Physiology Section, Department of ASEPAN, Institute of Agriculture, Visva-Bharati, Sriniketan – 731236, India

Bhawna Pant
Department of Plant Physiology, Institute of Agriculture Sciences, Banaras Hindu University, Varanasi – 221005, India, E-mail: bbosebhu@gmail.com

Pinky Raigond
ICAR-Central Potato Research Institute, Shimla – 171001, HP, India, E-mail: birju16@gmail.com

K. K. Rao
ICAR Research Complex for Eastern Region, Patna – 800014, India, E-mail: sharad.dwivedi9736@gmail.com

Mohd. Zahid Rizvi
Department of Botany, Shia Post Graduate College, Sitapur Road, Lucknow – 226020, Uttar Pradesh, India, E-mail: zahid682001@gmail.com

Anisha Rupashree
Centre for Life Sciences, Central University of Jharkhand, Brambe, Ranchi – 835205, Jharkhand, India, Tel.: +919470416688, +919437295915, E-mail: pallavi.sharma@cuj.ac.in, anm@cuj.ac.in

Mohammed Sabar
Department of Biochemistry, Université de Montréal, PO Box 6128, Station Centre-Ville, Montréal, Québec, H3C 3J7, Canada / McGill Univirsity Biology, Stewart Building 1205 Docteur Penfield, Montreal, Quebec Canada H3A 1B1, E-mail: mohammed.sabar@mcgill.ca

Abhijit Sarkar
Laboratory of Applied Stress Biology, Department of Botany, University of Gour Banga, Malda – 732 103, West Bengal, India

B. L. Sharma
UP Council of Sugarcane Research, Shahjahanpur – 242001, India

Pallavi Sharma
Centre for Life Sciences, Central University of Jharkhand, Brambe, Ranchi – 835205, Jharkhand, India, Tel.: +919470416688, +919437295915, E-mail: pallavi.sharma@cuj.ac.in, anm@cuj.ac.in

Achuit K. Singh
Crop Improvement Division, ICAR – Indian Institute of Vegetable Research, Varanasi, E-mail: bsinghiivr@gmail.com, achuits@gmail.com

Aneg Singh
U.P. Council of Sugarcane Research, Shahjahanpur – 242001 (U.P.), E-mail: ajju1985@gmail.com

B. P. Singh
ICAR-Central Potato Research Institute, Shimla – 171001, HP, India, E-mail: birju16@gmail.com

B. Singh
Crop Improvement Division, ICAR – Indian Institute of Vegetable Research, Varanasi, E-mail: bsinghiivr@gmail.com, achuits@gmail.com

Brajesh Singh
ICAR-Central Potato Research Institute, Shimla – 171001, HP, India, E-mail: birju16@gmail.com

Jagjeet Singh
PG. Department of Botany G.F. College (M.J.P. Ruhilkhand University) Shahjahanpur – 242001, India

Pooja Singh
Institute of Environment and Sustainable Development, Banaras Hindu University, Varanasi – 221 005, India, Tel.: 191-993-591-2997, E-mail: rajeevprataps@gmail.com

Priyanka Singh
UP Council of Sugarcane Research, Shahjahanpur – 242001, India

Rajani Singh
Centre for Life Sciences, Central University of Jharkhand, Brambe, Ranchi – 835205, Jharkhand, India, Tel.: +919470416688, +919437295915, E-mail: pallavi.sharma@cuj.ac.in, anm@cuj.ac.in

Rajeev Pratap Singh
Institute of Environment and Sustainable Development, Banaras Hindu University, Varanasi – 221
005, India, Tel.: +91-993-591-2997, E-mail: rajeevprataps@gmail.com

S. P. Singh
UP Council of Sugarcane Research, Shahjahanpur – 242001, India

Rajesh K. Singhal
Department of Plant Physiology, Institute of Agriculture Sciences, Banaras Hindu University, Varanasi
– 221005, India, E-mail: bbosebhu@gmail.com

Vaibhav Srivastava
Institute of Environment and Sustainable Development, Banaras Hindu University, Varanasi – 221
005, India, Tel.: +91-993-591-2997, E-mail: rajeevprataps@gmail.com

Vineet Srivastava
Centre for Life Sciences, Central University of Jharkhand, Brambe, Ranchi – 835205, Jharkhand,
India, Tel.: +919470416688, +919437295915, E-mail: pallavi.sharma@cuj.ac.in, anm@cuj.ac.in

Pierre Thibaut
Department of Chemistry, Université de Montréal, PO Box 6128, Station Centre-Ville, Montréal,
Québec, H3C 3J7, Canada / Institute for Research in Immunology and Cancer, Université de Montréal,
PO Box 6128, Station Centre-Ville, Montréal, Québec, H3C 3J7, Canada

A. K. Tiwari
U.P. Council of Sugarcane Research, Shahjahanpur – 242001 (U.P.), E-mail: ajju1985@gmail.com

Maharishi Tomar
ICAR-Central Potato Research Institute, Shimla – 171001, HP, India, E-mail: birju16@gmail.com

R. K. Upadhyay
Department of Botany, Haflong Government College (Assam University Affiliation), Haflong –
788891, Assam, India, E-mail: rku.univ@yahoo.com

Satya Pal Verma
PG. Department of Botany G.F. College (M.J.P. Ruhilkhand University) Shahjahanpur – 242001, India

LIST OF ABBREVIATIONS

ABA	abscisic acid
ADP-Se	adenosine phosphoselenate
AMP1	altered meristem program-1
AOS	active oxygen species
APX	ascorbate peroxidase
ARE	antioxidant responsive element
ARFs	auxin response factors
ARR4	arabidopsis response regulator 4
ATP	adenosine triphosphate
BA	boric acid
BL	blue light
BN-PAGE	blue native polyacrylamide gel electrophoresis
CBC	cap-binding complex
CBS	cystathionine β-synthase
CCC	chlorocholine chloride
CDC5	cell division cycle-5
CDK	cyclin dependent kinases
CDPK	calcium dependent protein kinase
CDPK	calmoduline dependent protein kinase
CGR	crop growth rate
CHS	chalcone synthase
CMT	cell membrane thermo-stability
CNSL	cashewnut shell liquid
COP-1	constitutive photomorphogenic 1
CPL1	C-terminal domain phosphatase-like-1
CRD	completely randomized design
CTD	canopy temperature depression
DAS	day after sowing
DSC	differential scanning calorimetric
EIGP	eastern Indo-Gangetic plains
ER	endoplasmic reticulum

FAD	flavin adenine dinucleotide
FMN	flavin mononucleotide
FR	far-red
FS	filterable solids
FT-IR	Fourier transform infrared
GB	glycine betaine
GFP	green fluorescent protein
GI	germination index
GIRK1	geminivirus rep interacting kinase
GPX	guiacol peroxidase
GR	glutathione reductase
GSK3	glycogen synthase kinase
HFR1	hypocotyle in FAR-Red-1
HIRs	high irradiance responses
HKRD	histidine kinase-related domain
HOS1	high expression of osmotically responsive genes 1
HOS5	high osmotic stress gene expression 5
HSP	heat shock proteins
HY5	hypocotyle 5
IGR	intergenic regions
INSS	integrated nutrient supply system
LAI	leaf area index
LD	long-day
LFRs	low fluence responses
MAPK	mitogen activated protein kinase
MDA	malondialdehyde
MF	melamine formaldehyde
MGT	mean germination time
MH	maleic hydrazide
ML	measuring light
MMT	methionine S-methyltransferase
MRE	metal responsive element
MSI	membrane stability index
MTF	metal-induced transcription factor
NAD	nicotinamide adenine dinucleotide
NADP	nicotinamide adenine dinucleotide phosphate

NAR	net assimilation rate
NDPK-2	nucleoside diphosphate kinase 2
NMR	nuclear magnetic resonance
NO	nitric oxide
NSERC	Natural Science and Engineering Research Council of Canada
NTE	N-terminal extension
NWPZ	North-West Plain Zone
PAL	phenylalanine ammonia-lyase
PAPP-5	phytochrome associated phosphatase 5
PARi	photosynthetically active radiation incidental
Pas	pyrrolizidine alkaloids
PC	phytochelatin
PCS	phytochelatin synthase
PIFs	phytochrome interacting factors
PIPS	phytochrome interacting proteins
PIPs	plasma membrane intrinsic proteins
PKA	protein kinase A
PKC	protein kinase C
PKG	protein kinase G
PSI	photosystem I
PSII	photosystem II
PTGS	post transcriptional gene silencing
PTK	protein tyrosine kinase
QTL	quantitative trait locus
RACK1	receptor for activated c kinase1
RBP	rubisco binding proteins
RG-II	rhamnogalacturonan II
RGR	relative growth rates
RISC	RNA induced silencing complex
ROS	reactive oxygen species
RWC	relative water content
SA	salicylic acid
SAM	shoot apical meristem
SAS	shade avoidance syndrome
SINE	short interspaced elements

SMT	selenocysteine methyltransferase
SOD	super oxide dismutase
SOS2	salt overly sensitive 2
SS	suspended solids
SUT	sulphate transporters
TBARS	thiobarbituric acid reactive substances
TBS	tris-buffered saline
TD	total solids
TDI	toluylene diisocyante
TDM	total dry matter
TEs	transposable elements
TFs	transcription factors
TOC1	timing of cab expression 1
TPP	thiaminepyrophosphate
UTR	untranslated region
VI	vigor index
VLFRs	very low fluence responses

PREFACE

Plant physiology is now considered as an essential ingredient for improving crop productivity. Since 1960s, Indian plant physiology has contributed significantly to the understanding of the basic parameters of crop productivity under Indian conditions.

Wheat, potato, rice, rapeseed, mustard, and pulses are some of the crops, that have received special attention. Among the cereals, wheat recorded an increase in yield by more than two-fold and the total production crossed the over 50-million ton mark. Although several factors have contributed to enhancing wheat productivity, the most important from the productivity point of view was the rise in the harvest index, which touched 50% growth. The world food production is expected to be doubled by 2050. India and China, the most densely populated countries of the world, will need to maintain at least a 4–5% annual growth rate in agriculture to keep pace with the growing population and sustain world food security. For example, India supports about 17% of human and 11% of livestock population of the world on just 2.8% land and with 4.2% of water resources. As per recent estimates, India will need to produce about 281 mt food grains, 53.7 mt oilseeds, 22 mt pulses, 127 mt vegetables, and 86 mt fruits by 2020–2021. The country will have to produce more quality food with a diminishing natural resource base and changing climate.

The issues of current concern to national food, nutritional, and environmental security include: diversion of agricultural land for nonagricultural use; decreasing land holding size; declining profit margin in agriculture; depletion of ground water, deterioration soil health and biodiversity; increased frequency of climate-related risks, such as cyclones/tsunami, drought and floods, cold and heat waves; contamination of soil and water with heavy metals like arsenic, selenium and fluoride; and their cycling in plant-animal-human-atmosphere chain.

Ever-increasing concentration of green house gases in the atmosphere, resulting in global warming, is likely to have serious repercussions for human beings, animals, plants, microbes, and environment. This book

deals with understanding the physiological basis of the various plant processes and their underlying mechanism under fluctuating environments. It
is therefore, is of great importance for sustainable crop production. Further advances in cellular and molecular biology hold promises to modify
the physiological processes, thereby improving the quality and quantity of
major food crops and ensuring stability in yield of the produce even under
severe abiotic stress.

This book covers the latest information in the present scenario on the
physiological basis of plant productivity, abiotic stress adaptation and
management, plant nutrition, climate change and plant productivity, transgenic and functional genomics, plant growth regulators and their applications. It is in this context that all chapters as a collective group shall help
in tackling some of these key issues of sustainable plant production and
help evolve future strategies in overcoming challenges faced by the agricultural sector as a whole. The topics covered in this book highlights the
general and provides an overview of some of the very important aspects of
physiological research by reputed scientists of the country and of abroad.
Keeping in view the vital nature of these emerging issues, this book will
provide a wide opportunity to readers to have complete information from
one source. Therefore, this book will be a useful resource for researchers
and extension workers involved in the plant physiology with other cognate
disciplines.

CHAPTER 1

POTATO PHYSIOLOGY IN RELATION TO CROP YIELD

BRAJESH SINGH, MAHARISHI TOMAR, SOM DUTT,
PINKY RAIGOND, and B. P. SINGH

*ICAR-Central Potato Research Institute, Shimla – 171001, HP, India,
E-mail: birju16@gmail.com*

CONTENTS

ABSTRACT

Potato is a highly versatile crop grown in varying environmental conditions *viz.* plains, hills, and plateau in India. It is grown under short days in the plains in autumn season and under long days during summers in hills. Therefore, the pattern of growth, duration of crop as well as the yield varies according to the environment. Besides, the potato cultivars differ from each other in relation to their morphological characters, response to the environmental factors and finally in yield and the dry matter accumulation. The present chapter deals with the physiological factors, which determine the growth, development and finally the yield of potato crop.

1.1   INTRODUCTION

The potato cultivars differ with each other in relation to their morphological characters, response to the environmental factors and finally in yield and dry matter accumulation. The biochemical reasons for such variations are not clear but it seems that genetically determined differences in the synthesis of growth regulators such as auxins, cytokinins, gibberellins; and abscisic acid play an important role and their balance at different growth stages affect the growth and development, and therefore, the yield of the cultivar. Because of the major influence of environment on the potato crop, the differential yield response may be obtained even in a single cultivar grown in different environment, due to which it is difficult to predict the actual yield of a particular crop grown over different locations and seasons. However, a general prediction of minimum and maximum yields and dry matter accumulation can be worked out for specific potato cultivar under given set of environment. The present chapter deals with the factors, which determine the growth and development and finally the yield of potato crop.

1.2   POTENTIAL YIELD OF POTATO

The potential yield of potato crop is influenced by many factors such as cultivar, morphological attributes of the crop, climatic factors (sunlight, photoperiod, temperature, and rainfall), and edaphic factors (soil type and availability of nutrients, etc.). The yield of any crop is dependent on the assimilation of carbohydrates by the plant through the process of photosynthesis (the process by which the atmospheric carbon is assimilated in the form of carbohydrate in the plant).

Based on climate data, the potential tuber yield for north Indian plains is 60t/ha for 110-days crop; however, the actual yield obtained is about 20t/ha. But, the potential yield is based on all the optimum conditions for the growth of potato crop in practice, which is indeed difficult, though these may be available for some period of the crop growth. This is the reason why the full potential of any potato cultivar is not exploited in the field. If the light intensity matches the photosynthetic saturation level throughout the day, temperatures are conducive for plant growth, there is no biotic and abiotic stress for the crop growth, the soil is fertile and all the required nutrients are available in sufficient quantity, then only we can get the maximum potential yield of the potato crop. Unfortunately, such growing conditions are not available and at every location there may be one or more physiological constraints, limiting the potential yield of the crop. These constraints are beyond the control of the cultivator, still there may be scope for improving the growth conditions through avoidance of any kind of nutrient, biotic and abiotic stresses and betterment of cultural practices for the superior stand and crop growth in particular location, so that the maximum achievable yield and dry matter is harvested from the potato crop.

1.3 GROWTH OF POTATO PLANT

Potato is a highly versatile crop and is grown in varying environmental conditions. Potato crop is cultivated both in plains and hills in India and the environmental conditions in these areas have immense differences. Mostly, the potato crop is grown under short days in plains and under long days in hills and therefore, the pattern of growth, duration of crop as well as the yield of crop varies in these areas. It is normally cultivated as a vegetatively propagated crop and the basic material used for multiplication, is potato tuber, which is used as seed for planting in the fields. It can also be grown through true seeds (TPS) obtained from fruits, but the area under TPS cultivation is still meager at present in India. The potato tuber seed has several eyes (growing points) on it and it may be planted as such (having all the eyes on one tuber) or as cut seed (cutting tuber in few pieces, where every piece contains one or more eyes) depending on the feasibility of the cultivator and economics involved in the cost of seed. At optimum physiological age these eyes develop into hairy structures called sprouts, which have apical meristem and are capable of growing into potato plant. The potato plant with its root and shoot (stem and leaves) system grows during the crop season and at the time of tuber initiation it produces stolons which grow underground and finally bulk at their terminal ends in the form of tuber (Figure 1.1). A major amount of photosynthate is translocated to these tubers at the time of bulking and the photoassimilate is stored in tubers in the form of starch and other biochemical constituents. Under appropriate conditions (mainly long days), potato plants produce flowers, fruits, and seed (TPS) as well. After full maturity the tubers are harvested and they may be either used for consumption or stored as seed. If they are to be used as seed, they are stored till the next crop season, during this period the stored photoassimilate in tuber helps to sustain its living processes (respiration and germination in the next season). Through these phases the life cycle of a potato plant is completed.

FIGURE 1.1 Dormant tuber, sprouted tuber, potato crop, stolon formation and tuberization, and mature tubers of potato.

1.3.1 SPROUT

Freshly harvested potato tubers do not sprout even if they are given the appropriate conditions, because they have dormancy duration or rest period of few weeks. Every potato cultivar has specific dormancy duration and in Indian cultivars generally this duration varies from 6 to 8 weeks. The dormancy duration is determined by the physiological age and the balance of endogenous growth regulators, which are further influenced by the environmental interactions. As soon as the tubers attain proper physiological age and the dormancy period is terminated, the buds on the tubers develop into sprouts. The temperature and humidity at which the tubers are stored play a key role in sprouting behavior and it has been well documented that the temperatures ranging from 18–20°C, high humidity and darkness are most appropriate conditions for sprout growth.

Each sprout consists of three parts viz. base, middle portion, and apical tip. The roots arise from the base of the sprouts and at the time of tuber initiation the stolons are also formed from the base. The middle portion of the sprout gives rise to the stem and the apical part contains the meristematic tissue, which provides growth and differentiation in the potato plant. Morphologically, the pigmentation of sprout may vary from white to green, pink, red, violet or purple depending on the cultivar. This pigmentation is caused by anthocyanin and if the sprouts are exposed to light they may also develop chlorophyll pigmentation.

The sprouting behavior of Indian cultivars has been extensively studied. The temperature, humidity and gaseous environment under which the sprout growth takes place and the soil and climatic conditions in which the potato crop is raised have good bearing on the sprouting behavior. Very low (5°C or below) and high (30°C and above) temperatures are known to reduce sprout growth. Similarly low levels of humidity (30% or below) reduce the sprout growth. However, increase in the CO_2 concentration upto 15% has been found to reduce the dormancy duration and enhance the sprout growth to a great extent at 20–25°C temperature. A potato cultivar when grown in short days enhances sprout growth early in comparison to when grown in long days. The levels of fertilizers applied to potato crop also influence the sprout growth. Based on the studies related to hormonal control of sprout growth it was observed that ABA (abscisic acid) inhibits the sprout growth and initiation and maintains tuber dormancy (Figure 1.2). While on the other hand Cytokinins are associated with dormancy break and sprout growth initiation. Seed potatoes are sometimes treated with Gibbberellin (GA), which is a growth stimulating substance and has an important role in regulating sprout growth and tuber dormancy (Figure 1.2). Thus, the sprout growth pattern besides being a cultivar characteristic is dependent on various factors, which determine the potato plant stand at later stages.

1.3.2   SHOOT

It is a well-known fact that large sized tubers produce more stems than small sized tubers, though the cultivar differences exist due to their genetic constitution and environmental interactions. The potato crop when grown in hills produces taller stems in comparison to when grown in plains. Similarly, the number of leaves per plant are more in hills under long days than under short days. The leaf abscission is also initiated early under short days than long days. The number of stem and leaves per plant are important as they influence the final yield of the potato crop and good amount of work has been done in India for establishing their role in crop productivity.

The leaves are sites for photosynthetic activity and more number of photosynthetically active leaves are desirable for better productivity of the crop. Defoliation at different levels has shown that the lower and upper leaves contribute more towards the dry matter production. The leaf insertion angle is also important for light reception on to the leaves and it has been found that full grown potato plants intercept as much as 95% of the incident radiation and most of it is intercepted by the top half of the plant.

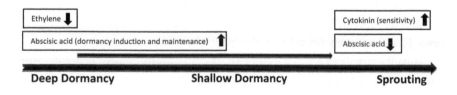

Dormancy progression time

FIGURE 1.2   Hormonal regulation of tuber dormancy: An overview.

If the light interception is increased towards the lower layers as well, it is likely that more photosynthate will be translocated to the sink, as it is closer to the lower leaves.

The increased leaf area index (LAI) has been found to increase the yield but increasing LAI beyond a certain level has not been found beneficial because of the increase in the area of unproductive leaves. It is the early stage of plant growth when LAI plays an important role in increasing the photosynthetic efficiency and if plant achieves higher LAI at initial stage and also maintains it for the longer duration, the final yield is positively influenced. Thus the potato plants should have more number of photosynthetically active leaves for a longer duration of crop growth for achieving higher productivity.

1.3.3 ROOT

The root development is initiated early and before plant emergence from the soil. The potato root system is mainly confined to the top 20–30 cm of soil and sometimes goes beyond this layer. Adventitious roots develop at the base of developing sprouts. In general potatoes have a highly branched and fibrous root system and is mainly concentrated in the upper soil profile. The roots have been classified into basal roots (arising from stem), stolon roots (from stolons), junction roots (from the joint of stem and stolon), and tuber roots (from the tuber bud base). As far as water and nutrient absorption is concerned the basal and junction roots are more important. It has generally been found that the early maturing cultivars have shallow root system, whereas the late maturing cultivars have deeper roots. For the better productivity of the potato crop it is desirable that the plants have good root system for proper absorption of water and nutrients from the soil.

1.3.4 STOLON

The stolons are modified lateral shoots that arise from the nodes at the base of the shoots below the soil. They have elongated internodes, small-scale leaves, hooked tips and chlorophyll is absent in them. The first stolons form at the basal nodes on the stem (near the mother tuber) and then develop with an upward progression. The terminal part of the stolons accumulates partitioned dry matter and swells in form of bulbous structure called tuber. Its initiation is favored by darkness and high humidity.

1.3.5   TUBER

A potato tuber is a modified stem possessing leaves and axillary buds. The end of the tuber where it joins stolon is called as the heel end or stem end while the other end is called the rose or bud end. The eyes present on tubers are the nodes of modified stem and contain axillary buds and scale leaf. Potato tubers originate from stolons and its initiation is referred as tuberization when the stolon starts expanding radially due to cell division and cell enlargement. During this growth process dry matter is translocated from the foliage to the growing tubers.

The number of tubers per hill is a varietal characteristic and the tuberization is influenced by several environmental factors like moisture, temperature and nutrient availability. The moisture stress is known to affect the growth of the tubers adversely and it decreases the yield of the potato plant. The night temperature of about 18–20°C has been found optimum for balanced growth of the tubers and temperatures beyond it adversely affect the process of tuberization. Application of nitrogen fertilizer has been found effective in increasing the number and size of the tubers up to optimal level.

Tuberization is majorly controlled by photoreceptors, which includes phytochromes (PHYA and PHYB). The action of PHYB has been found to have a central role in controlling tuber formation. At molecular and genetic levels a number of proteins and transcription factors are involved in controlling tuberization by a complex signaling mechanism. Among these proteins (*StSP3D and StSP6A*) which have a very important role in floral and tuberization transition. StCO (CONSTANS) is found to repress the tuberization process in a photoperiod dependent manner. All these proteins work together in a specific consortium like manner for regulating the tuberization process.

Figure 1.3a shows regulation of tuberization by sucrose transporter StSUT4. Accumulation of StCO and StSP6A mRNA is affected by StSUT4 in a photoperiod-dependent manner. StSUT4 induces StCO accumulation under long day condition which in turn inhibits StSP6A accumulation and

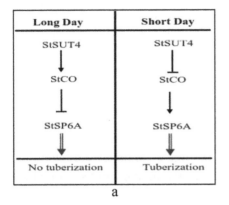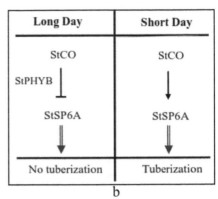

a   b

FIGURE 1.3   (a) Regulation of tuberization by sucrose transporter StSUT4 [4], (b) StSP6A mediated regulation of tuberization in potato [4].

hence results in no tuberization. On the other hand under short day conditions StSUT4 inhibits StCO accumulation and hence StCO mediated inhibition of StSP6A is prevented thus leading to tuberization.

Figure 1.3b shows StSP6A mediated regulation of tuberization in potato. StCO represses StSP6A under long day conditions. PHYB is suggested to have a role in modulation of this repressor activity. When transferred to short days StCO activates StSP6A gene expression in the leaves. This signal is amplified by an auto-relay mechanism during transport which is partially mediated by StCO. StSP6A activation in leaves and stolons promotes tuber formation.

1.4   PHOTOSYNTHESIS

Potato is majorly grown as a short day crop in India as short duration crop (80–110 days) and the yields are low. In some parts the yields are high, where the potato is grown as a long duration crop. The longer duration of crop enables the production of more total dry matter per unit land area by the photosynthetic apparatus of the plant and this is the reason for higher productivity of the potato crop under long duration. Besides the crop duration, appropriate growth conditions are required for the proper growth and development of the potato plant that is available for only a short duration. Therefore, the cultivation of potato crop for a longer duration in the plains is a limitation and it has to be grown only for a short duration, which finally results in the low productivity potential of the potato crop.

1.4.1 UNIT RATE

The unit rate of photosynthesis may be described as the amount of carbon dioxide fixation per unit leaf area, per unit time under saturated light intensity by the potato leaf. It is often termed as the light-saturated net photosynthetic rate (P_{max}).

More than 90% of the dry weight of potato plant is derived from the photosynthetically fixed carbon dioxide. The potato crop has C3 photosynthetic metabolism and the P_{max} is greatly influenced by the genetic and climatic factors. The peak photosynthesis occurs during the forenoon (10 a.m. to 1 p.m.) whereas, during the mid afternoon a depression in P_{max} is observed even when the conditions are favorable. This phenomenon is known as the diurnal variation in the net photosynthetic rate. The rate of photosynthesis is generally higher in tuberized potato plants, which is believed to be due to the presence of an active sink. However, it is generally observed that the P_{max} is more in young plants and it decreases with the age of the crop.

1.4.2 GENETIC VARIATION

The potato genotypes differ with respect to their photosynthetic efficiency and this information can be utilized during potato breeding for a higher photosynthetic and productivity potential in potato crop. The P_{max} though varies among the genotypes, it does not correlate well with the potato yield. This shows that the photosynthetic rate per unit leaf area is not the sole determinant of final yields. The rate of photosynthesis per unit ground area by the whole canopy for the full season as well as the partitioning of the assimilate to the tubers is more important for the estimation of the yield. However, higher P_{max} is an essential feature towards higher yields and therefore, breeders must identify the parental lines with higher P_{max} and other desirable traits and then breed to combine these attributes for better productivity of the potato crop.

1.4.3   CO₂ REQUIREMENT

Potato like other C3 plants is influenced by the availability of CO_2 in the atmosphere. With the increased CO_2 concentration in the atmosphere the entry of primary substrate for photosynthesis into the sub-stomatal cavity is enhanced and it results in increased biomass productivity and yield. At present the CO_2 concentration in atmosphere has reached to the level of approximately 400 ppm from the pre-industrial level of about 280 ppm. It has been established that production of C3 plants is increased by 10–40% by doubling the CO_2 concentration. Work on potato has not shown such a great increase in yield on increasing atmospheric CO_2 but root zone CO_2 enhancement in potatoes has been found to increase the dry matter upto maximum of 25%. The CO_2 thus can be termed as a limiting factor for photosynthesis in potato plant and it is likely that the increase in atmospheric CO_2 will positively effect the potato yield, provided the other factors do not become limiting.

1.4.4   WATER REQUIREMENT

The efficiency of conversion of light into dry matter is not much influenced by water stress. However, prolonged drought results in a decline in minimum and maximum fluorescence, indication premature senescence and a decline in the chlorophyll light-harvesting complex. Therefore, regular supply of irrigation is desirable for effective photosynthetic rate of the potato crop. The work on Indian cultivars has shown that removal of irrigation for two consecutive days reduced the P_{max} by 32–84% as compared to the control plants watered daily. The leaves of stressed plants had 0.3 to 0.5 MPa lower water potential and 61–86% lower stomatal conductance than control.

1.4.5   TEMPERATURE REQUIREMENT

The temperature is known to influence the process of photosynthesis and the optimum range of temperature varies between 15 to 30°C for P_{max}. Beyond 30°C temperature the P_{max} declines substantially. The stomatal conductance has been found to reach at a maximum level by 24°C temperature and remains at this level for up to 30–35°C temperature. The temperatures lower than 15°C result in low P_{max}, which may be due to the lower activity of the photosynthetic apparatus involved in the process.

1.4.6   LIGHT REQUIREMENT

Most of the Indian potato cultivars achieve maximum photosynthetic rate between 500 and 800 photosynthetically active radiation incidental (PARi). The reduction in light intensity may reduce the tuber yields adversely in potato plant. The interception of light by the canopy plays major role in achieving P_{max}, and penetration of light to the whole canopy if increased, may result in higher P_{max}. The stomatal conductance is also influenced by the light intensity and is directly correlated with it. The leaves on the upper canopy, which receive more light, have greater stomatal conductance at a given light intensity than the shaded lower leaves and also have higher P_{max} than the lower ones.

1.5   DRY MATTER DISTRIBUTION

The dry matter produced as a result of photosynthetic process in utilized by the plant for its growth and developmental processes. For better tuber yields it is important that a substantial part of the assimilated carbon is translocated to the tubers and rest of it is utilized by the other parts. It has been found that under short days more dry matter is partitioned to the tubers whereas under long days more dry matter is partitioned to the foliage. The proportion of the total dry matter in the tubers increase with maturity and it has been documented that accumulation of dry matter is more in shoots till the initiation of tuber enlargement phase and it declines afterwards due to accumulation of more dry matter in the tubers.

The dry matter distribution in plant parts is dependent on the source-sink interaction. This sink may be either haulms or tubers and the capacity to attract the photoassimilate by the sink is known as sink strength. The stronger sink thus will attract more photosynthate than the weak sink. The sink strength also influences the rate of photoassimilation and it has been found that after tuberization the sink strength of the tuber increases and as a result the rate of photosynthesis also increases. Generally, before tuber initiation the major part of the photoassimilate is translocated to the haulms for growth and development processes and after the tuber initiation a substantial amount of assimilate is partitioned to the tubers. It has been demonstrated that at 52 days the dry matter accumulation in leaf of a potato cultivar was 54% and decreased to about 10% by 80 days of crop growth, thereby showing greater partitioning of dry matter to the tubers rather than the haulms at maturity.

1.6   YIELD COMPONENTS

The yield in potato plant is dependent on genotypic, phenotypic and environmental factors. The morphological factors are important as most of the physiological processes involved in yield formation of plant depend on them. Therefore, it is pertinent to mention the desirable traits of the plant parts in the yield formation.

1.6.1   STEM NUMBER

The number of stem produced per tuber is influenced by the genetic constitution, tuber size and environment. It is desirable that a potato tuber seed gives rise to more number of stem per tuber so that more foliage is supported on it, thereby producing more photoassimilate for growth, development, and yield. It is a well-known fact that large size tubers bearing more number of eyes give rise to more number of stems. The environmental factors influence the growth of stem and the taller stem are produced in hills than in plains. The temperature also influences the growth of stem and these influences have been found to vary with the cultivars.

1.6.2   LEAF AREA

For better productivity of the potato plant it is desirable that the plant should have higher leaf area per unit land area, so that higher photosynthetic activity can be taken up in the given area and more dry matter is

produced. The increase in leaf area at initial growth stages and its maintenance over a longer duration of crop growth has been found to be favorable
for larger yields in Indian potato cultivars. The potato cultivars attaining
higher LAI at early stages have shown higher harvest index and yield. It
has been widely proven that leaf area has positive correlation with the
yield in potato crop, and thus, quick achievement of high LAI and its con-
tinuation for longer duration is beneficial for increased productivity.

1.6.3   LEAF AREA DURATION

The leaf area duration is another parameter for estimation of plant growth
and expected yield. It has been documented that effective participation
of leaves for longer duration in the process of photosynthesis results in
higher productivity. The leaves having low efficiency act as sink rather
than source for the dry matter accumulation and are detrimental for the
productivity. Hence, the more number of photosynthetically active leaves
for the overall duration of the crop are desirable for increased productivity.

1.6.4   TUBER NUMBER AND TUBER SIZE

For better productivity of the potato plant more number of tubers with
bigger size is important. The tuber development is influenced by various environmental factors and after initiation, tubers require appropriate
growth conditions for bulking. When the potato plant is healthy and active
a large amount of photosynthate produced in foliage is translocated to the
tubers and the bulking takes place with the growth in size of the tuber. The
strong tuber sink influences the photosynthetic accumulation positively
and thus more number of tubers cause an increase in the accumulation
process. The ultimate result is higher yields with more number of tubers
and bigger size.

1.7   TEMPERATURE

All the growth processes of a plant are influenced one way or the other
by the prevailing temperatures under the provided conditions. The

temperature if is not optimum (low or high) to the growth requirement of the potato plant it hampers the processes of plant growth, development and yield production. The effect of temperature on the plant growth and yield are discussed here.

1.7.1   PLANT GROWTH

The potato plant growth initiates with the sprouting in seed potatoes and ends with the production of tubers in the next crop. The dormancy and sprout growth of seed tubers are temperature dependent to some extent and it is established that dormancy duration is shortened and the sprout growth is enhanced between 18–20°C storage temperature. The temperatures above or below this level are not that favorable for the dormancy breaking and sprout growth.

Once the sprouting takes place the temperature of soil becomes important factor for the further growth of sprouts and root system. It has been found that soil temperature of about 20°C is favorable for the plant emergence and development of root system in potato. The shoot growth is slow at lower temperatures (10°C or so) and is optimum at 20°C in Indian potatoes. The number of stems per hill increase with increase of temperature from 20°C day/10°C night temperature to 30°C day/20°C night temperature. The leaf area is similarly affected by temperature as the number of stems. The life span of the potato crop increases with temperature under long days but under short days it is adversely affected. The optimum temperature for rate of stem elongation is above 25°C whereas, for leaf characteristics it is relatively low (20°C). The specific leaf area is positively correlated with temperature upto 30°C, but for the individual leaf size and longevity of leaves the optimum temperature is around 20°C. For induction of flowering, the optimum temperature range is 16–28°C.

1.7.2   YIELD

The yield of the potato crop is a result of photoassimilate production in the plant through the process of photosynthesis. The optimum temperature for gross photosynthesis in potato plant ranges between 24 to 30°C whereas for net photosynthesis it is less than 25°C. Though the plant response to photosynthesis is a genetic character and plants have adaptive capacity, but higher temperatures reduce rate of photosynthesis as a result of accelerated senescence, chlorophyll loss, reduced stomatal conductance and inhibition of dark reaction. The partitioning of dry matter is greatly influenced by the temperatures and it has been found that in heat sensitive cultivars high temperatures result in more starch accumulation in the leaves and low rate of its translocation to the tubers than in tolerant genotypes. High temperature strongly reduces the harvest index in potatoes. Thus, as temperature increases the tubers contribute proportionally less to the total dry matter yield of the potato plant while the stem and leaves contribute more.

The stolon initiation is strongly influenced by temperature. It has been documented that the stolon development is delayed though the numbers of stolons increase by high temperature. Higher temperatures stimulated the stolon initiation, its growth and branching but delay tuberization resulting in more stolons and more potential tuber sites, but less number of sizeable tubers. The tuber initiation is delayed, impeded or even inhibited at higher temperatures, particularly at high night temperatures (above 22°C). Higher temperatures also reduce the tuber bulking and thus the final yield of the potato plant is decreased at higher temperatures.

1.8   LIGHT

The light is one of the most crucial factors influencing the potato plant growth and the basic production system (photosynthesis). The light intensity though effects only the process of photosynthesis and thereby final yield, the photoperiod (light availability period) greatly influences almost all the growth and developmental phases of potato plant. In India potato crop is majorly grown in the North Indo-Gangetic plains under short days (less than 12-hour photoperiod), whereas, in some parts it is grown during long days (more than 12-hour photoperiod). Under such circumstances the growth and development of potato plant shows difference as discussed further.

1.0.1 PLANT GROWTH UNDER DIFFERENT PHOTOPERIODS

Like temperature, the photoperiod also influences most of the growth and developmental processes in potato plant (Table 1.1). The effect of photoperiod during storage of seed tuber is small, but after the dormancy is broken, photoperiod may have a significant effect. At lower temperatures the long photoperiods (above 12 hours) increase the number of stems per hill while at higher temperatures both the long and short photoperiods stimulate formation of more number of stems per hill. The number of leaves is generally more under long days that under short days in any specific cultivar. The life span of the entire shoot is increased by long days in comparison to short days and therefore, the crop duration of a particular cultivar is less under short days and more under the long days. The leaf area increases rapidly under short days and the specific leaf area is generally more in comparison to the long days. The overall influence of photoperiod on growth shows that long photoperiods stimulate the development of haulm by increasing branching and number of leaves per stem, but reduce leaf size and specific leaf area.

TABLE 1.1   Effect of Increased Photoperiods on Selected Crop Characteristics [12]

Effect on	Response
Sprout length	Reduced at pre sprouting
Stem branching	Increased
Number of leaves	Increased in first level of main stems only
Life span of shoots	Increased
Leaf characteristics	Specific leaf area and leaf size decreased
Photosynthesis	Delayed
Stolon initiation	Delayed
Tuber initiation	Small and inconsistent
Tuber growth	Decreased
Tuber number	Increases when season is long enough as growth cycle increases
Tuber yield	

1.8.2 YIELD

Short photoperiods increase the rate of dry matter production per unit of intercepted light and the rate of photosynthesis per unit leaf dry wt. The production of total dry matter is influenced as per the influence of photoperiod on already described processes. The short day conditions are known to increase the daytime assimilate transport with higher harvest index whereas, the harvest index is reduced under long days. During the early phases of plant development stolon initiation is stimulated by short days, while is delayed by long days. The overall influence of photoperiod on stolonization shows that under long days more number of long stolons with more branches per stolon are formed, thus having more potential sites for tuber development. Tuber initiation, however, is favored by short photoperiods, though it is a variety dependent character. Under Indian conditions four cultivars when planted under both SD and LD conditions showed tuberization in all the four cultivars under SD, while only one cultivar tuberized under LD. Some other experiments also proved that the potato has a short day requirement for tuberization. The leaves are the sites for photoperiodic reception and the tuberization stimulus is transferred from here. It has been found that several days SD cycle is required to produce the tuberization stimulus. Once this requirement is fulfilled the plant exposed to long days produces higher yields. The number of tubers is usually higher under long days, though the growth rate of individual tubers is slow in comparison to short days.

1.9   RESPIRATION

Respiration is a physiological process that sustains cellular life in which the organic substitute is broken down to produce energy, which is used in the biosynthetic events needed for growth and development. It is a gaseous exchange process where the tissue takes up the oxygen and CO_2 is evolved out in the atmosphere. The respiration in potato takes place both in tubers and plants.

The respiratory rate of dormant potato tuber is extremely low. The thick periderm of the tuber acts as a barrier for gaseous exchange and the gaseous concentration inside the tuber is normally different from outside environment. The rate of respiration in potato tuber increases after wounding or slicing which may be a result of increase in the number of mitochondria in the cells.

Just after the harvesting the respiratory rate of potato tuber is generally high and then falls off rapidly. The rate of respiration is high in immature tubers than the mature tubers. Following this initial decrease, respiratory activity during storage at constant temperature changes little with time until sprouting begins, when there is an increase in respiration rate. At the whole plant level the rate of respiration in potato is like any other plant and the plant organs utilize the accumulated carbohydrate as substrate for generating energy to meet their requirement of growth and metabolism.

1.10 PLANT GROWTH REGULATORS

The growth and developmental processes of potato like any other plant are directed by the plant growth substances in one way or the other. These plant growth regulators either act as hormones or they effect the biosynthesis or the action of plant hormones. The first kind of PGR's are called primary regulators, whereas, the second kind as secondary regulators. The primary growth regulators include auxins, gibberrellins, kinins and abscissic acid and examples of secondary growth regulators are chlorocholine chloride (CCC), maleic hydrazide (MH), dormins, etc. These regulators do not operate in isolation but their ratios at particular stage influence the growth stimulus and accordingly the plants respond to these stimuli. Thus the interactions between these primary and secondary growth regulators (be it antagonistic, additive or synergestic) results in growth and development of the plant parts individually or as a system. The internal hormonal balance can be influenced by external factors and this is how the environment interferes with the expression of genetic information into metabolic activity and, by that, in the form and function of the plant. Apart from nutritive effects and from direct influences on enzymatic functions, this influence upon the hormonal balance is one of the major effects of the environment in the growth and development of the plant, on which a great deal of our cultural techniques are unconsciously being based. In potato plant various growth and developmental processes depend on such hormonal balance and are influenced by the surrounding environment. Here we shall deal with a general profile of balance of these regulators during some important stages of potato plant growth and tuberization.

1.10.1   PLANT GROWTH

Once the tuberization is over in potatoes, the plants are harvested after maturity and the seed tubers are stored till the next season planting. During this transition period various activities take place in tubers. The tubers remain dormant for quite some time depending on their genetic constitution and the environment under which they are stored. The dormancy break is induced by the hormonal regulation and it has been found that the content of ABA decreases in the peels of potato tubers at the termination of their dormancy period. On the other hand gibberellic acid is known to induce the dormancy break in potatoes, which means that at the time of dormancy break GA/ABA ratio is increased. Auxins however, have not been shown to influence dormancy itself but influence only sprout growth after dormancy break. Cytokinins like gibberellins are generally considered to be growth promoters, and are capable of breaking dormancy when used as external treatment. After the dormancy period is over the sprouts emerge and elongate and this process is thought to be affected positively by gibberellins. The roots are formed from the base of sprouts and auxins play an important role in the development of root system in potatoes. Gibberrellins, auxins, and cytokinins in combination influence the growth of potato plant and each process requires a particular set of balance of these growth regulators. The flowering in potatoes in induced by 6 weeks GA application.

1.10.2   YIELD

It is generally believed that the specific tuber forming stimulus responsible for tuberization may be related to a cytokinin since application of cytokinins have shown tuber induction and starch accumulation in *invitro* grown potato stolons. This is further strengthened by the presence of high levels of cytokinins in the stolon tips and newly initiated tubers. GA on the other hand inhibit tuber formation and cause elongation of stolons. Natural tuberization is inhibited by ethylene as well, and it does not even allow stolons to elongate. ABA stimulates the tuberization process. Studies in India on application of exogenous ABA could not induce tuberization in potato. However, it is widely believed that higher ABA/GA ratio is important for tuberization stimulus and once this stimulus is obtained, higher cytokinin activity helps in rapid cell division for tuber growth in potato. The overall sequence for the role of PGR's may be that GA promotes stolon formation, IAA helps in cell enlargement in the subapical region, ABA causes cessation of apical growth and cytokinin helps in cell division at later stages of development of the young tubers. Jasmonic acid also plays a role during tuber initiation in potatoes. Application of various plant growth substances has been found to increase the yield components in potato. The number of tubers per plant has been found to increase with the application of auxins (IBA at 10 ppm concentration). Plant growth regulators (such as auxins and cytokinins) are also known to increase the proportion of medium and large sized tubers in potatoes and consequently the number of small sized tubers is less. However, such results are not uniform and in general the attempts to increase the potato yield through application of growth regulators have not yielded consistent results.

1.11 CONCLUSION

The yield in potato plant is dependent on genotypic, phenotypic, and environmental factors. The morphological factors are important as most of the physiological processes responsible for tuber yield depend on them. Therefore, it is pertinent to mention the desirable traits of the plant parts in the yield formation.

The number of stems produced per tuber is influenced by the genetic factors, seed tuber size, and environment. It is desirable that a potato seed tuber gives rise to more number of stem so that more foliage is supported by it, thereby producing more photoassimilate for growth, development and yield. It is a well-known fact that large size tubers bearing more eyes give rise to more number of stems. For better productivity of the potato plant it is desirable that the plant should have higher leaf area per unit land area, so that higher photosynthetic activity can be taken up in the given area and more dry matter is produced. The increase in leaf area at initial growth stages and its maintenance over a longer period of crop growth is favorable for higher yields in potato. The cultivars attaining higher LAI at early stages have shown higher harvest index and yield. The leaf area duration is another parameter for estimation of plant growth and expected yield. It has been established that effective participation of leaves for longer duration in the process of photosynthesis results in higher productivity. The leaves having low efficiency act as sink rather than source for the dry matter accumulation and are detrimental for the productivity. Hence, the more number of photosynthetically active leaves for the overall duration of the crop are desirable for increased productivity. Similarly, for better productivity of the potato plant, more number of tubers with bigger size is important. When the potato plant is healthy and active, a large amount of photosynthate produced in foliage is translocated to the tubers leading to growth in size of the tuber. The strong tuber sink influences the photosynthetic accumulation positively and thus more number of tubers cause an increase in the accumulation process. The ultimate result is higher yields with more number of tubers and bigger size.

KEYWORDS

- **physiology**
- **potato**
- **tuberization**
- **yield**
- **photosynthesis**
- **growth**
- **productivity**

REFERENCES

1. Basu, P. S., & Minhas, J. S. (1991). Heat tolerance and assimilate transport in different potato genotypes. *Journal of Experimental Botany. 42*, 861–866.
2. Burton, W. G. (1989). *The Potato.* Longman Scientific and Technical, New York, USA. 742p.
3. Turnbull, C. G. N. & Hanke, D. E. (1985). The control of bud dormancy in potato tubers. Measurement of the seasonal pattern of changing concentrations of zeatin-cytokinins, *Planta. 165,* 366–376.
4. Dutt, S., Manjul, A. S., Raigond, P., Singh, B., Siddappa, S., Bhardwaj, V., Kawar, P. G., Patil, V. U., & Kardile, H. B.(2017). Key players associated with tuberization in potato: potential candidates for genetic engineering. *Crit Rev Biotechnol. 37,* 942–957.
5. Ewing, E. E. (1981). Heat stress and tuberization stimulus. *American Potato Journal, 58,* 31–49.
6. Ezekiel, R., & Bhargava, S. C. (1993). The influence of photoperiod on the growth and development of potato: a review. *Plant Physiology and Biochemistry. 20,* 63–72.
7. Harris, P. M. (1992). *The Potato Crop.* Chapman and Hall, London, UK, 909p.
8. Jeffrey C. Suttle, & Julie F. Hultstrand (1994). Role of Endogenous Abscisic Acid in Potato Microtuber Dormancy. *Plant Physiol. 105,* 891–896.
9. Jeffrey C. Suttle. (2004). Physiological Regulation of Potato Tuber Dormancy. *Amer J of Potato Res. 81,* 253–262.
10. Li, P. H. (1985). *Potato Physiology.* Academic Press, London. 586p
11. Minhas, J. S., & Singh, B. (2003). Physiology of crop growth, development and yield. In: *The Potato-Production and Utilization in Sub-Tropics.* Khurana, S. M. P., Minhas, J. S., Pandey, S. K. (eds.), Mehta Publishers, New Delhi, pp. 292–300
12. Struik, P.C. & Ewing, E. E. (1995). Crop physiology of potato *(Solanum tuberosum):* responses to photoperiod and temperature relevant to crop modelling. In: Haverkort, A. J., & MacKerron, D. K. L. (Eds.), Potato Ecology and Modelling of Crops under Conditions Limiting Growth. Kluwer Academic Publishers, Dordrecht, pp. 19–40.
13. Sukumaran, N. P. (1993). Studies on photosynthesis in relation to potato yields. In: *Advances in Horticulture,* Vol. 7. Chadha, K. L., & Grewal, J. S. (eds.), Malhotra Publishing House, New Delhi, India, pp. 375–382.
14. Van Loon, C. D. (1986). Drought, a major constraint in potato production and possibilities of screening for drought resistance. In: *Potato Research of Tomorrow.* Pudoc, Wageningen, The Netherlands, pp. 3–55.

CHAPTER 2

PHYTOCHROME: PHYSIOLOGY, MOLECULAR ASPECTS, AND SUSTAINABLE CROP PRODUCTION

BANDANA BOSE,[1] BHAWNA PANT,[1] RAJESH K. SINGHAL,[1] MAHESH KUMAR,[1] and SANANDA MONDAL[2]

[1]Department of Plant Physiology, Institute of Agriculture Sciences, Banaras Hindu University, Varanasi – 221005, India, E-mail: bbosebhu@gmail.com

[2]Plant Physiology Section, Department of ASEPAN, Institute of Agriculture, Visva-Bharati, Sriniketan – 731236, India

CONTENTS

ABSTRACT

Plants utilize several families of the photoreceptors to fine-tune growth and development over a large range of environmental conditions. Among them Phytochrome is one of the families of photoreceptors or pigment sensitive to light in the red and far-red region of the visible spectrum. Phytochrome display both unique and overlapping roles throughout the life cycle of plants, regulating a range of the developmental processes from seed germination to maturation. The chapter summarizes the structure, interconvertable forms, phosphorylation, functional domain, interacting proteins, and physiological responses of phytochrome. Furthermore, it describes the role of phytochrome in various physiological processes occurring in plants such as flowering, shade avoidance, and sustainable crop production.

2.1 INTRODUCTION

Plants are particularly sensitive to the environmental changes because they are sessile in nature. Light is one of the most important environmental cues, controlling plant development throughout their life cycle, from seed germination to floral induction. The importance of light signals in regulation of plant growth has been documented before centuries. Indeed, Darwin himself provided a detailed observations of the developmental processes occurring following emergence of a dark-grown (etiolated) seedling into the light, in a book written with his son, Francis, 'the power of movement in plants' [1]. In this chapter, the authors recorded important roles of light throughout the development of plant. Phytochromes are red- and far-red-sensing photoreceptors that sense the quantity, quality, and duration of light all over the entire life cycle of plants.

The detection of physiological response such as the germination of lettuce seeds, promoted by the red (R) light and repressed by subsequent far-red (FR) light lead to the discovery of phytochrome [2]. Phytochromes were first detected in 1959 by investigators at the USDA Plant Industry Station in Beltsville, MD [3]. Phytochrome is a blue protein pigment with a molecular mass of about 250 KDa. It occurs as a dimmer made up of two subunits. Each subunit consists of two components: a light absorbing pigment – referred as chromophore; and a polypeptide chain – referred as apoprotein. Both chromophore and apoprotein are combined and formed it as holoprotein, which is functional. For sensing red light–depleted (shade) and red light–enriched (full sun) conditions, plants use phytochromes, large (120 kD) proteins that posses the covalently linked linear tetrapyrrole (bilin) chromophore phytochromobilin.

Plant phytochromes (Phy family) are members of a more widespread family of photo-sensors that are found in cyanobacteria (cyanobacterial phytochromes Cph1 and Cph2) as well as in purple and nonphotosynthetic bacteria (bacteriophytochromes; BphP) and even in fungi (fungal phytochromes; Fph family) [4–6]. Three discrete apoprotein-encoding genes (PHYA–PHYC) are conserved within angiosperms [7]. In the model species *Arabidopsis thaliana*, five genes (PHYA-E) encoding phytochrome apoproteins have been sequenced and characterized [8, 9]. The isolation of separate null mutants in all five phytochromes has facilitated the construction of multiple, higher-order mutants, deficient in a variety of phytochrome combinations. An analysis of both individual and multiple phytochrome-deficient mutants has subsequently provided a refined picture of phytochrome functions throughout plant development. Phytochrome physiology, biochemistry, molecular aspect, and role in sustainable crop production are discussed in this chapter.

2.2 PHYTOCHROME: STRUCTURE, FUNCTIONAL DOMAIN, PHYSIOLOGICAL RESPONSE, GENE FAMILY, PIPS, AND ACTIVATION OF LIGHT RESPONSE

2.2.1 PHYTOCHROME INTERCONVERSION

Phytochrome, present in a red light absorbing form is referred to as Pr; it is blue in color (inactive form) and it is converted by red light to far red light absorbing form Pfr(active form), which is blue-green in color. The phytochrome chromobilin alone cannot absorb red and far red light. Light can be absorbed only when the polypeptide is covalently linked with phytochromobilin to form holoprotein. Phytochromobilin is synthesized inside plastids and exported to cytosol where it attaches to apoprotein through thio-eather linkage to a cystein residue. Its assembly is autocatalytic.

Phytochromes have two relatively stable, spectrally distinct, and inter-convertible conformers: an R-absorbing Pr form and a FR-absorbing Pfr form [10]. The Pfr form is considered to be the active form because many physiological responses are promoted by R light. The ability of a given phytochrome is absorbed red and far-red light stems from its bound phyto-chromobilin, which undergoes a reversible photoisomerization at the C15-C16 double bond in response to red light (666 nm) and far-red light (730 nm) [11, 12]. Isomers of phytochromobilin are presented in Figure 2.1.

2.2.2 *LIGHT INDUCED CONFORMATIONAL CHANGES IN PHYTOCHROME*

Members of the extended family of phytochrome proteins share a common N-terminal photosensory core, with three blocks of homology sometimes referred to as P2, P3, and P4 [13]. Phytochrome C-terminal domains medi-ate the transmission of photo sensory signals perceived by the N-terminal region to signal transduction pathways within the cell. This region typically contains a histidine kinase–related domain that has been shown to confer ATP-dependent protein phosphotransferase activity in several cases [14–21].

While bacterial and cyanobacterial phytochromes typically employ classical two-component phosphotransferrylase, the mechanism of

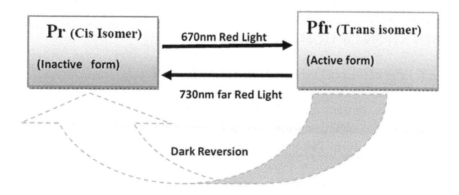

FIGURE 2.1   Photochemical property of phytochromes, two isomers of phytochromobilin; (Pr chromophore) and (Pfr chromophore) (Adapted and modified from Ref. [12]).

plant phytochrome signaling appears considerably more complex, involving light mediated nuclear translocation and regulation of transcription factor function [22–24]. The C-terminal regulatory domains of plant phytochromes have been shown to mediate homo dimerization and light-modulated nuclear targeting, both of which are required for signal transmission [25]. Signaling by plant phytochromes also involves protein–protein interactions with the N-terminal part of the protein [26, 27].

2.2.3. PHYTOCHROME: AUTOPHOSPHORYLATING PROTEIN KINASE

Phytochromes are dimeric chromo-proteins covalently linked to tetrapyrrole chromophore phytochromobilin, and exist as two photo-inter convertible species, red-light absorbing Pr and far-red-light absorbing Pfr forms. Phytochromes are biosynthesized as the Pr form in the dark, and are transformed to the Pfr form upon exposure to red light. This photo activation of phytochromes induces a highly regulated signaling network for photomorphogenesis in plants [28, 29]. In recent times, phosphorylation and dephosphorylation have been suggested to play significant roles in phytochrome- mediated light signaling [30, 31], for example; a few phytochrome-associated protein phosphatases have been shown to act as positive regulators of phytochrome signaling [32, 33]. However, functional role of phytochrome phosphorylation stay behind to be explored. Phytochromes are known as auto phosphorylating serine/threonine protein kinases [15, 34]. The N-terminal extension (NTE) of phytochrome is functionally important for providing autophosphorylation sites. The NTE is required for the full biological activity of phyA [35], and serine-to-alanine substitutions in the NTE region have been shown to result in increased biological activity [36]. In bacterial phytochrome, the input domain receives the red light signal and transmitted to sensor protein which autophosphorylate the histidine (Figure 2.2). This phosphorylated sensor protein phosphorylates the response regulator protein at an as parted. The phosphorylated response regulator stimulates the process.

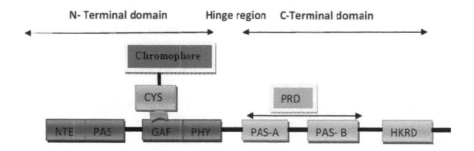

FIGURE 2.2   General overview of phytochrome structure in plants. (NTE: N-terminal extension; HKRD: Histidine kinase related domain; PAS, GAF: cGMP stimulated phosphodiesterases; PHY: Phytochrome; PRD: Phytochrome related domain) [Adapted and modified from Ref. [29]).

2.2.4   FUNCTIONAL DOMAINS IN PLANT PHYTOCHROMES

All plant phytochromes can be divided into an N-terminal photosensory (signal input) domain and a C-terminal dimerization and signal output domain. The N-terminal photosensory domain further divided into four consecutive sub domains called P1(NTE), P2(PAS), P3(GAF) (cGMP stimulated phosphodiesterases), and P4(PHY) (Phytochrome) (named sequentially from the N terminus). The P4 domain has been suggested to directly interact with the D ring of the chromophore to maintain its extended linear conformation in the Pr form and to stabilize the Pfr form. The PAS, GAF, and PHY domains share a common core fold defined by a central, antiparallel β sheet with the strands in the order of 2–1–5–4–3 and a connecting element between strands 2 and 3 containing an α helix. Although the overall spatial arrangement of these three domains is linear in a beads-on-a-string fashion and they are closely integrated via the NTE of the PAS domain, a highly unusual knot [37], and an arm extending from the PHY core domain [38–39]. Whereas, the C-terminal domain may be divided into two subdomains: the PAS-A and PAS-B domains, and the histidine kinase-related domain (HKRD), which belongs to the ATPase/kinase GHKL (gyrase, Hsp90, histidine kinase, MutL) super family [40]. The PAS domain is named after three proteins in which it occurs: Per (period circadian protein), Arn (Ah receptor nuclear translocator protein), and

Sim (single-minded protein). The HKRD lacks a critical histidine residue, and thus may be an evolutionary remainder rather than an active histidine kinase [41].

Among the N-terminal subdomains, the P1 domain is uniquely present in plant phytochromes, whereas the P2(PAS), P3(GAF), and P4(PHY) domains are also found in phytochrome-like proteins of various origins. Among the C-terminal subdomains, the PAS-A and PAS-B domains are unique to plant phytochromes, whereas HKRDs are also found in phytochrome-like proteins (Figure 2.2).

2.2.5 PHYSIOLOGICAL RESPONSES MEDIATED BY PHYTOCHROMES

Phytochrome responses have been subdivided into different classes based on the radiation energy of light that is required to obtain the response. These include low fluence responses (LFRs), very low fluence responses (VLFRs), and high irradiance responses (HIRs). LFRs are the classical phytochrome responses with R/FR reversibility. VLFRs are not reversible and are sensitive to a broad spectrum of light between 300 to 780 nm. HIRs require prolonged or high-frequency intermittent illumination and usually are dependent on the fluence rate of light. The light signals can be perceived by different plant photoreceptors: phytochromes (red light (R)/far-red light (FR) photoreceptors), cryptochromes (blue light/UV-A photoreceptors), and UV-B photoreceptor(s). These pigments convert light signals into biochemical signals later transduce through largely unidentified pathways into molecular and physiological changes that modulate growth and development. Usually there are three 'modes of action' of phytochrome have been described: (i) very-low-fluence responses (VLFR); (ii) low-fluence responses (LFR) and (iii) high-irradiance responses (HIR). Very-low proportions of phytochrome as Pfr (Pfr/P) are sufficient to induce VLFR whereas higher levels of Pfr/P are necessary to induce LFR. Since VLFR are very sensitive to Pfr, exposure to long-wavelength far red' is sufficient to obtain and even saturate this type of effect. LFR occur between Pfr/P typical of FR and R and are therefore R/FR reversible. Dual responses to Pfr/P are observed when both VLFR and LFR are available.

Dual responses have been observed for seed germination [42–44], promotion of coleoptile growth [45], inhibition of mesocotyl growth [45–46], cotyledon unfolding [47], anthocyanin synthesis [48], glutamate synthase activity [49], and greening [50–52]. With the exception of seed germination and the shade-avoidance response, which are controlled solely by phytochromes in *Arabidopsis* [53], other physiological processes, including seedling development and floral induction, are controlled by interconnected networks of both phytochromes and cryptochromes (Figure 2.3) [54–57].

2.2.6   PHYTOCHROME GENE FAMILY STRUCTURE AND FUNCTIONS

In the model species *Arabidopsis* thaliana, five discrete phytochrome-encoding genes, *PHYA-PHYE*, have been isolated and sequenced [58] and Phytochromes A, B and C are conserved among angiosperms. Phytochromes A, B, C and E are evolutionarily divergent proteins, sharing near 46–53% sequence identity, while *PHYD* encodes an apoprotein that shares 80% sequence identity with PHYB. Molecular phylogenetic analysis supports the occurrence of four major duplication events in the evolution of phytochrome genes (Figure 2.4). An initial duplication is believed to have separated *PHYA* (light labile in the far-red light absorbing (Pfr) form) and *PHYC* (light stable in the Pfr form) from *PHYB/D/E* (all light stable in the Pfr form). The subsequent separation of *PHYA* from *PHYC* and *PHYB/D* from *PHYE* resulted in three sub families: *A/C, B/D* and *E* [59].

2.2.7   PHYTOCHROME-INTERACTING PROTEINS

It is established that phytochrome molecules are in fact induced to translocate from the cytoplasm to the nucleus for photo activation, thereby potentially rendering a second messenger pathway. From several laboratories data, using phy–β-glucuronidase (GUS), phy–green fluorescent protein (GFP) fusions, and immune cytochemical localization procedures, point

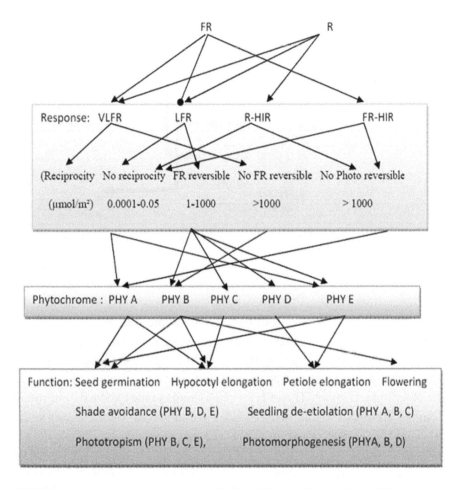

FIGURE 2.3 Relationship between red light (R) and far red light (FR) response, reciprocity, reversibility, phytochrome family and their functions (Adapted and modified from Ref. [55]).

toward that full-length phytochromes are initially present in the cytosol in their inactive Pr conformer, but subsequent to photo conversion to the active Pfr conformer, the photoreceptor molecules are induced to translocate into the nucleus. Some phytochrome-interacting proteins are needed for the nuclear/cytoplasmic partitioning of phytochromes. In eukaryotes, proteins larger than 40 kD must be actively transported into or out of the nucleus through nuclear pore complexes, with the help of transport proteins such

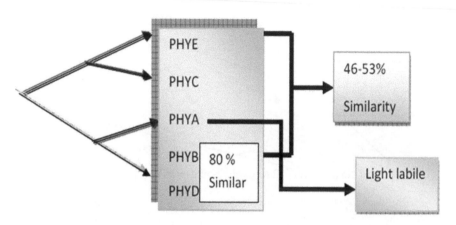

FIGURE 2.4 Evolutionary relationship among phytochrome genes in dicotyledonous plants. Phylogenetic analysis has revealed *PHYA* and *PHYC* to share a common ancestry, with *PHYB*, *PHYD* and *PHYE* forming a distinct subgroup (Adapted and modified from Ref. [6]).

as importins and exportins [60, 61]. Phytochrome dimmers are approximately 240 kD in size so they require interacting proteins for their nuclear localization. At first the absorption of light by phytochromobilin occurs which photoisomerizes and accompanying structural changes that expose the N-terminal domain. Then FHY or FHL binds to the exposed N-terminal domain. Translocation of PHYA-FHY-1 complex into the nucleus via the NLS of FHY1 occurs. Binding of FHY-1 is required only for PHYA, PHYB does not require the function of FHY1 or FHL because the C-terminal of PHYB contains NLS (nuclear-localization signal). The Pfr form of PHYB might be functionally equivalent to the PHY-FHY1/FHL complex [62, 63].

Under certain light condition some phytochrome-interacting proteins modulate the signaling output of phytochrome. The Pfr form is the biologically active form of phytochrome, it is expected that some proteins are to interact with the Pfr form and control its output activity by either changing its concentration or by modulating its ability to transmit signals to downstream components. The phytochrome output activity can also be modulated either by changing its affinity for its downstream component or by altering its transmitting activity.

Arabidopsis response regulator 4 (ARR4) specifically inhibits the dark reversion of PHYB, whereas constitutive photomorphogenic 1 (COP-1) ubiquitinates the Pfr form of PHYA. ARR-4, which binds to PHYB and stabilizes the Pfr form, was the first identified phytochrome interacting protein that modulates the output activity of phytochrome. ARR-4 binds to the N-terminal end of PHYB. The binding of ARR-4 inhibits the dark reversion of PHY B in both yeast and plants. COP-1 binds to PHYA and decreases PHYA output activity by decreasing the total PHY-A concentration, phytochrome associated phosphatase 5 (PAPP-5) binds to both PHYA and B and preferentially dephosphorylate Pfr form. The dephosphorylation of phytochromes by PAPP 5 increases their affinity for the interacting proteins nucleoside diphosphate kinase 2 (NDPK-2) and PIF-3 increases the stability of PHYA (Figure 2.5).

2.2.8   ACTIVATION OF LIGHT RESPONSES BY PHYTOCHROMES (PHY) AND THEIR INTERACTING PROTEINS

The light responses are repressed in the dark, because negative components such as phytochrome interacting factors (PIFs)/PIF-3 like proteins inhibit light responses where as positive components such as long hypocotyle in FAR-Red-1 (HFR1), long hypocotyle 5 (HY5) and long after FAR-red light (LAF-1) are degraded by the nuclear localization constitutive photomorphogenic-1 (COP-1). Upon irradiation, the Pfr forms of phytochrome initiate cytosolic light responses by binding cytosolic interacting protein such as phytochrome kinase substrates and enter the nucleus with PHYA

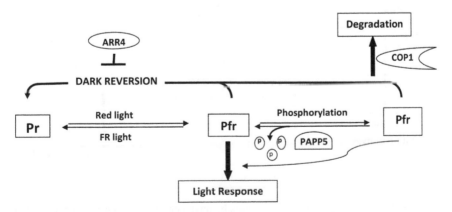

FIGURE 2.5   Modulation of phytochrome output activity (Adapted from Ref. [29]).

(with FHY-1 and FHL) and without PHYB. In the nucleus the Pfr form activates the degradation of PIFs/PILs through unidentified E_3 ubiquitin ligase and inhibits COP-1 by the exclusion of it from the nucleus. Owing to decrease levels of negative components and increase levels of positive components, light responses are initiated (Figure 2.6).

2.3   PHYTOCHROME IN CROP SUSTAINABILITY

2.3.1   RESOURCE ALLOCATION AND BIOMASS PRODUCTION

It was found that phytochrome controls carbon allocation and biomass production in the developing plants [64]. Phytochrome mutants have a reduced CO_2 uptake, yet over accumulate daytime sucrose and starch. Phytochrome depletion alters the proportion of day-night growth. In addition, phytochrome loss leads to sizeable reductions in overall growth,

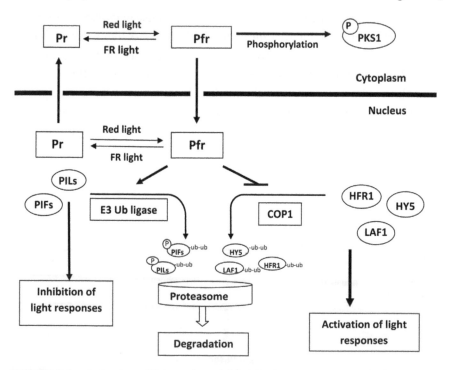

FIGURE 2.6   Activation of light responses by phytochrome and their interacting protein (Adapted from Ref. [29].

dry weight, total protein levels, and the expression of *cellulose synthase-like* genes. Furthermore, the growth-retarded phytochrome mutants are reported to be less responsive to growth-inhibiting abiotic stresses and have elevated expression of stress marker genes. In nature, this strategy may be activated in phytochrome-disabling, vegetation-dense habitats to enhance survival in potentially resource-limiting conditions.

2.3.2   FLOWERING

Phytochromes sense light signals and mediate diverse developmental processes. In many plants, the timing of floral initiation is regulated by the daily cycle of photoperiod. Plants in which flowering is accelerated by short days (short-day plants, SDP) generally flower in the autumn to finish reproduction before the adverse temperatures of winter. Plants in which flowering is accelerated by long days (long-day plants, LDP) generally flower in late spring, thus promoting seed set in a more favorable climate. The roles of individual photoreceptors in mediating flowering responses have been largely inferred from the mutant analyses. Two common experimental approaches for studying the regulation of floral initiation are manipulation of day length using day extension and night break light treatments [65]. In day extension experiments, light of low fluence rate is applied at the end of a short-day photoperiod. In night break experiments, a light exposure is given in the middle of a long night, thus mimicking long-day (LD) conditions. In rice, all three phytochrome genes—*OsphyA*, *OsphyB*, and *OsphyC*—are involved in regulating flowering time. It was investigated that role of *OsPhyA* by comparing the *osphyA:osphyB* double mutant to an *osphyB* single mutant. Results indicated that *OsPhyA* influences flowering time mainly by affecting the expression of *OsGI* under SD and *Ghd7* under LD when phytochrome B is absent. They also demonstrated that far-red light delays flowering time via both OsPhyA and OsPhyB [66].

In *Arabidopsis* plant, genes of five pathways control both flowering and time of flowering. Autonomous pathway and vernalization pathway, gene inhibit FLC (flowering locus C, a floral repressor), and the FT (flowering locus T, it alters flowering time) gene, which can activate AP1 (APETALA 1) gene. Perhaps, both diurnal and photoperiodic pathways can activate

FT directly. GA pathway activates AP1 gene directly or by the activation of SOC1gene. Finally, AP1 and LFY genes are interchangeable form and regulates flowering in plant systems. Recently, analysis of FT gene is found to be quite encouraging in inducing flowering in transgenic *Populous* species [67]. Role of Diurnal clock, automated pathway, vernalization pathway/photoperiodic pathway and GA pathway in respect to flowering are summarized in Figure 2.7.

2.3.3   SEED GERMINATION STIMULANTS

Karrikins (KAR1) are a class of seed germination stimulants identified in smoke from wildfires. Microarray analysis of imbibed *Arabidopsis*

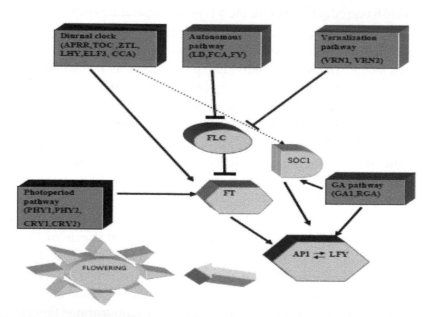

FIGURE 2.7 Representation of genetic pathways (photoperiodic, light quality, vernalization, autonomous, and GA), regulating flowering time in *Arabidopsis*. (LHY: late elongated hypocotyl, ZTL: zeitlupe, ELF3: early flowering, CCA1: circadian clock associated 1, FLC: flowering locus C, AP1: APETALA 1, FT: flowering locus T, timing of cab expression 1 (TOC1), APRR: arabidopsis pseudo-response regulator, LFY: leafy, LD: luminidependence; FCA: flowering time control protein, SOC1: suppressor of overexpression of constants, VRN1&2: vernalization) (Adapted and modified from Ref. [67]).

thaliana seeds was performed to identify transcriptional responses to KAR1 before germination. Karrikins enhance light responses during germination and seedling development in *Arabidopsis thaliana [68]*. It was showed that *Arabidopsis* DET1 (de-etiolated 1) degrades HFR1 (hypocotyl in far red 1) but stabilizes PIF1 to precisely regulate seed germination [69].

2.3.4 SHADE AVOIDANCE

In recent years, the concept of shade avoidance has provided a functional meaning to the role of the phytochrome photoreceptor family in mature plants in their natural environment, and the question of which of these phytochromes is responsible for shade avoidance reactions has inevitably been raised. It is now fully accepted that shade avoidance reactions are all initiated by a single environmental signal, the reduction in the ratio of red (R) to far-red (FR) radiation (i.e., R: FR) that

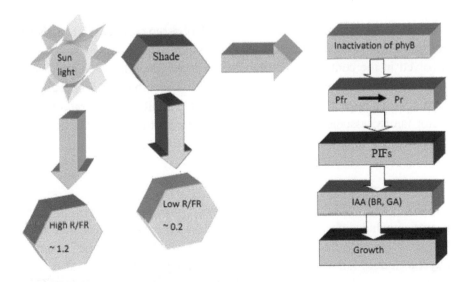

FIGURE 2.8 Schematic representation of the shade-avoidance response in seedlings of dicotyledonous plants, such as Thale cress (*Arabidopsis thaliana*). In sunlight, at a high R:FR, photo morphogenesis is mediated by phytochromes [Adapted and modified from Ref. [70]).

occurs within crowded plant communities [70]. Figure 2.8 showed that in presence of sunlight R/FR ratio is near to 1.2 while in shade this ratio is very less (0.2). In shade condition inactivation of phyB gene convert pfr to pr, which can further activate PIF. These PIFs regulate phytohormones like IAA, BR and GA, which in turn regulate the plant growth and development.

2.4   BIOTIC AND ABIOTIC STRESS TOLERANCE

It is well-documented that phytochromes can control plant growth and development from germination to flowering. Additionally, these photoreceptors have been shown to modulate both biotic and abiotic stresses. Role of phytochrome in abiotic stress are discussed in the following subsections.

2.4.1   DROUGHT TOLERANCE

Some observations have lead to a possible relationship between phytochromes and drought stress being raised, since there is evidence that phytochromes are involved in the control of leaf transpiration, germination of dormant seed and ABA metabolism. Ref. [71] showed that a member of C-repeat/dehydration responsive element-binding factor (CBF/DREB1) of transcription factor family, associated with promoters of drought-responsive genes, was repressed by a phytochrome-interacting factor (PIF7) in *A. thaliana*.

2.4.2   HIGH AND LOW TEMPERATURE STRESS

It was concluded that phyA and phyB function antagonistically to regulate cold tolerance that essentially involves FR light-induced activation of phyA to induce ABA signaling and, subsequently, JA signaling, leading to an activation of the CBF pathway and a cold response in tomato plants [72]. It was reported that a family of *Small Auxin UP RNA* (*SAUR*) genes that are expressed at high temperature in a *PIF4*-dependent manner and promote elongation growth [73].

2.4.3 HIGH LIGHT STRESS

In recent studies, PIFs are found to be involved in the prevention of photo-oxidation in *A. thaliana* [74]. It is suggest that PIF1 and other PIFs in transducing light signals to regulate *PSY* (phytotine synthetase) gene expression and carotenoid accumulation during daily cycles of light and dark in mature plants [75]. An important strategy allowing seedling survival during early development is the photo protection promoted by anthocyanin accumulation in hypocotyls and cotyledons [76]. It is probable that phytochrome plays a role in these responses either through the induction of phenylalanine ammonia-lyase (PAL) or chalcone synthase (CHS), both of which play a crucial role in the anthocyanin biosynthesis pathway [77].

2.4.4 HERBIVORE DEFENSE

It has been pointed out that phyB inactivation triggers shade avoidance syndrome (SAS) responses and results in lower levels of constitutive defenses; phyB inactivation represses JA-induced direct defense; *phyB1* and *phyB2* plants support greater herbivore damage than WT plants; phyB inactivation changes the blend of VOCs (volatile organic compounds) emitted by JA-induced plants; phyB inactivation increases the attractiveness of JA-induced plants to the mirid predatory bug *M. pygmaeus* [78].

2.5 PHOTO MORPHOGENESIS AND TROPISM

Phytochrome has an important role in phototropism and gravitropism in plant. Interaction of different photoreceptors like phytochrome, crypto-chrome, phototropin regulate photomorphogenesis and tropism in plant [79]. The role of phytochrome A (phyA) and phytochrome B (phyB) in phototropism was investigated by using the phytochrome-deficient mutants *phyA-101*, *phyB-1* and a *phyA/phyB* double mutant [80]. Study on *Arabidopsis thaliana* (L.) showed that phy C, D, E plays an important role in tropisms in light grown seedlings and inflorescence stems [81–84].

2.6 CONCLUSION

This chapter summarizes the structure, interconvertable forms, light induced changes, activation, functional domain, and gene family of phytochrome interacting proteins thoroughly. It also includes role of phytochrome gene family in various developmental processes of plants starting from seed germination, hypocotyls elongation, seedling de-etiolation, photo morphogenesis, and shade avoidance to flowering. Finally the role of phytochrome in crop sustainability and various abiotic/biotic stresses is also taken into consideration. However, literature suggested that this vast area of stress effect has been managed by the plants through the cross talking between various hormones, metabolic pathways and phytochrome action via up and down regulation of various genes in plants. Therefore, future challenge in phytochrome research will be to find out more detailed information about various phytochrome genes network and the related proteins, which can be helpful to Plant Scientists in plant improvement programme, in this climate change scenario.

KEYWORDS

- de-etiolation
- functional domain
- photo morphogenesis
- phytochrome
- phytochrome interacting proteins (PIPS)
- shade avoidance
- sustainable crop production

REFERENCES

1. Darwin, C. (1880). *The Power of Movement in Plants. 187*, 957–567.
2. Kendrick, R. E., & Kronenberg, G. H. M. (1994). *Photomorphogenesis in Plants.* Kluwer Academic, Dordrecht, The Netherlands.

3. Butler, W. L., Norris, K. H., Siegelman, H. W., & Hendricks, S. B. (1959). Detection, assay, and preliminary purification of the pigment controlling photoresponsive development of plants. *Proceedings of the National Academy of Sciences, 45*(12), 1703–1708.

4. Blumenstein, A., Vienken, K., Tasler, R., Purschwitz, J., Veith, D., Frankenberg-Dinkel, N., & Fischer, R. (2005). The Aspergillus nidulans phytochrome FphA represses sexual development in red light. *Current Biology, 15*(20), 1833–1838.

5. Froehlich, A. C., Noh, B., Vierstra, R. D., Loros, J., & Dunlap, J. C. (2005). Genetic and molecular analysis of phytochromes from the filamentous fungus *Neurospora crassa*. *Eukaryotic Cell. 4*(12), 2140–2152.

6. Karniol, B., Wagner, J. R., Walker, J. M., & Vierstra, R. D. (2005). Phylogenetic analysis of the phytochrome superfamily reveals distinct microbial subfamilies of photoreceptors. *Biochemical Journal. 392*(1), 103–116.

7. Mathews, S., Lavin, M., & Sharrock, R. A. (1995). Evolution of the phytochrome gene family and its utility for phylogenetic analyses of angiosperms. *Annals of the Missouri Botanic Garden. 82,* 296–232.

8. Sharrock, R. A., & Quail, P. H. (1989). Novel phytochrome sequences in Arabidopsis thaliana: structure, evolution and differential expression of a plant photoreceptor family. *Genes and Development. 3,* 1745–1757.

9. Clack, T., Mathews, S., & Sharrock, R. A. (1994). The phytochrome apoprotein family in Arabidopsis is encoded by positive genes: the sequences and expression of PHYD and PHYE. *Plant Molecular Biology. 25,* 413–427.

10. Quail, P. H. (1997). An emerging molecular map of the phytochromes. *Plant Cell Environ. 20,* 657–666.

11. Abe, H., Yamamoto, K. T., Nagatani, A., & Furuya, M. (1985). Characterization of green tissue-specific phytochrome isolated immunochemically from pea seedlings. *Plant and Cell Physiology. 26*(7), 1387–1399.

12. Chen, M., Chory, J., & Fankhauser, C. (2004). Light signal transduction in higher plants. *Genet. 38,* 87–117.

13. Montgomery, B. L., & Lagarias, J. C. (2002). Phytochrome ancestry: sensors of bilins and light. *Trends in Plant Science. 7*(8), 357–366.

14. Yeh, K. C., Wu, SH., Murphy, J. T., & Lagarias, J. C. (1997). A cyanobacterial phytochrome two-component light sensory system. *Science. 277*(5331), 1505–1508.

15. Yeh, K. C., & Lagarias, J. C. (1998). Eukaryotic phytochromes: light-regulated serine/threonine protein kinases with histidine kinase ancestry. *Proc Natl Acad Sci USA. 95,* 13976–3981.

16. Bhoo, S. H., Davis, S. J., Walker, J., Karniol, B., & Vierstra, R. D. (2001). Bacteriophytochromes are photochromic histidine kinases using a biliverdin chromophore. *Nature. 6865,* 776–779.

17. Hübschmann, T., Jorissen, H. J., Börner, T., Gärtner, W., & de Marsac, N. T. (2001). Phosphorylation of proteins in the light-dependent signaling pathway of a filamentous cyanobacterium. *European Journal of Biochemistry. 268*(12), 3383–3389.

18. Lamparter, T., Esteban, B., & Hughes, J. (2001). Phytochrome Cph1 from the cyanobacterium Synechocystis PCC6803. *European Journal of Biochemistry. 268*(17), 4720–4730.

19. Karniol, B., & Vierstra, R. D. (2003). The pair of bacteriophytochromes from Agrobacterium tumefaciens are histidine kinases with opposing photobiological properties. *Proceedings of the National Academy of Sciences. 100*(5), 2807–2812.

20. Giraud, E., Zappa, S., Vuillet, L., Adriano, J. M., Hannibal, L., Fardoux, J. Berthomieu, C., Bouyer, P., Pignol, D., & Verméglio, A. (2005). A new type of bacteriophytochrome acts in tandem with a classical bacteriophytochrome to control the antennae synthesis in Rhodopseudomonas palustris. *Journal of Biological Chemistry. 280*(37), 32389–2397.

21. Tasler, R., Moises, T., & Frankenberg-Dinkel, N. (2005). Biochemical and spectroscopic characterization of the bacterial phytochrome of *Pseudomonas aeruginosa. FEBS Journal.* 272(8), 1927–1936.

22. Nagatani, A. (2004). Light-regulated nuclear localization of phytochromes. *Current Opinion in Plant Biology.* 7(6), 708–711.

23. Hiltbrunner, A., Viczián, A., Bury, E., Tscheuschler, A., Kircher, S., Tóth, R., Honsberger, A., Nagy, F., Fankhauser, C., & Schäfer, E. (2005). Nuclear accumulation of the phytochrome A photoreceptor requires FHY1. *Current Biology.* 15(23), 2125–2130.

24. Huq, E., & Quail, P. H. (2005). Phytochrome signaling. *Handbook of Photosensory Receptors. 15*, 151–170.

25. Matsushita, T., Mochizuki, N., & Nagatani, A. (2003). Dimers of the N-terminal domain of phytochrome B are functional in the nucleus. *Nature. 424*(6948), 571–574.

26. Ni, M., Tepperman, J. M., & Quail, P. H. (1999). Binding of phytochrome B to its nuclear signalling partner PIF3 is reversibly induced by light. *Nature. 400*(6746), 781–784.

27. Oka, Y., Matsushita, T., Mochizuki, N., Suzuki, T., Tokutomi, S., & Nagatani, A. (2004). Functional analysis of a 450–amino acid N-terminal fragment of phytochrome B in Arabidopsis. *The Plant Cell. 16*(8), 2104–2116.

28. Jiao, Y., Lau, O. S., & Deng, X. W. (2007). Light-regulated transcriptional networks in higher plants. *Nat Rev Genet. 8,* 217–230.

29. Bae, G., & Choi, G. (2008). Decoding of light signals by plant phytochromes and their interacting proteins. *Annu Rev Plant Biol.* 59, 281–311.

30. Kim, J. I., Shen, Y., Han, Y. J., Park, J. E., Kirchenbauer, D., & Soh, M. S. (2004). Phytochrome phosphorylation modulates light signaling by influencing the protein-protein interaction. *Plant Cell. 16*, 2629–2640.

31. Kim, J. I., Park, J. E., Zarate, X., & Song, P. S. (2005). Phytochrome phosphorylation in plant light signaling. *Photochem Photobiol Sci., 4*, 681–687.

32. Kim, D. H., Kang, J. G., Yang, S. S., Chung, K. S., Song, P. S., & Park, C. M. (2002). A phytochrome-associated protein phosphatase 2A modulates light signals in flowering time control in Arabidopsis. *Plant Cell. 14*, 3043–3056.

33. Phee, B. K., Kim, J. I., Shin, D. H., Yoo, J., Park, K. J., & Han, Y. J. (2008). A novel protein phosphatase indirectly regulates phytochrome-interacting factor 3 via phytochrome. *Biochem J. 415,* 247–255.

34. Fankhauser, C. (2000). Phytochromes as light-modulated protein kinases. *Semin Cell Dev Biol. 11*, 467–473.

35. Cherry, J. R., Hondred, D., Walker, J. M., & Vierstra, R. D. (1992). Phytochrome requires the 6-kDa N-terminal domain for full biological activity. *Proc Natl Acad Sci USA, 89*, 5039–5043.

36. Stockhaus, J., Nagatani, A., Halfter, U., Kay, S., Furuya, M., & Chua, N. H. (1992). Serine-to-alanine substitutions at the amino-terminal region of phytochrome A result in an increase in biological activity. *Genes Dev. 6*, 2364–2372.

37. Wagner, J. R., Brunzelle, J. S., Forest, K. T., & Vierstra, R. D. (2005). A light-sensing knot revealed by the structure of the chromophore-binding domain of phytochrome. *Nature. 438*, 325–331.

38. Essen, L. O., Mailliet, J., & Hughes, J. (2008). The structure of a complete phytochrome sensory module in the Pr ground state. *Proc. Natl. Acad. Sci. USA, 105*, 14709–4714.

39. Yang, X., Kuk, J., & Moffat, K. (2008). Crystal structure of *Pseudomonas aeruginosa* bacteriophytochrome: photoconversion and signal transduction. *Proc. Natl. Acad. Sci. USA, 105*, 14715–14720.

40. Wu, S. H., & Lagarias, J. C. (2000). Defining the bilinlyase domain: lessons from the extended phytochrome superfamily. *Biochemistry. 39*, 13487–3495.

41. Boylan, M., & Quail, P. H. (1996). Are the phytochromes protein kinase? *Protoplasma. 195*, 12– 17.

42. Blaauw-Jansen, G., & Blaauw, O. H. (1975). A shift of the response threshold to red irradiation in dormant lettuce seeds. *Acta botanica neerlandica. 24*(2), 199–202.

43. Cone, J. W., Jaspers, P. A., & Kendrick, R. E. (1985). Biphasic fluence-response curves for light induced germination of Arabidopsis thaliana seeds. *Plant, Cell and Environment, 8*(8), 605–612.

44. Van Der Woude, W. J. (1985). A dimeric mechanism for the action of phytochrome: evidence from photothermal interactions in lettuce seed germination. *Photochemistry and Photobiology. 42*(6), 655–661.

45. Mandoli, D. F., & Briggs, W. R. (1981). Phytochrome control of two low-irradiance responses in etiolated oat seedlings. *Plant Physiology 67*(4), 733–739.

46. Clough, R. C., Casal, J. J., Jordan, E. T., Christou, P., & Vierstra, R. D. (1995). Expression of functional oat phytochrome A in transgenic rice. *Plant Physiology. 109*(3), 1039–1045.

47. Casal, J. J., & Sanchez, R. A. (1998). Phytochromes and seed germination. *Seed Sci. 8*, 317–329.

48. Drumm, H., & Mohr, H. (1974). The dose response curve in phytochrome-mediated anthocyanin synthesis in the mustard seedling. *Photochemistry and Photobiology, 20*(2), 151–157.

49. Hecht, U., & Mohr, H. (1990). Relationship between phytochrome of photoconversion and response. *Photochemistry Photobiology, 51*, 369–373.

50. Raven, C. W., & Shropshire, Jr. W. (1975). Photoregulation of logarith-micuence-response curves for phytochrome control of chlorophyll formation in Pisum sativum L. *Photochemistry and Photobiology, 21*, 423–432.

51. Horwitz, B. A., Thompson, W. F., & Briggs, W. R. (1988). Phytochrome regulation of greening in Pisum. Chlorophyll accumulation and abundance of mRNA for light harvesting chlorophyll a/b binding proteins. *Plant Physiology. 86*, 299–305.

52. Mösinger, E., Batschauer, A., Apel, K., Schäfer, E., & Briggs, W. R. (1988). Phytochrome regulation of greening in Barley effects on mRNA abundance and on transcriptional activity of isolated nuclei. *Plant Physiology. 86*(3), 706–710.

53. Neff, M. M., Fankhauser, C., & Chory, J. (2000). Light: an indicator of time and place. *Genes Dev. 14*, 257–271.

54. Hennig, L., Funk, M., Whitelam, G. C., & Schafer, E. (1999). Functional interaction of cryptochrome 1 and phytochrome D. *Plant J. 20*, 289–294.

55. Neff, M. M., & Chory, J. (1998). Genetic interactions between phytochrome A, phytochrome B, and cryptochrome 1 during Arabidopsis development. *Plant Physiol. 118*, 27–35.

56. Mockler, T. C., Guo, H., Yang, H., Duong, H., & Lin, C. (1999). Antagonistic actions of Arabidopsis cryptochromes and phytochrome B in the regulation of floral induction. *Development. 126*, 2073–2082.

57. Mazzella, M. A., Cerdan, P. D., Staneloni, R. J., & Casal, J. J. (2001). Hierarchical coupling of phytochromes and cryptochromes reconciles stability and light modulation of Arabidopsis development. *Development, 128*, 2291–2299.

58. Mathews, S., & Sharrock, R. A. (1997). Phytochrome gene diversity. *Plant, Cell and Environment. 20*, 666–671.

59. Smith, H. (2000). Phytochromes and light signal perception by plants—an emerging synthesis. *Nature. 407*(6804), 585–591.

60. Stewart, M. (2007). Molecular mechanism of the nuclear protein import cycle. *Nat. Rev. Mol. Cell. Biol. 8*, 195–208.

61. Stewart, M. (2007). Ratcheting mRNA out of the nucleus. *Mol Cell. 25*, 327–330.

62. Casal, J. J., Davis, S. J., Kirchenbauer, D., Viczian, A., & Yanovsky, M. J. (2002). The serine-rich N-terminal domain of oat phytochrome A helps regulate light responses and subnuclear localization of the photoreceptor. *Plant Physio. 129*, 1127–1137.

63. Nagy, F., & Schafer, E. (2002). Phytochromes control photomorphogenesis by differentially regulated, interacting signaling pathways in higher plants. *Annu. Rev. Plant Bio. 53*, 329–55

64. Yang, D., Seaton, D. D., Krahmer, J., & Halliday, K. J. (2016). Photoreceptor effects on plant biomass, resource allocation, and metabolic state. *Proc Natl Acad Sci USA. 113*, 7667–7672.

65. Thomas, B., Vince-Prue, D. (1996). *Photoperiodism in Plants*. London: Academic Press, 2nd edition 143–179.

66. Lee, Y. S., Yi, J., & An, G. (2016). OsPhyA modulates rice flowering time mainly through OsGI under short days and Ghd7 under long days in the absence of phytochrome B. *Plant Molecular Biology, 91*(4–5), 1–15.

67. Boss, P. K., Bastow, R. M., Mylne, J. S., & Dean, C. (2004). Multiple pathways in the decision to flower: enabling, promoting, and resetting. *The Plant Cell. 16*, S18–S31.

68. Nelson, D. C., Flematti, G. R., Riseborough, J. A., Ghisalberti, E. L., Dixon, K. W., & Smith, S. M. (2010). Karrikins enhance light responses during germination and seedling development in Arabidopsis thaliana. *Proceedings of the National Academy of Sciences, 107*(15), 7095–7100.

69. Shi, H., Wang. X. M. X., Tang, C., Zhong, S., & Deng, X. W. (2015). Arabidopsis DET1 degrades HFR1 but stabilizes PIF1 to precisely regulate seed germination. *Proceedings of the National Academy of Sciences, 112*(12), 3817–3822.

70. Smith, H., & Whitelam, G. C. (1997). The shade avoidance syndrome: multiple responses mediated by multiple phytochromes. *Plant Cell Environ. 20*, 840–844.

71. Kidokoro, S., Maruyama, K., Nakashima, K., Imura, Y., Narusaka, Y., Shinwari, Z. K., Osakabe, Y., Fujita, Y., Mizoi, J., Shinozaki, K., & Yamaguchi-Shinozaki, K. (2009). The phytochrome-interacting factor PIF7 negatively regulates DREB1 expression under circadian control in *Arabidopsis. Plant Physio. 151*, 2046–2057.

72. Wang, F., Guo, Z. Li. H., Wang, M., Onac, E., Zhou, J., Xia, X., Shi, K., Yu, J., & Zhou, Y. (2016). Phytochrome A and B Function Antagonistically to Regulate Cold Tolerance via Abscisic Acid-Dependent Jasmonate Signaling. *Plant Physiology. 170*(1), 459–471.

73. Franklin, K. A., Lee, S. H., Patel, D., Kumar, S. V., Spartz, A. K., Gu, C., Ye, S., Yu, P., Breen, G., Cohen, J. D., & Wigge, P. A. (2011). Phytochrome-interacting factor 4 (PIF4) regulates auxin biosynthesis at high temperature. *Proceedings of the National Academy of Sciences, 108*(50), 20231–20235.

74. Zhong, S., Zhao, M., Shi, T., Shi, H., An, F., Zhao, Q., & Guo, H. (2009). EIN3/EIL1 cooperate with PIF1 to prevent photo-oxidation and to promote greening of *Arabidopsis* seedlings. *Proc. Natl. Acad. Sci. USA. 106*, 21431–21436.

75. Toledo-Ortiz, G., Huq, E., & Rodríguez-Concepción, M. (2010). Direct regulation of phytoene synthase gene expression and carotenoid biosynthesis by phytochrome-interacting factors. *Proceedings of the National Academy of Sciences. 107*(25), 11626–11631.

76. Chalker-Scott, L. (1999). Environmental significance of anthocyanins in plant stress responses. Photochem. *Photobiol. 70*, 1–9.

77. Brödenfeldt, R., & Mohr, H. (1988). Time courses for phytochrome-induced enzyme levels in phenylpropanoid metabolism (phenylalanine ammonia-lyase, naringenin-chalcone synthase) compared with time courses for phytochrome-mediated end-product accumulation (anthocyanin, quercetin) *Planta. 176*, 383–390.

78. Cortés, L. E., Weldegergis, B. T., Boccalandro, H. E., Dicke, M., & Ballaré, C. L. (2016). Trading direct for indirect defense? Phytochrome B inactivation in tomato attenuates direct anti-herbivore defenses whilst enhancing volatile-mediated attraction of predators. *New Phytologist.* doi: 10.1111/nph.14210.

79. Batschauer, A. (1998). Photoreceptors of higher plants. *Planta. 4*, 479–492.

80. Janoudi, A. K., Konjevic, R., Whitelam, G., Gordon, W., & Poff, K. L. (1997). Both phytochrome A and phytochrome B are required for the normal expression of phototropism in Arabidopsis thaliana seedlings. *Physiologia Plantarum 101*(2), 278–282.

81. Kumar, P., & Kiss, J. Z. (2006). Modulation of phototropism by phytochrome E and attenuation of gravitropism by phytochromes B and E in inflorescence stems. *Physiologia Plantarum. 127*(2), 304–311.

82. Kumar, P., Montgomery, C. E., & Kiss, J. Z. (2008). The role of phytochrome C in gravitropism and phototropism in Arabidopsis thaliana. *Functional Plant Biology. 35*(4), 298–305.

83. Casal, J. J., Luccioni, L. G., Oliverio, K. A., & Boccalandro, H. E. (2003). Light, phytochrome signalling and photomorphogenesis in Arabidopsis. *Photochem. Photobiol. Sci 2*, 625–636.

84. Shinomura, T., Uchida, K., & Furuya, M. (2000). Elementary processes of photoperception by phytochrome A for high-irradiance response of hypocotyl elongation in Arabidopsis. *Plant Physio. 122*, 147–156.

CHAPTER 3

SIGNAL TRANSDUCTION IN PLANTS WITH RESPONSE TO HEAVY METALS TOXICITY: AN OVERVIEW

PRASANN KUMAR

Department of Plant Physiology, Institute of Agricultural Sciences, Banaras Hindu University, Varanasi, 221005, U.P., India, E-mail: prasann0659@gmail.com

CONTENTS

ABSTRACT

Heavy metal toxicity is one of the abiotic stresses that cause deleterious health effects both in animals and plants. Owing to their higher reactivity they directly affect growth, development, senescence and energy synthesis process. The signaling network eventually leads to transcriptional activation of metal-responsive gene to combat the stress. *Sorghum* is one of the most important staple foods for the world's poorest and most food-unsecured people across the semi-arid tropics. Root system of *Sorghum vulgare* is fibrous in nature, which leads to accumulation of metals more in the root. But interestingly, grains of the plant remain free from contamination because transfer and transportation ratios remain low for the metals within the plants under field condition.

3.1   INTRODUCTION

Due to rapid industrial development and urbanization, there is high emission of toxic ions into environment [28]. The soil is primary recipient of a myriad of waste products and chemicals. Heavy metals contamination of soils results from the application of phosphate fertilizers, sewage, sludge and wastewater effluents. The signaling responses of heavy metals stress comprises of a signal transduction network which is activated by sensing of heavy metal and synthesis of stress related proteins and signaling molecules. This signaling network eventually leads to transcriptional activation of metal-responsive gene to combat the stress [25]. There are a number of signal transduction pathways that converge on synthesis of stress-related genes and their cross talk, ROS signaling and mitogen activated protein kinase (MAPK) phosphorylation cascade. The metal-dependent gene expression, particularly to Cd, Pb, Al, and Zn may be regulated on multiple levels which involve direct action on DNA, chromatin modifications, transcription, RNA processing, translation and post-translational events. The metal- dependent modulation of gene expression has not received much attention of investigation, through most of such study been focused on the gene or the protein products that confer plant to tolerance to metals. The recent advancements in the field of transcriptomics, proteomics and metabolism have widened over knowledge on metal response in plants. Genes induced by metal stress are classified into two classes: (a) gene encoding for protein gives direct protection against metal stress such as detoxifying enzymes and other functional proteins;

and (b) regulating gene expression and signal transduction i.e., transcription factors and kinases.

3.2 METAL RECEPTORS

Receptor like protein kinase mediates perception of extracellular signals. For instance, in barley, gene coding for lysine motif receptor like kinase (HvLysMR1) was shown to be induced by Cr, Cd and Cu during leaf senescence [31]. Similarly, a putative receptor protein kinase was involved in signaling of Cd by rice roots.

3.2.1 CA-CALMODULIN SYSTEM

The expression of HvLysMR1 is also mediated by Ca level. Ca^{+2} signaling features in responses to a number of abiotic stress signaling. Heavy metals modify the stability of Ca channels thereby increasing calcium flux into the cell. Studies have demonstrated that certain metals like Ni, Cd causes disbalance in intracellular calcium level and thus interfere with calcium signaling by substituting Ca in calmodulin regulation [11]. Cadmium and Cu induce calcium accumulation in rice roots [45]. Ca^{+2} acts as a secondary messenger interacting with calmodulin to initiate the signal leading to down regulation of genes associated with heavy metal transport, metabolosim and tolerance. Similarly, Ca calmodulin is also involved in the response to other heavy metal toxicity, like Ni and Pb; transgenes tobacco expressing NtCBP4 (*Nicotiana tabacum* calmodulin-binding protein) tolerate higher levels of Ni^{+2} but are hypersensitive to Pb^{+2}, indicating foe exclusion of Ni^{+2} but the accumulation of more Pb^{+2} than wild type plants [3].

3.2.2 ROLE OF ROS AND MAPK CASCADE

Cadmium produces reactive oxygen species (ROS) directly via the fenton and Haber-weiss reaction, while indirectly by inhibiting anti-oxidant enzymes. H_2O_2 levels increased in response to Cd and Cu treatment in

Arabidopsis [2, 9, 16, 41], besides Hg treatment in tomato and in response to Mn toxicity in barley [5]. Similarly, the treatment of tobacco cells and Scots Pine Mots with Cd and lupine roots with lead caused H_2O_2 production [30, 33]. The oxidative brust in tobacco due to Cd was shown to be mediated by calmoduline and/or calmoduline-dependent protein. ROS are able to induce anti-oxidant defence mechanism directing via antioxidant responsive element (ARE) found in promoter region of such gene: one such gene is from CAT1 (CATALASE1) whose gene expression was upregulated during Cd toxicity [38]. Cytosolic ascorbate perioxidase such as APX1 & APX2 as well as plastidic iron superoxide dismutase (FSD1) are activated by H_2O_2 and affected by metal stress [38]. ROS are known to activate scavenging mechanism via redox-sensitive transcription factor or via activation of kinase cascades which subsequently activate transcription factors that trigger gene transcription [37]. MAPKs are one of the largest families of serine-threonine kinases in higher plants that transduce extracellular signals regulating cellular processes. The MAPK pathways are activated in response to Cd [15], Cu [45] and iron stress. MAPK cascade is a well-known response in plant subjected to both biotic and abiotic stress. At the end of the cascade of phosphorylation, MAPKS phosphorylated different substrate in different cellular compartments, thus it allows transduction of information to downstream targets. Cd and Cu activated four different MAP kinase (S1MK, MMK2, MMK3 & SAMK) in alfalfa [15]. Cd also induced ATMEKK1 in kinase in Arab [40] and OsMAPK2 in rice [45]. Scientist [45] reported that Cd and Cu-inducedd MAPK activation required involvement of Calmoduline dependent protein kinase (CDPK) and phosphatidyl-inositol 3-kinase (PI3 kinase); this reflects that Cd and Cu signaling pathway induce ROS production as well as calcium accumulation. These signaling pathways eventually converge in the regulation of transcription factors that activate genes needed for stress adoption particularly the genes that activate metal transporters and biosynthesis of chelating compounds.

3.2.3   PHYTOHORMONE SIGNALING

Plant hormones assume a critical role in the adaptation to abiotic stress since there is a regulation of hormone (plant growth regulator) synthesis in the presence of heavy metals. The signaling pathways involving ABA, SA and auxin participate in response to metals; cis-DNA regulatory elements were detected in metal-induced genes [40]. Similarly, auxin-responsive mRNA was detected in Cd-treated *Brassica* plants [10]. Cd and Cu activate synthesis of ethylene by up-regulating ACC synthase expression and activity. These heavy metals also shown to induce accumulation of Jasmonic acid in *Phaseolus* plant [29, 32, 35, 36]. SA protected the barley roots from lipid perioxidaion caused by Cd tocxicity [27]. The transcription activation of SAMT gene associated with SA synthesis was detected in pea plants subjected to Hg.

3.2.4 NITRIC OXIDE SIGNALING

Nitric oxide (NO) is an important signaling molecule influencing plant development process and defense response [6]. Cd-treated soyabean cells have been shown to release NO, thus demonstrating role of NO in alleviating heavy metal induced stress [23]. There was NO accumulation observed in plastids of *Arabidopsis* cells exposed to Fe; NO acts denovostream of Fe and upstream of P2A-type phosphatase to promote the increase of mRNA coding for Ferritin (*AtFer1*) [4] (Table 3.1).

3.3 GENES CONTROLLING METAL-DEPENDENT TRANSCRIPTION

The transcription profiling of plats treated with Cd, Pb and Zn indicate for the existence of metal-induced transcription factors [24]. Cd-induced transcripts for bZIP (basic region leucne zipper), Myb and zinc finger

TABLE 3.1 Examples of Genes and Proteins Induced in Plants by Heavy Metals

Genes/proteins induced	Metal involved	References
Ish, gsh2, gr1-glutathione metabolic genes	Cu, Cd	[42]
PR-1 protein caused by SA, TMV	Cu	[44]
OsMSRMK2-multiple-responsive MAPKgene induced by elicitor, drought, H_2O_2, ABA	Cu, Hg, Cd	[1]
CSAP-Cd-stress associated 51kDa protein	Cd	[28]
HSP	Hg	[7]

transcription factors were detected in *Arabidopsis thaliana,* and *Brassica juncea* [10, 39]. The multiple MYB genes of *Arabidopsis* not only expressed in response to Cd but also salinity stress and phytohormones like ABA, GA, IAA, JA, SA and ethylene. The Cd-induced bZIP transcription factor-OBF5) in *Arabidopsis* binds to promoter region of glutathione transferase gene-GST6, which is induced by auxin, SA and oxidative stress [39]. In another study, it was demonstrated that Zn treatment of *Arabidopsis* induced transcription fators-bHLH, while the expression of other transcription factor –WRKY and GATA-type decreased in the presence of excess of Zn [9, 16, 26, 29, 36]. It is likely that ROS plays significant role in activating metal-induced transcription factor in plants. Organism like yeast and animals posses specific metal-induced transcription factor (MTF), which bind to metal responsive element (MRE) present in promoters of metal-responsive genes [13]. The cis-acting elements associated with MRE are found within promoters of a few plants genes inducing metallothionein-like genes [31, 33]. There are two cis-DNA elements reported in plants to be functional in metal response: the iron-dependent regulatory sequences (IDRS) mediates iron-regulated transcription of genes related to iron acquisition, and the other one identified within the promoter region of PvSR2 gene from *Phaseolus vulgaries* [33], PvSR2 gene encodes a heavy metal stress related protein whose expression is strongly stimulated by Cd, As, Hg and Cu [46]. Apart from the effect of heavy metal on transcription of specific genes, there also affect the transcription process indirectly by substitution of Zn in RNA polymerase or in Zn-requising transcription factors [34]. Several metals such as Mn, Ni, Mg, Cu, Co, Cd, Pb bind to DNA directly causing conformational changes [8]; these changes impair the genes present within the DNA region affecting the transcription related events. For instance, Cr stress altered gene expression in animal cells by forming chromium-DNA adducts and Cr-DNA crosses links, thereby disturbing the transcription activator complexes [40]. Such metal induced to damage to DNA have been reported in Cd-treated soyabean, tobacco, broad bean [12, 38], Pb-treated lupine and Al-treated barley. Heavy metals are also known to alter DNA repair processes; process; Cd interaction with with DNA inhibiting mismatch repair in yeast and human cells. This inactivation of DNA repair machinery restricts DNA to exhibit its basic coding function in plants.

3.4 CHROMATIN MODIFICATION

The chromatin structure can be influenced by position of nucleosomes on DNA through ATP dependent remodeling complex. Such re-positioning of nucleosomes to other part of DNA exposes the binding site resulting in gene expression. Induction of transcripts homologous to coding for ATP-dependent chromatin remodeling (SWI/SNF) and putative histone deacetylase (PDAC) was detected in Cd-treated Brassica plant [10] as well as Al-treated Saccharum species, indicating for the metal-induced chromatin modifications in plants. The chromation modification can also be manifested by post-translational modification of histone protein such as acetylation, methylation, phosphorylation and ubiquitination (Figure 3.1). Instances are known whereby chromation modification was effected by replacing major histone proteins with their variants [14]. Ni and Cu inhibit histone acetylation via action of HAT (histone acetyltransferases) enzyme, which is linked to silencing of gene expression [17, 43]. Besides, the post-translational process involving microRNA plays a key role in

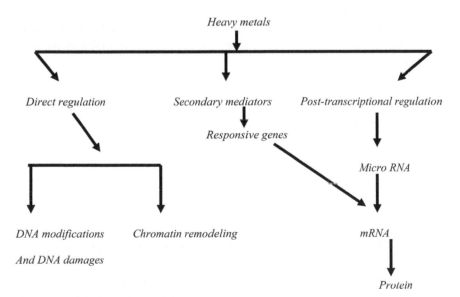

FIGURE 3.1 Schematic representation of modulation of gene expression in plants subjected to heavy metal (*Source*: drawn by the author, unpublished, 2016).

regulation of at least one plant metal-dependent gene, i.e., Cu, Zn-SOD [42] (Figure 3.1).

3.5   HEAVY METAL CHELATION IN THE CYTOSOL

Morton and Drew [47] gives the word "chelate," subsequent from the Greek term chela, meaning "great claw" of the lobster or other crustaceans. The word chelate suggests the approach in which an organic compounds "clamps" onto the cationic elements, which it chelates. In order for a compound to be called a true chelating agent, it must have certain chemical characteristics. This chelating compound must consist of at least two sites capable of donating electrons (coordinate covalent bond) to the metal it chelates. For true chelation to occur the donating atoms(s) must also be in a position within the chelating molecule so that a formation of a ring with the metal ion can occur. The term sequestered deals additional with the accomplishment of chelation or complexing, not the actual chemical understanding or meaning. The term "complexed" originates from combinations of minerals and organic compounds that do not gather the guidelines of a true chelate.

There are five categories of compounds that are usually diverse with minerals and used in agricultural foliar and soil applied applications:
1. Synthetic chelates
2. Ligno sulfonates
3. Humic or fulvic acids
4. Organic acids
5. Protein (amino acids)

3.5.1   SYNTHETIC CHELATING AGENTS

The most ordinary form is EDTA that is typically worn in agricultural mineral formulations as the disodium salt of EDTA. Synthetically chelated minerals among the strongest forms of chelation used in commercial agricultural applications. Usually used forms of minerals are: calcium EDTA, magnesium EDTA, zinc EDTA, manganese EDTA, and copper EDTA. Forms of synthetic iron chelates diverge rather, according to the circumstance in which they are utilized. One of the accepted forms is iron EDDHA (ethylenediamine [di(o-hydroxyphenyacetic) acid]). The efficiency of artificial chelating agents for foliar application, it is recognized that in a nutrient solution it is necessary to have an surplus of overall mineral in relation to chelating agent so that the chelating agent does not participate for the mineral with the plant (Table 3.2).

3.5.2   LIGNO SULFONATES

Ligno sulfonates are measured to be a water soluble and non-toxic polymer. Polymer typically means that the molecule is reasonably large. Lignin, a most important polymeric constituent of woody tissues in higher plants, is composed of repeating phenyl propane units and usually amounts to 20–30% of the dry weight of wood. It is a consequence of the pulp and paper industry. The lignin resulting from soft woods is dissimilar that from hardwoods, consequently, their reactivity differs fairly. Ligno sulfonates and mineral combinations are fairly frequently referred to as sequestered or complexed. With all the obtainable sites on the lingo sulfonate molecule, it is very probable that there would be 2 or 3 obtainable to make a true chelate; though, the size of molecule

TABLE 3.2   Advantages and Disadvantages of EDTA

Advantages	Disadvantages
EDTA chelates are extremely stable and can be speckled with just about any phosphate containing fertilizer at basically any pH.	EDTA is a synthetic compound, not formed by plants.
EDTA is incredibly resistant to microbial degradation; consequently it remains quite constant in soils.	Synthetic chelating agents can compete with the plant for a mineral. EDTA is known for its strong affinity toward calcium and although debatable, some think that because of its insistence in a plant, it is probable that a zinc EDTA foliar applied, could be winning for zinc, but the EDTA could fight for calcium once in the plant.

Source: Refs. [18–20].

to number of potential chelate sites makes these molecule incompetent chelators. Even though it is likely to make extremely effectual mineral formulations from lingo sulfonate, the majority of commercially utilized sources of this material are not formulated for effectual foliar uptake (Table 3.3).

3.5.3 HUMIC OR FULVIC ACIDS

In foliar applications, the fulvic acids in NUE amino minerals may not play a straight role in accessibility and movement of minerals in the plant. Though, they accomplish act as plant stimulants and perhaps precursors to plant hormones (Table 3.4).

TABLE 3.3   Advantages and Disadvantages of Ligno Sulfonate

Advantages	Disadvantages
The best advantage of lingo based minerals is its lower cost.	The main disadvantage of lingo sulfonate for foliar application is the size of the molecule. There are many different grades of lignin and thus different molecular weights. Some are as high as 21,000.
Another advantage is the polymer is biodegradable and supportive of soil microorganism.	With proper sulfonation, lower MW lignin sulfonates do exist.
Ligno sulfonate also has surfactant properties.	

Source: Ref. [22].

TABLE 3.4   Advantages and Disadvantages of Humic Acid and Fulvic Acid

Advantages	Disadvantages
Humic and fulvic acids are price effectual compounds to adjoin to foliar and soil applied minerals	In foliar applications humic has a definite disadvantage of being too large in size
When they are "small" adequate to efficiently move into the plant, they can provide a valuable resource of precursors to significant plant chemicals	In trace mineral mixtures humic molecules be inclined to resolve out in the container, mostly due to their molecular size and of their reduced suspension of these complex colloids in water at low pH

Source: Refs. [18–20].

3.5.4   ORGANIC ACIDS

These compounds are one of the two groups of compounds that are essential for the transportation and solubility of divalent elements in plants. The organic acids can be called "anionic organic acids" because of their negative charge. One such organic acid in agriculture is citric acid. Although organic acids do not directly chelate monovalent elements, they are associated in plant systems by loosely held ionic attractions.

In foliar applications, these anionic organic acids play an important role in converting cationic minerals into nonionic compounds for increased leaf penetration and movement into plant. Plants have the ability to manufacture many types of organic acids such as: citric, malic, fumaric, succinic, and others. Some of these acids are utilized to transport minerals in the plant. Other are excreted the roots to solublize and take needed minerals into the plant system. NUE amine minerals contain a group of these natural organic acids, derived from a fermentation process (Table 3.5).

3.5.5   PROTEIN (AMINO ACIDS)

Amino acids are the second type of natural compounds that plants produce to solublize and translocate minerals. Plants manufacture these

TABLE 3.5   Advantages and Disadvantages of Organic Acid

Advantages	Disadvantages
Organic acids are one of the usual systems a plant utilizes to solublizing and translocate minerals.	Commercially produced individuals organic acids are quite expensive compared to some other chelating and complexing agents
The groups of organic acids in NUE Amino minerals have been proven in field demonstrations to be more effective in correcting mineral deficiencies on some crops than mineral applications without these compounds.	Compared to synthetic chelating agents and some amino acids chelating compounds organic acids chelates form weaker bonds with minerals, particularly at rising pH levels.
By neutralizing the charge of minerals, organic acids create mineral uptake additional effective.	

Source: Refs. [18–21].

compounds to make mineral biologically available in the cell. Even when unchelated or uncomplexed minerals are sprayed on plants, the mineral must combine with an organic compounds such as an amino acids before the plant can effectively utilize it. As with organic acids, amino acid compounds also play a role in mineral uptake into plant tissue, because of the increase permeability effect of the amino acid on the cuticle (Table 3.6).

Inside the cell, heavy metal ions that are not instantly necessary metabolically may attain toxic concentrations and plant cells have evolved various mechanisms to store excess metals to stop their participation in unwanted toxic reactions. If the toxic metal concentration exceeds a certain threshold inside the cells, an active metabolic process contributes to the production of chelating compounds. Specific peptides such as PCs and MTs are used to chelate metals in the cytosol and to sequester them in specific sub-cellular compartments. A large number of small molecules are also involved in metal chelation inside the cells, including organic acids, amino acids and phosphate derivatives.

3.5.6 PHYTOCHELATINS

PCs are the best-characterized heavy metal chelators in plants, particularly in the context of Cd tolerance. PCs are a family of metal-binding

TABLE 3.6 Advantages and Disadvantages of Amino Acids

Advantages	Disadvantages
Amino acids are one of the natural systems a plant utilizes to translocate and utilize minerals.	Enzymatically hydrolyzed protein is expensive to produce.
The groups of amino acids in NUE Amino minerals have been proven in field demonstrations to be more effective in correcting mineral deficiencies on some crops than mineral applications without these compounds.	The actual combination of properly prepared amino acid/peptide compounds and minerals is not 100% complete with simple tank mix procedures.
By neutralizing the charge of minerals, amino acids make mineral uptake more effective.	

Source: Refs. [18–21].

peptides with the general structure (γ-Glu-Cys)$_n$Gly ($n = 2 - 11$). They are present in plant and fungi. The cysteine thiol groups allow PCs to chelate metals and form complexes with a molecular weight of 2.5–3.6 kDa. PCs are synthesized in the cytosol and then transported as complexes to vacuole. Their synthesis is fast activated in the presence of heavy metals such as Cd, Cu, Zn, Ag, Hg and Pb. Synthesis involves the chain extension of GS by PCS a constitutively expressed cytosolic enzyme whose activity is controlled post-translationally because the metal ions chelated by PCs are required for enzyme activity. Due to their metal likeness, PCs are thinking to be involved in the homoeostasis and trafficking of essential metals such as Cu and Zn and in the detoxification of heavy metals, but they do not seem to be involved in Hyperaccumulation. Confirmation has been reported for the role of PCs in heavy metal tolerance. They have a clear function in the reply of plants and yeast to Cd, e.g., they are induced rapidly in *Brassica juncea* following the intracellular accumulation of Cd, thus protecting the photosynthetic apparatus despite a decline in transpiration and leaf expansion. In addition, the Cd compassion of various *A. thaliana* mutants correlated with their ability to accumulate PCs. Cd and Cu treatment also induces the transcription of genes involved in the synthesis of GS, the precursor of PCs.

3.5.7   METALLOTHIONEINS AND FERRITINS

MTs have been established in many organisms, even though the MTs in plants differ considerably from those found in mammals and fungi. As they contain mercaptide groups they are able to bind metal ions. In plants, MTs are induced by a variety of abiotic stresses but are also expressed during development. In wheat and in rice, MTs are induced by metal ions, such as Cu and Cd and by abiotic stresses such as temperature extremes and nutrient deficiency. Plant MTs sequester surplus of metals by coordinating metal ions with the multiple cysteine thiol groups and have particular affinity for Zn^{+2} and Cu^{+2}. MTs possibly have diverse functions in response to dissimilar heavy metals and could also contribute in additional antioxidant defense mechanism and plasma membrane repair. Ferritins are ever-present multimeric proteins that can store up to 4500 iron atoms in a central cavity. Animal ferritins can also store other metals, including Cu, Zn, Cd, and Al, whereas plant Ferritins has only been exposed to store Fe. Plants ferritins are synthesized in responses to various environmental stresses, including photoinhibition and iron overloading. Ferritins are therefore a front-line defense mechanism against free iron-induced oxidative stress. The chief function of plant ferritins is not to store and release iron, as previously reported, but to scavenge free reactive iron and prevent oxidative damage.

3.5.8   ORGANIC ACIDS, AMINO ACIDS, AND PHOSPHATE DERIVATIVES

Organic acids and amino acids can attach heavy metals and may consequently be deployed in reaction to metal toxicity. Organic acids such as malate, citrate and oxalate confer metal tolerance by transporting metals through the conducting tissue xylem and sequestrating ions in the vacuole, but they have multiple additional roles in the cell. Citrate, which is synthesized in plans by enzymes citrate synthase, has an elevated ability for metal ions than malate and oxalate and even though its principle role in chelate Fe^{+2}. It also has a strong affinity also for Ni^{+2} and Cd^{+2}. Malate is a cytosolic Zn-chelator in zinc-tolerant plants. Amino acids and derivatives are capable to chelate metals conferring to plants resistance to toxic levels of metal ions. Histidine is measured the most important free amino acids in heavy metal metabolism. Due to existence of carboxyl, amino and imidazole groups, it is a flexible metal chelator, which can confer Ni tolerance and enhance Ni transport in plants when supplied in the growth medium, possibly reflecting its normal role as a chelator in root exudates. Phytate (myo-inositol hexabisphosphate) is the standard form of preserve phosphorous in plants and is frequently localized in the roots and seeds. The molecule comprises six phosphate groups, which permit the chelation of multiple cations, including Ca^{+2}, Mg^{+2}, and K^+, but also Fe^{+2}, Zn^{+2}, and Mn^{+2}. The allocation of phytates and its capacity to chelate multiple metal species propose it could be mobilized as a detoxification strategy. Extraction methods are mainly based on the ratio of a given reagent (extractants) to a given weight of soil sample, and on the chemical feature of the extractants, i.e. with or without chelating agent. Some extractants commonly extract a wide range of soil elements while others are applied for single elements. Amount of Pb extracted were poorly correlated with soil physicochemical properties, irrespective of the type of extractants. The multiple regression analysis established for EDTA gave the best fit. The accessibility to plants of soil elements is ruled by more than a few natural factors such as soil reaction, soil physicochemical properties, temperature and climate. Redundant side reactions can be made more selective for specific elements using complexions.

3.6 CONCLUSION

Plants have evolved miscellaneous resources to support physiological concentrations of essential metal ions and to decrease contact to non-essential heavy metals. A number of method are ubiquitously since they are also essential for universal metal homeostasis and they reduce injure caused by eminent concentration of heavy metals in plants by detoxifications, consequently conferring tolerance to heavy metals stress. Many plants naked to lethal concentrations of metal ions effort to avoid or decrease uptake into root cells by restricting metal ions apoplast, binding them to the cell wall or to cellular exudates, or by inhibiting long distance convey. If this fails, metals before in the cell are addressed using a choice of storage space and detoxification strategies, counting metal transport, chelation, trafficking and sequestration into the vacuole. When these tackle are worn-out, plants turn on oxidative stress protection mechanisms and the amalgamation of stress-related proteins and signaling molecules, such as heat shock proteins, hormones and reactive oxygen species. When metal ions are accumulated in surplus within the cytosol, plants have to take away them in order to diminish their toxic belongings. Plants react to high intracellular concentrations of metal ions by means of efflux pumps either to export the ions to the apoplast or to compartmentalize them within the cells. The major storage compartment for metal ions is the vacuole, which in plants accounts for up to 90% of the volume. A number of families of intracellular transporters concerned in this procedure have been recognized in plants and yeast and they come into view to be extremely discriminating. The reply to heavy metal stress involves a complex signal transduction system that is activated by sensing the heavy metal and is characterized by the production of stress-related proteins and signaling molecules and lastly the transcriptional commencement of precise metal-responsive genes to offset the stress. The relevant signal transduction passageway includes the Ca-calmodulin system, hormone, ROS signaling and the mitogen-activated protein kinase (MAPK) phosphorylation cascade. Dissimilar signaling pathways may be used to respond to dissimilar heavy metals.

ACKNOWLEDGMENT

Author is thankful to the University Grants Commission for providing Senior Research Fellowship to the author.

KEYWORDS

- **agriculture**
- **basic**
- **cadmium**
- **DNA**
- **element**
- **ferritins**
- **G-protein**
- **histidine**
- **insoluble**
- **junction**

REFERENCES

1. Agarwal, M. (2003). Molecular characterization of rice hsp101: complementation of yeast hsp104 mutation by disaggregation of protein granules and differential expression in indica and japonica rice types. *Plant Molecular Biology, 151*(4), 543–553.

2. Agarwal, M., Sahi, C., Katiyar-Agarwal, S., Agarwal, S., Young, T., Gallie, D. R., Sharma, V. M., Ganesan, K., & Grover A. Journal Article, Research Support, Non-U.S. Gov't, Research Support, U.S. Gov't, Non-P.H.S.

3. Arazi, T., Sunker R., & Kalpan B., (1999). A tobacco plasma membrane calmodulin-binding transporter confers Ni^{+2} tolerances and Pb^{+2} hyypersensitivity in transgenic plants. *Plant Journal. 20*, 171–182.

4. Arnaud, N., Murgia, I., Boucherez, J., Briat, J. F., Cellier, F., & Gaymard, F. (2006). An iron-induced nitric oxide burst precedes ubiquitin-dependent protein degradation for *Arabidopsis AtFer1* ferritin gene expression. *J. Bio. Chem. 281*, 23579–23588.

5. Cho. U. H., & Par J. O. (2000). Mercury-induced oxidative stress in tomato seedlings. *Plant Sci., 156*, 1–9.

6. Crawford, N. M., & Guo, F. Q. (2005). New insights into nitric oxide metabolism and regulatory functions. Trends Plant Sci., *10*, 195–120.

7. Didierjean, L., Frendo, P., Nasser, W., Genot, G., Marivet, J., & Burkard, G. (1996). Heavy-metal-responsive genes in maize: identification and comparison of their expression upon various forms of abiotic stress. *Planta. 199*(1), 1–8.

8. Duguid, J., Bloomfield, V. A., Benevides, J., & Thomas, Jr., G. J. (1993). Raman spectroscopy of DNA metal complexesI, interaction and conformational effects of the divalent cation: Mg, Ca, Sr, Ba, Mn, Co, Ni, Cu, Pd and Cd. *Biophysics J., 65*, 1916–1928.

9. Fidalgo, F., Freitas, R., Ferreira, R., & Pessoa, A. M. J. (2011). Teixeira. *Environ. Exp. Bot., 72*, 312–319.

10. Fusco, N., Micheletto, L., Dalcorso, G., Borgato, L., & Furin, A. (2005). Identification of cadmium-regulated genes by cDNA-AFLP in the heavy metal accumulor *Brassicas juncea* L. *J. Exp. Bot., 56*, 3017–3027.

11. Ghelis, T., Dellis, O., Jeannette, E., Bardat, F., Miginiac, E., & Sotta B. (2000). Abscisic acid plasmalemma perception triggers a calcium influx essential for RAB18 gene expression in Arabidopsis thaliana suspension cells. *FEBS Lett. 483*(1), 67–70.

12. Gichner, T., Ptácek, O., Stavreva, D. A., Plewa, M. J. (1999). Comparison of DNA damage in plants as measured by single cell gel electrophoresis and somatic leaf mutations induced by monofunctional alkylating agents. *Environ. Mol. Mutag., 33*, 279–286.

13. Giedroc, D. P., Chen, X., Apuy, J. L. *(2001)*. Metal response element (MRE)-binding transcription factor-1 (MTF-1): structure, function, and regulation. *Antioxid. Redox. Signal, 3*(4), 577–596.

14. Hsieh, T. F., & Fischer, R. L. (2005).Biology of chromatin dynamics. *Annu. Rev. Plant Biol., 56*, 327–51.

15 Jonak, C., Nakagami, H., & Hirt, H. (2004). Heavy metal stress: activation of distinct mitogen-activated protein kinase pathways by copper and cadmium. *Plant Physiol., 136*(2), 3276–3283.

16. Judith, E., van de Mortel, Ric C. H., de Vos, Ester Dekkers, Ana Pineda, Leandre Guillod, Klaas Bouwmeester, Joop J. A., van Loon, Marcel Dicke, & Jos M. Raaijmakers. (2012). Metabolic and Transcriptomic Changes Induced in Arabidopsis by the Rhizobacterium *Pseudomonas fluorescens* SS101. *Plant Physiol., 160*(4), 2173–2188.

17. Kang, S. A., Marjavaara, P. J., & Crane, B. R. (2004). Electron transfer between cytochrome c and cytochome c peroxidase in single crystals. *J Ame. Chem. Soc., 126*(35), 10836–10837.

18. Kumar, P. (2012a). Empowerment of the Indian dry land agriculture in the face of climate change. *Agri. Update.* 7(3 & 4), 453–460.
19. Kumar, P. (2013). Cultivation of traditional crops: an overlooked answer. *Agri. Update.* 8(3), 504–508.
20. Kumar, P., & Dwivedi, P. (2014). Phytoremediation of cadmium through sorghum. In *Recent Advances in Crop Physiology* (Ed. Amrit Lal Singh). Daya Publishing House, pp. 311–342.
21. Kumar, P. (2012b). Feeding the future: crop protection today. *Acta Chem. & Farm.* 2(4), 231–236.
22. Kumar, P., & Dwivedi, P. (2011). Land use policy is the key driver for biodiversity management. Intl. J. Agric. Environ. Biotech. 4(4), 291–297.
23. Kopyra, M., Stachoń-Wilk, M., & Gwóźdź, E. A. (2006). Effects of exogenous nitric oxide on the antioxidant capacity of cadmium-treated soybean cell suspension. *Acta Physiol. Plant, 28,* 525–536.
24. Kovalchuk, I., Abramov, V., Pogrbhy, I., & Kovalchuk, O. (2004). Molecular aspects of plant adaptation to life in the Chernobyl zone. *Plant Physiol., 135,* 357–363.
25. Maksymiec, W., Wójcik, M., & Krupa, Z. (2007). Variation in oxidative stress and photochemical activity in *Arabidopsis thaliana* leaves subjected to cadmium and excess copper in the presence or absence of jasmonate and ascorbate. *Chemo., 66*(3), 421–427.
26. Metwally, A., Finkermeier, I., Georgi, M., Dietz, K. J. (2003). Salicylic acid alleviates the cadmium toxicity in barley seedlings, *Plant Physiol., 132,* 272–281.
27. Mitra, B., Ghosh, P., Henry, S. L., Mishra, J., Das, T. K., Ghosh, S., & Mohanty, P. (2004). Novel mode of resistance to Fusarium infection by a mild dose pre-exposure of cadmium in wheat. *Plant J., 48*(4), 485–498.
28. Nriagu, J. O., & Pacyna, J. M. (1998). Quantitative assessment of worldwide contamination of air, water and soils by trace metals. *Nature, 333*(6169), 134–139
29. Olmos, E., Reiss, B., & Dekker, K. (2003). The ekeko mutant demonstrates a role for tetraspanin-like protein in plant developments. *Biochem. Biophy. Res. Communi., 310,* 1054–1061.
30. Ouelhadj, D., & Petrovic, S. (2009). A survey of dynamic scheduling in manufacturing systems. *J. Sche., 12*(4), 417–431.
31. Perry, N. B., Anderson, R. E., Brennan, N. J., Douglas, M. H., Heaney, A. J., & McGimpsey, J. A., (1999). Small field, B. M. Essential Oils from Dalmatian Sage (*Salvia officinalis* L.): Variations among Individuals, Plant Parts, Seasons, and Sites. *J. Agric. Food Chem., 47,* 2048–2054.
32. Przymusinski, R., & Gzyl. J. (2013). Immunohisto- and cytochemical localization of PR-10 proteins induced by heavy metals in lupine roots. Acta Physiol. *Plant, 35,* 1707–1711.
33. Qi, X., Zhang, Y., & Chai, T. (2007). Characterization of novel plant promoter specifically induced by heavy metal and identification of the promoters regions conferring heavy metal responsiveness. *Plant Physiology, 143,* 50–59.
34. Robinson, B. H., Brooks, R. R., Howes, A. W., Kirkman, J. H., & Gregg, P. E. H. (1997b). The potential of the high biomass *Berkheya coddii* for phytoremediation and phytomining. *J Geochem. Expl., 60,* 115–126.

35. Rodríguez-Serrano, M., Romero-Puertas, M. C., &Pazmiño, D. M. (2009). Cellular response of pea plants to cadmium toxicity: cross talk between reactive oxygen species, nitric oxide, and calcium. *Plant Physiology, 150*(1), 229–243.

36. Scandalios, J. G. (2005). Oxidative stress: molecular perception and transduction of signals triggering antioxidant gene defenses. Brazallian. *J Med. Biol. Res., 38*(7), 995–1014.

37. Smeets (2008). Cadmium-induced transcriptional and enzymatic alterations related to oxidative stress. *Environment Experiment Botany, 63*(1–3), 1–8.

38. Sobkowiak, R., Rymer, K., Rucińska, R., & Deckert, J. Cadmium-induced changes in antioxidant enzymes in suspension culture of soybean cells. *Acta Biochim Pol. 51*(1), 219–222.

39. Suzuki, N., Buechner, M., Nishiwaki, K., Hall, D. H., Nakanishi, H., Takai, Y., Hisamoto, N., & Matsumoto, K. (2001). A putative GDP-GTP exchange factor is required for development of the excretory cell in Caenorhabditis elegans. *EMBO Report, 2,* 530–535.

40. Wei, F., Chen, J., Wu, Y. (1991). Study on the soil background value in China. *Environmental Science, 12*(4), 12–19.

41. Xiang, C., Werner, B. L., Christensen E. M., & Oliver D. J. (2001). The biological functions of glutathione revisited in Arabidopsis transgenic plants with altered glutathione levels. *Plant Physiology, 126,* 564–574.

42. Yamasaki, M., Wright, S. I., & McMullen, M. D. (2007). Genomic screening for artificial selection during domestication and improvement in maize. *Annals of Botany, 100,* 967–973.

43. Yang, P., Chen, C., Wang, Z., Fan, B., & Chen, Z. (1999). A pathogen- and salicylic acid-induced WRKY DNA-binding activity recognizes the elicitor response element to the tobacco class I chitinase gene promoter. *The Plant Journal, 18,* 141–149.

44. Yeh, C. M., Chien, P. S., & Huang H. J. (2007). Distinct signaling pathways for induction of MAP kinase activities by cadmium and copper in rice roots. *Journal of Experimental Botany, 58*(3), 659–667.

45. Zaitao, P., Raymond, W., Arritt, Eugene, S., Takle, William, J., Gutowski, Jr. Christopher, J., & Moti, S. Anderson, (2004). Altered hydrologic feedback in a warming climate introduces a "warming hole." *Geophysical Research Letter. 31*(17), 1-4.

46. Zhang, B., Egli, D., Georgien, O., & Schaffner, W. (2001b). The drosophila homology of mammalian zinc finger factors MTF-1 activates transcription in response to heavy metals. *Molecular Cell Biology, 21,* 4505–4514.

CHAPTER 4

VEGETABLE OIL AND ITS SIGNIFICANCE IN SUSTAINABLE DEVELOPMENT

A. HASNAT

Natural Products and Polymer Research Laboratory, Department of Chemistry, G.F. College (Affiliated to MJP Rohilkhand University), Shahjahanpur – 242001, U.P., India

CONTENTS

ABSTRACT

Renewable resources provide a suitable platform to many sustainable developments as they have ability to grow again and again. The utilization of renewable raw materials is taking the advantage of synthetic potential of nature built-in-design for biodegradability and low toxicity towards human being. Today, due to interdisciplinary approaches through research and technological innovation in synthetic chemistry, biosciences and engineering, it is possible to architect the specialty biomaterials from the spectrum of nature. Among different renewable resources vegetable oils obtained from the seeds of various plants represents a promising class of raw materials for chemical industries, owing to their universal availability, sustainability, biodegradability and excellent environmental credential. The vegetable oils are triglyceride of fatty acids. The fatty acids constitute about 94–96% of the total weight of the one mole of the vegetable oils and largely govern the properties. The fatty acid compositions of vegetable oils mainly vary with the plants; however, climatic conditions, environments and growing conditions are also responsible for notable variations. In present chapter efforts have been made to discuss the various physicochemical characterizations of vegetable oils and inference of the results. The spectral characterizations of vegetable oils are highlighted. Characteristic IR bands for the different functional groups present in triglyceride oils and its derivatives are summarized in tabular form. Position of signals, their integral values and splitting patterns in ^{1}H-NMR spectra are very useful to assign the fatty acids in the particular vegetable oil. The δ values in ppm for different types of protons present in the vegetable oils and their derivatives are provided. Several processes for the epoxidation of vegetable oils including enzymatic have been highlighted in view to optimize the various parameters like type of catalyst, temperature, molar ratios of the reactants. The direct and indirect utilizations of epoxidized oils in development of different practicable materials as a precursor of renewable resource for various industrial arenas are provided. In addition to these, materials derived from vegetable oils with antimicrobial properties also enlightened at the end of the chapter.

4.1 INTRODUCTION

The use of materials derived from the spectrum of nature is becoming important due to both environmental and sustainability. The sustainable development means development that meets the need of present without compromising the ability of future generations to meet their own needs [1,

2]. The conservation and management of renewable resources for the development through environment friendly routes is the main focus of interest [3, 4]. Renewable resources can provide a suitable platform to many sustainable developments as they have ability to grow again and again. Furthermore their productions can increases whenever required by more cropping of particular species. The utilization of renewable raw materials is taking the advantage of synthetic prospective of nature built-in-design for biodegradability and low toxicity towards human being. From the time of immemorial plants have been wieldy used for a variety of industrial developments, curing of ailments in addition to their non-ignorable contribution to environmental balance [5, 6]. Copious agriculture based materials have been utilized to design the bio-based recipes of colossal utility. Some common examples of renewable raw materials used as precursors of different valuable bio-based materials are starch, cellulose, cashew nut shell liquid (CNSL), lignin, chitosen, lactic acid, bagasses, wool fiber, proteins, vegetable oil, and many others [4, 6–9]. Among these bio-based resources vegetable oil is chiefly focused by the academicians and scientist due to their universal abundance at low cost, superb environmental credentials and functionalities [10–12]. Now-a-days oil bearing plantation is being raised and modified genetically for use in different areas like folk medicines, biodiesel, cosmetics, composites and polymeric resins. Vegetable oils are triglycerides of fatty acids basically extracted from the different parts of plants and have been utilized by human being for the ancient time in medicines, edible and non-edible purposes [5, 13, 14]. The fatty acid compositions of vegetable oils mainly vary with the plants; however, climatic conditions, environments and growing conditions are also responsible for variations [1, 5, 15]. The fatty acids constitute about 94–96% of the total weight of the one mole of the vegetable oils. Among the different fatty acids some of them are saturated and some are unsaturated. The saturated fatty acids do not have the double bonds between carbon-carbon. Unsaturated fatty acids have one or more than one double bonds between carbon-carbon. Furthermore, double bonds are conjugated or isolated. If double bonds and single bonds are arrange in alternate ways then systems are called conjugated where as if double bonds are separated by two or more than carbons then systems are called isolated (Figure 4.1).

The nature of fatty acids present in the vegetable oils largely governs the properties and plays the pivotal role in the different applications

$\sim\sim$-CH$_2$-CH$_2$-CH$_2$-CH$_2$-CH$_2$-CH$_2$-CH$_2$-$\sim\sim$

(a)

$\sim\sim$-CH$_2$-CH=CH-CH=CH-CH$_2$-CH$_2$-$\sim\sim$

(b)

$\sim\sim$-CH$_2$-CH=CH$_2$-CH$_2$-CH=CH$_2$-CH$_2$-CH$_2$-CH$_2$-$\sim\sim$

(c)

FIGURE 4.1 Hydrocarbon chain of different fatty acids: (a) saturated, (b) conjugated, and (c) isolated.

[16]. A general structure of the triglyceride oil and common fatty acids comprising different vegetable oils depicted in Figure 4.2(a and b). Vegetable oils have been used as a binder or additive in paints and coatings for the many centuries, even in the days of cave paintings [2]. Oils alone do not fulfill the desirable properties and hence to value addition they are further modified according to requirements and service conditions [17]. Triglyceride oil has built in various functionalities and reactive site

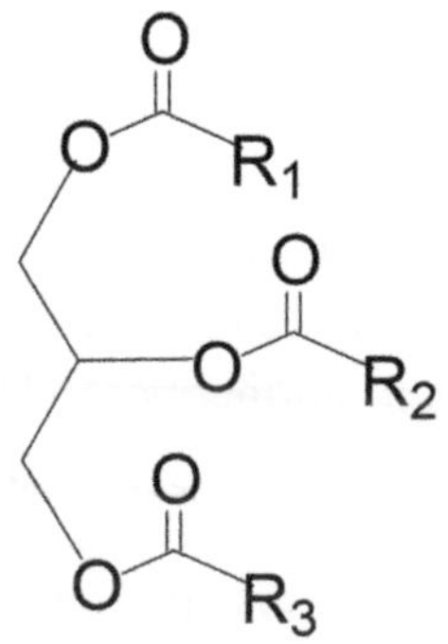

FIGURE 4.2A General structure of triglyceride R$_1$, R$_2$, R$_3$ are fatty acid hydrocarbons either same or different.

-CH$_2$-CH$_2$-CH$_2$-CH$_2$-CH$_2$-CH$_2$-CH$_2$-CH$_2$-CH$_2$-CH$_2$-CH$_2$-CH$_2$-CH$_2$-CH$_2$-CH$_2$-CH$_2$-CH$_2$-COOH

Stearic acid

(Octadecanoic acid)

-CH$_2$-CH$_2$-CH$_2$-CH$_2$-CH$_2$-CH$_2$-CH$_2$-CH$_2$-CH=CH-CH$_2$-CH$_2$-CH$_2$-CH$_2$-CH$_2$-CH$_2$-CH$_2$-COOH

Oleic acid

(Octadec-9-enoic acid)

-CH$_2$-CH$_2$-CH$_2$-CH$_2$-CH$_2$-CH=CH-CH$_2$-CH=CH-CH$_2$-CH$_2$-CH$_2$-CH$_2$-CH$_2$-CH$_2$-CH$_2$-COOH

Linoleic acid

(Octadec-9,12-dienoic acid)

-CH$_2$-CH$_2$-CH=CH-CH$_2$-CH=CH-CH$_2$-CH=CH-CH$_2$-CH$_2$-CH$_2$-CH$_2$-CH$_2$-CH$_2$-CH$_2$-COOH

Linolenic acid

(Octadec-9,12,15-trienoic acid)

-CH$_2$-CH$_2$-CH$_2$-CH$_2$-CH$_2$-CH$_2$-CH$_2$-CH$_2$-CH$_2$-CH$_2$-CH$_2$-CH=CH-CH$_2$-CH$_2$-CH$_2$-CH$_2$-COOH

Petroselinic acid

(Octadec-6-enoic acid)

$$\overset{\text{OH}}{|}$$
-CH$_2$-CH$_2$-CH$_2$-CH$_2$-CH$_2$-CH$_2$-CH-CH$_2$-CH=CH-CH$_2$-CH$_2$-CH$_2$-CH$_2$-CH$_2$-CH$_2$-CH$_2$-COOH

Ricinoleic acid

(12-Hydroxyoctadec-9-enoic acid)

$$\overset{\text{OH}}{|}$$
-CH$_2$-CH$_2$-CH$_2$-CH$_2$-CH$_2$-CH$_2$-CH-CH$_2$-CH=CH-CH$_2$-CH$_2$-CH$_2$-CH$_2$-CH$_2$-CH$_2$-CH$_2$-CH$_2$-CH$_2$-COOH

Lesquerolic acid

(14-Hydroxyeicosa-11-enoic acid)

$$\overset{\text{O}}{\diagup \diagdown}$$
-CH$_2$-CH$_2$-CH$_2$-CH$_2$-CH$_2$-CH$_2$-CH$_2$-CH$_2$-CH=CH-CH$_2$-CH$_2$-CH$_2$-CH$_2$-CH$_2$-CH$_2$-CH$_2$-COOH

Vernolic acid

12,13-Epoxyoctadec-9-enoic acid

FIGURE 4.2B Common fatty acids present in different vegetable oil.

for derivativzations. These include carbon-carbon double bonds (C=C), allylic carbons, ester linkages, α-carbons with respect to the ester group [7, 18]. Some common modifications such as vinylation, acrylation, male-inization, epoxidation, urethanation and by using the vegetable oils as precursors of many others bio-based polymers [1, 16, 19]. The bio-based materials derived from vegetable oils have got enormous applications not only in industrial arena like paints, coatings, adhesives, laminates, packaging but also used in pharmaceuticals and bio-medicals [20].

4.2   PHYSICOCHEMICAL AND SPECTRAL CHARACTERIZATION

Vegetable oils can be characterized by physicochemical analyses like specific gravity, refractive index, solubility, viscosity, acid value, iodine value, soponification value, and peroxide values. Iodine value of the oil is defined as the mg of I_2 consume for the double bonds of 100 gram of the sample under specified conditions. It can be easily measured analytically by treating the oil sample with an excess of ICl, followed by addition of KI to ensure the liberation of equivalent amount of I_2. The librated I_2 can be easily measured analytically by titrating against standard solution of hypo, using freshly prepared starch as an indicator. The iodine value directly provides information about the degree of unsaturation. Higher the iodine value means higher the unsaturation, whereas lower iodine value indicates the lower unsaturation in the oil samples. Triglycerides oils are alienated into three categories on the basis of iodine values; non-drying, semi-drying and drying. Triglyceride oils with iodine values less than 90 are generally known as non-drying, the triglycerides with iodine values in between 90–130 termed as semi-drying, whereas vegetable oil with iodine value more than 130 are commonly known as drying oil [3, 16]. Iodine values of some vegetable oils are provided in Table 4.1 [5, 11, 20, 21].

The acid value of the vegetable oil defined as mg KOH/NaOH consume by one gram of oil sample to complete neutralization. It can be easily measured analytically by the titration of the oil samples against the standard solution of alcoholic KOH/NaOH using acid base indicator like phenolphthalein. An acid value provides the information about the acidity or free fatty acids presents in the vegetable oil under study [20].

TABLE 4.1   Iodine Values of Some Common Vegetable Oils

S. No.	Vegetable oil	Iodine value
1.	Linseed oil	182.0
2.	Soybean oil	128.8
3.	Safflower oil	134.7
5.	Sunflower oil	133.0
6.	Pogamia glabra seed oil	82.0
7.	Castor oil	84.2

Soponification value of the vegetable oil defined as the number of mg KOH or NaOH required to completely soponify the one gram of sample. It can be measured by soponification of vegetable oil with the known excess amount of KOH or NaOH and then excess of KOH/NaOH titrated back with the standard solution of oxalic acid to evaluate the KOH/NaOH consume. It is useful to point out the average molecular weight or chain length of fatty acids constituting the vegetable oil. Higher the soponification value means short chain fatty acids, whereas lower the soponification value indicates the fatty acids of longer chain length [20].

Peroxide value requires for detecting the freshness or rancidity of the vegetable oils. The densities of vegetable oils are lower than water, ranges from 0.8–0.96 gm cm^{-1}, they float over water layer. These characteristics of vegetable oils directly influence the properties of end products obtained from them. For example alkyd resins formulated using vegetable oils of high iodine value have the air drying properties and can form film lonely at ambient temperature, whereas alkyds derived from vegetable oils of low iodine values cannot cure at room temperature.

Fourier transform infrared (FT-IR) spectroscopy is a power full tool to find out the functional groups present in the organic compounds within very short time by using a minute quantity of the samples. The different functional groups like hydroxyl, epoxy, double bonds, carbonyls, and many others show the characteristic absorption bands in the FT-IR spectrum. The curing behavior or the functional group responsible for the cross-linking can also be investigated by taking the spectrum before and after the curing [22]. Characteristic absorption bands for some common functional groups present in the vegetable oils and materials derived from them are listed in Table 4.2 [16, 20, 23].

TABLE 4.2   Characteristic Absorption Bands for Common Functional Groups in FT-IR Spectra

S. No.	Absorption bands (cm⁻¹)	Functional groups
1.	3680–3100	O-H, stretching
2.	3400–3250	N-H, stretching
3.	2960–2850	C-H, aliphatic ethyl, methylene, methane (with a weak shoulder about 2960 cm⁻¹ due to terminal methyl group
4.	1870–1600	Carbonyl group (of ester, aldehyde, ketone, carboxylic acid, amide)
5.	1100–1300	-C(=O)-O-C Stretching vibration of ester
6.	1640–1670	-C=C- (uncojugated), stretching
7.	970–990	Conjugated unsaturated double bonds in eleostearic acid
8.	1342–1266	C-N (stretching vibration)
9.	1375–1465	C-H bending vibration

Nuclear magnetic resonance (NMR) is another useful technique to investigate the structural feature of the compounds. Position of signals, their integral values and splitting patterns are very useful to assign the fatty acids in the particular vegetable oil. The δ values in ppm for different types of protons present in the vegetable oils and their derivatives are provided in the Table 4.3 [23, 24, 25]. Fatty acids composition of linseed

TABLE 4.3   ¹H-NMR Chemical Shift Values for Different Types of Protons

S. No.	¹H chemical shifts (δ-ppm)	Functional groups
1.	5.3–5.6	-CH=CH-CH$_2$-R
2.	6.4–8.0	Ar-H (protons attached to aromatic rings)
4.	4.0–5.3	-CH_2-OCH$_2$COR, -CH-OCH$_2$COR
5.	1.5–3.0	-CH_2-CH=CH-
6.	3.0–4.2	-CH_2-O-CH$_2$-
7.	2.0–2.5	-CH_2-C=O
8.	1.2–1.4	Chain -CH_2-
9.	0.7–1.0	-CH_3
10.	2.5–3.1	Protons attached to epoxy carbons

oil was investigated by using the characteristic peaks of different methyls and methylene groups and their ratios [24, 25].

Gas chromatography is also used to investigate the fatty acids compositions of various vegetable oils. The triglyceride oils are transformed to monomethyl ester through the trans-estrification reaction with methanol using KOH as a catalyst. Fatty acids identification was made by comparing the retention time for each peak with the standard samples of fatty acid methyl esters.

4.3 MODIFICATION OF VEGETABLE OIL

4.3.1 EPOXIDATION

The vegetable oil can be chemically modified to more reactive and versatile through the epoxidation (Figure 4.3). As the epoxy groups are more reactive due to ring strain, consequently can be utilized as precursors for many practicable recipes [26, 27]. Numerous processes have reported for the epoxidation of vegetable oils such as epoxidation with peracids, enzymatic epoxidation, metallic catalyzed epoxidation and many others [16, 20, 28]. Peracid epoxidation performed industrially at large scale. In this process peracid synthesized in-*situ* by reacting carboxylic acid with concentrated hydrogen peroxide (H_2O_2) in presence of certain inorganic acids like HCl, H_2SO_4, HNO_3, H_3PO_4, etc. It has been reported that epoxidation using acetic acid as oxygen carrier is more suitable than formic acid and H_2SO_4 is the more effective inorganic acid [29].

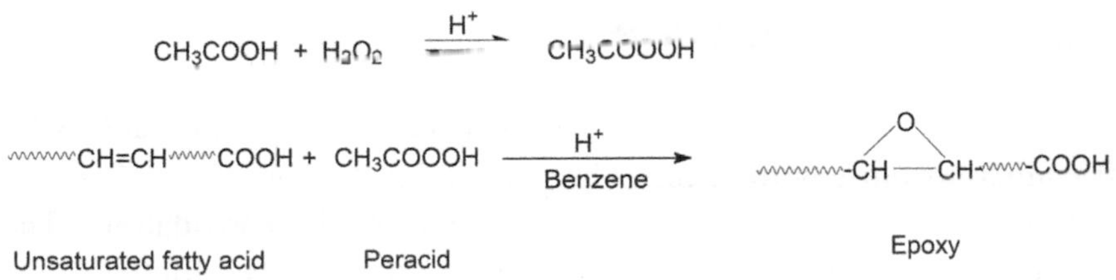

FIGURE 4.3 Reaction scheme for epoxidation.

Epoxidation of soybean oil containing about 52% linoleic acid using H_2O_2 as oxygen carrier and catalytic amount of formic acid was reported by Saremi et al. [30]. Soybean oil is used as polymerizable monomers in a radiation curable system due to its environmental friendly character and lower cost as compared to petrochemical based chemicals. They claimed that high molar ratio of H_2O_2 compared to unsaturation (H_2O_2: C=C) about 1:1.7 is required to achieve to highest percentage of epoxy contents.

The progress of reaction was monitored by determination of iodine values and epoxy equivalents at regular intervals. The iodine value and epoxy equivalent both decreases on progress of reaction. The lower value of epoxy equivalent indicates that higher the number oxirane rings, whereas low epoxy equivalent means lower the epoxy groups in the given compound [31].

The epoxy equivalent can be analytically measured by titrating the samples against the standard solution of HBr or by pyridinium chloride methods [32, 33]. Aerts and Jacobs [34] use the ^{1}HNMR spectroscopy to quantify the yield of epoxidation of methyl oleate, methyl linoleate, sunflower, and safflower oils. Investigated results were further verified by comparing with GC. Reaction yield was determined by evaluating the integral values of disappearing and appearing groups before and after the epoxidation. They found that peak area at $\delta = 2.01$ ppm due to $-CH_2-CH = CH-CH_2-$ decreases where as new signals appears at $\delta = 1.50$ ppm due to CH_2-CHOCH-CH_2 and at $\delta = 2.90$ ppm due to $-CHOCH-$.

Goud et al. [35] reported the epoxidation of Mahua oil using H_2O_2 and glacial acetic acids and also claimed for the improvement in economic value by the epoxidation. They also optimized the various parameters like type of catalyst, temperature, molar ratios of the reactants.

4.3.2 ENZYMATIC METHOD

Enzymatic method for epoxidtion of vegetable oil is supposed to be friendlier to the environment. The immobilized lipase from *Candida antartica* was use as catalyzed chemo-enzymatic epoxidation. The parameters affecting the lipase activity and operational lifetime during the chemo-enzymatic epoxidation of fatty acids were investigated

[28, 36]. Optimization of chemo-enzymatic epoxidation of soybean oil was investigated [37]. It has been reported that the rate of reaction was affected by the amount of lipase biocatalyst in addition to other parameters like temperature, concentrations of the components [38]. The lipase is remarkably stable under reaction conditions and can be reused about 15 times effectively [38].

4.3.3 UTILIZATION OF EPOXIDIZED OILS

Now-a-days epoxidized vegetable oils are used as natural plasticizers and are also claimed for eco-friendly due to their low toxicity and low migration. Universally accepted definition of a plasticizer is the substance incorporated in a material to increase its flexibility, workability, practical utility or extensibility. Numerous epoxidized vegetable oils like castor oil, sunflower oil, linseed oil, soybean oil and many others have been successfully utilized as plasticizers in many industrial formulations [39,40].

Epoxidized vegetable oils are used as precursors for many practicable materials, such as epoxy groups easily react with the carboxyl groups of acrylic acid to yield acrylated epoxidized oil (Figure 4.4). The acrylated epoxidized oil has several prospective of polymerizations. They can polymerized itself or copolymerize with other vinyl monomers like styrene.

Linseed oil epoxy undergoes trans hydroxylation and used a polyol for the polyurethane syntheses. Polyol obtained in the previous step reacted with toluylene diisocyante (TDI) in different ratios to yield a series of polyurethane resins. The resulting polymers were characterized by physicochemical analyses. The structural elucidation and formation of new

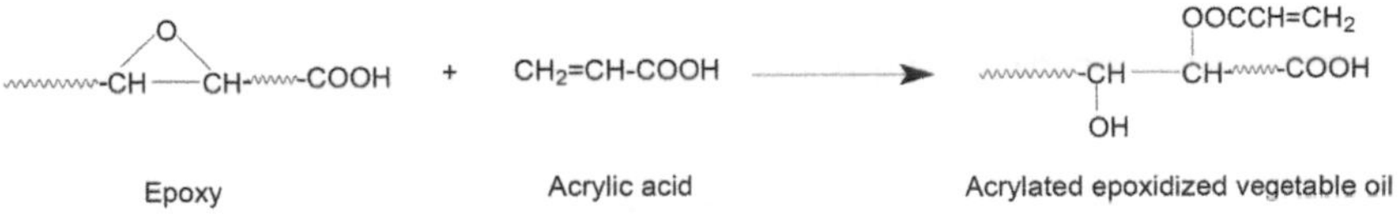

| Epoxy | Acrylic acid | Acrylated epoxidized vegetable oil |

FIGURE 4.4 Acrylation through epoxy.

moieties while the synthesis of the polymer was carried out by using IR and ^{1}HNMR spectral data's. Polyurethane obtained through this route used as a corrosion protective coating materials cured at ambient temperature. They claimed that the coated samples show good physic-mechanical and weather resistance performances [41].

Vegetable oil based epoxies embedded with long hydrophobic hydrocarbon chain and hence shows the good flexibility and corrosion protection abilities especially in aqueous and acidic environment. However, films of these aliphatic epoxies show the poor toughness and low load bearing performances. To improve these properties of linseed oil based epoxy blended with acrylic polymers like polystyrene and poly (methyl methacrylate) in different ratios. These polymers are classified as hard and brittle polymers [42]. The polyblends of linseed oil epoxy with

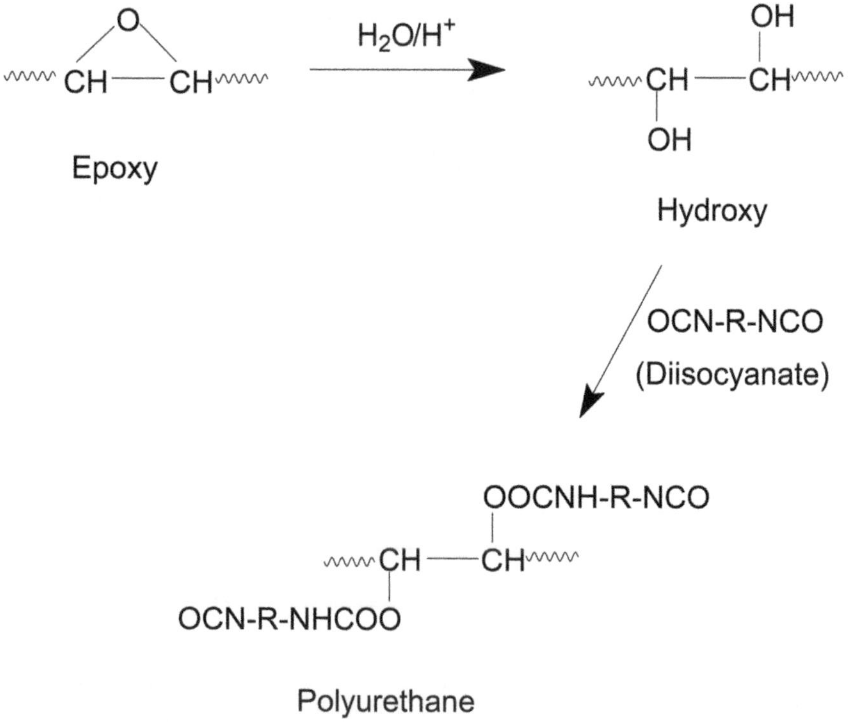

FIGURE 4.5 Urethanation through epoxy.

polystyrene and poly(methyl methacrylate) were found to homogenous and no phase separation was reported in cold and hot conditions. These are further claimed that no chemical changes occur during the blending as functional groups like epoxy groups, double bonds remain the constant for the same amount of linseed oil epoxy used in polyblends. This was also confirmed by the authors' using differential scanning calorimetric (DSC) and IR analyses. They found that polyblends systems show the single gloss transition temperature (Tg) a characteristic single-phase system. They reported that linseed oil epoxy turned into rigid mass by addition of only a small amount of these polymers (about 16.0% w/w) [31]. In another report it has been estimated that paints developed from these materials cured with melamine formaldehyde (MF) resin reduce the cost of epoxy paints about 20%. Furthermore linseed oil epoxy synthesized from sustainable resource cut down the use of raw materials derived from petrochemicals [43].

4.4 VEGETABLE OIL BASED POLYMERS WITH ANTI-MICROBIAL ACTIVITY

The vegetable oils and vegetable oil based polymers show the potential applications in making coatings, paints, adhesives, packaging materials and other industrial appliances. In addition to these several vegetable oils and numerous polymers derived from them reported to show the remarkable antimicrobial activities [44]. Antimicrobial properties of *Moringa oleifera* was overviewed and concluded that oil extract of *Moringa oleifera* resist the growth of *P. Aeruginosa* [45]. Antimicrobial activities of *Moringa oleifera* seed oil, pre-polymer *N,N,*-bis (2-hydroxyethyl) *Moringa oleifera* oil fatty amide (HEMA) and final polyesteramide (MOPEA) were investigated against Gram-positive organisms and Gram-negative organisms by Siyanbola et al. [46]. They claimed that HEMA and MOPEA both sow good inhibitive properties MOPEA in particular. Urethane modified boron filled polyesteramide of soybean oil was developed by reacting boron filled polyesteramide of respective vegetable oil with tolylene-2,4-di-isocyanate (TDI) in variable wt% ratios. The developed urethane modified polymers were characterized by physic-chemical analyses like hydroxyl value, iodine value, acid value, refractive index, viscosity, structural elucidation of the polymers was made by FT-IR, ^{1}H-NMR and ^{13}C-NMR spectral data's. The anti-bacterial and anti-fungal activities of these polymers were investigated by the measurement of the size of clarity of zone of the inhibition in nutrient media. The boron containing urethane modified polyesteramide resins were screened for their antibacterial activity against *E. coli*, *Pseudomonas* sp. and *Staphylo-coccus* sp. in nutrient agar media. Antifungal activities of the polymers were investigated on Candida *albicans* and *Mucor* sp. It has been reported that urethane modified boron filled Polyesteramide resins were inhibit the microbial growth. Furthermore polymer obtained by 8 wt% loading of TDI showed the highest anti-microbial activity [47]. Cadmium incorporated castor and soybean oil based polyesteramides were synthesized using poly(condensation) polymerization between N,N -bis-2-hydroxy ethyl fatty amides of these vegetable oils, $Cd(OH)_2$ and sebasic acid. The developed polymers were tested for biological activities in vitro against several pathogenic fungi and bacteria to evaluate their inhibiting potential. It was reported that the tested polymers much more effective against fungi than bacteria, which are responsible for causing candidiasis (a disease that varies from superficial mucosal to life threatening systematic disorders) [48].

ACKNOWLEDGMENT

The author would like to thanks the authorities of G. F. College, Shahjahanpur, U.P., India, for providing necessary facilities to carry out this study and Dr. Z. Abbas, HOD, Botany for valuable suggestions and encouragements.

KEYWORDS

- anti-microbial activity
- expoxidized oils
- Fourier transform infrared
- NMR
- soponification
- vegetable oil

REFFRFNCES

1. Samarth, N. B., & Mahanwar, P. A. (2015). Modified vegetable oil based additives as a future polymeric materials. *Open J. Org. Polym. Mater. 5*, 1–22.
2. Lligadas, G., Ronda, J. C., Galia, M., & Cadiz, V. (2013). Renewable polymeric materials from vegetable oils: a perspective. *Materials Today 16*, 337–343.
3. Sharmin, E., Zafar, F., Akram, D., & Alam, M. (2015). Recent advances in vegetable oil based environment friendly coatings: a review. *Industrial Crops and Products 76*, 215–229.
4. Lochab, B., Varma, I. K., & Bijwe, J. (2012). Sustainable polymers derived from naturally occurring materials. *Advances in Mater. Physics and Chem. 2*, 221–225.
5. Meier, M. A. R., Metzger, J. O., & Schubert, U. S. (2007). Plant oil renewable resources as green alternatives in polymer science. *Chem. Soc. Rev. 36*, 1788–1802.
6. Sangwan, S., Rao, D. V., & Sharma R. A. (2010). A review on *Pongamia Pinnata* (L.) Pierre: A great versatile leguminous plant. *Nature and Science 8*, 130–139.
7. Narine, S. S., & Kong, X. (2004). Vegetable oils in production of polymers and plastics. In: *Bailey's Industrial Oil and Fat Products*; Shahidi, F. (ed.). John Wiley and Sons Inc., pp. 1–28.
8. Ahamad, S., Ahmad, S. A., & Hasnat, A. (2016). Synthesis and characterization of methyl methacrylate modified poly(ester-amide) resins from melia azedarach seed oil as coating materials. *Mater. Sci. Res. India 13*, 50–56.
9. Gallegos, A. M. A., Carrera, S. H., Parra, R., Keshavarz T., & Iqbal, H. M. N. (2016). Bacterial cellulose: A sustainable source to develop value-added products: a review. *BioResources 11*, 5641–5655.
10. Zafar, F., Sharmin, E., Ashraf, S. M., & Ahmad, S., (2004). Studies on poly (styrene-co-maleic anhydride) polyesteramide based anticorrosive coatings synthesized from sustainable resource, *J. Appl. Polym. Sci. 92*, 2538–2544.
11. Gultekin, M., Beker, U., Guner, F. S., Erciyes, A. T., & Yagci, Y. (2000). Styrenation of castor oil and linseed oil by macromer method. *Macromol. Mater. Eng. 283*, 15–20.
12. Yadav, S., Zafar, F., Hasnat, A., & Ahmad, S. (2009). Poly(urethane fatty amide) resin from linseed oil-A renewable resource. *Prog. Org. Coat. 64*, 27–32.
13. Petrovic, Z. S. (2010). Polymers from biological oils. *Contem. Mater. 1*, 39–50.
14. Babu, R., Connor, K. O., & Seeram, R. (2013). Current progress on bio-based polymers and future trends. *Prog. Biomater. 2*, 1–16.
15. Abdurrahman, M., Kumar, A., Kishor, B., & Abbas, Z., (2015). Genotypic characterization of Jatropha curcas L. Germplasm by RAPD analysis. *J. Biol. Chem. Res. 32*, 353–360.
16. Guner, F. S., Yagci, Y., & Erciyes, A. T. (2006). Polymers from triglyceride oil, *Prog. Polym. Sci. 31*, 633–670.
17. Majumdar, S., Kumar, D., & Nirvan, Y. P. S. (1999). Acrylic grafted dehydrated castor oil alkyd, a binder exterior paints, *Paintindia 60*, 57–65.
18. Datta, N., Karak, N., & Dauli, S. K. (2007). Stoving paint from *Mesuaferrea* L. seed oil based short oil polyester and MF resins blend. *Prog. Org. Coat. 58*, 40–45.
19. Sharma, V. P. (2006). Addition polymers from natural oils: a review. *Prog. Polym. Sci. 31*, 983–1008.

20. Islam, M. R., Beg, M. D. H., & Jamari. S. S. (2014). Development of vegetable oil based polymers. *J. Appl. Polym. Sci. 131*, 40787–40799.

21. Ahmad, S., Ashraf, S. M., Naqvi, F., Yadav, S., & Hasnat, A. (2003). A polyesteramide from *Pongamia glabra* oil for biologically safe anticorrosive coating. *Prog. Org. Coat. 47*, 95–102.

22. Zafar, F., Ashraf, S. M., & Ahmad, S. (2008). Self-cured polymers from non-drying oil. *Chemist. Chem. Tech. 2*, 286–293.

23. Silverstein, R. M., Bassler, G. C., & Morril, T. C. (1991). *Spectroscopic Identification of Organic Compounds,* 5th edn., John Wile y & Sons, New York. pp. 1–430.

24. Cavallo, A. S., Senouci, H., Jerry, L., Klein, A., Bouquey, M., & Terrisse, J. (2003). Linseed oil and mixture with maleic anhydride: ^{1}H and ^{13}C NMR. *J. Am. Oil. Chem. Soc. 80*, 311–314.

25. Balanuca, B., Stan, R., & Hanganu, A. (2014). Novel linseed oil-based monomers: synthesis and characterization. *U. P. B. Sci. Bull., Series B, 76*, 129–139.

26. Schmitz, W. R., & Wallace, J. G. (1954). Epoxidation of methyl oleate with hydrogen peroxide. *J. Am. Oil Chem. Soc. 31*, 363–365.

27. Fiser, S. S., Jankovic, M., & Petrovic, Z. S. (2001). Kinetics of *in situ* epoxidation of soybean oil in bulk catalyzed by ion exchange resin. *J. Am. Oil Chem. Soc. 78*, 725–731.

28. Saurabh, T., Patnaik, M., Bhagt, S. L., & Renge, V. C. (2011). Epoxidation of vegetable oils: A review. *Int. J. Adv. Eng. Tech. 2*, 491–501.

29. Dinda, S., Patwardhan, A. V., Goud, V. V., & Pradhan, N. C. (2008). Epoxidation of cotton seed oil by aqueous oxygen peroxide catalyzed by liquid inorganic acids, *Biores. Tech. 99,* 3737–3744.

30. Saremi, K., Tabarsa, T., Shakeri, A., & Babanalbandi, A. (2012). Epoxidation of soybean oil. *Annals Biol. Res. 3*, 4254–4258.

31. Ahmad, S., Ashraf, S. M., Hasnat, A., & Noor, (2001). A. Studies on epoxidised oil and its blend with polystyrene and poly(methyl methacrylate), *Indian J. Chem. Tech., 8*, 176–180.

32. Lee, H., & Neville, K. (1967). *Handbook of Epoxy Resins*, McGraw-Hill, New York, pp. 83.

33. Pandey, S., & Srivastava, A. K. (1999). Synthesis and characterization of epoxy resins containing zinc chloride. *Indian J. Chem. Tech. 6*, 313–316.

34. Aerts, H. A. J., & Jacobs, P. A. (2004). Epoxide yield determination of oils and fatty acid methyl esters using ^{1}H NMR. *J. Am. Oil Chem. Soc. 81*, 841–846.

35. Goud, V. V., Patwardhan, A. V., & Pradhan, N. C. (2006). Studies on Mahua oil (Madhumica Indica) by hydrogen peroxide, *Biores. Tech. 97*, 1365–1371.

36. Orellana-Coca, C., Camocho, S., Adlercreutz, D., Mattiasson, B., & Hatti-Kaul, R. (2005). Chemo- enzymatic epoxidation of linoleic acid, parameters influencing the reaction. *Eur. J. lipid Sci. Tech. 107*, 864–870.

37. Vlcek, T., & Petrovic, Z. S. (2006). Optimization of chemoenzymatic epoxidation of soybean oil. *J. Am. Oil Chem. Soc. 83*, 247–252.

38. Charlie Scrimgeour, (2005). *Bailey's Industrial Oil and Fat Product*, Sixth edition, Six volume set. Edited by Shahidi, F., John Wiley & Sons Inc., pp. 1–43.

39. Baltacioglu, H., & Balkose, D. (1999). Effect of zinc stearate and/or epoxidized soybean oil on gelation and thermal stability of PVC-DOP plastigels. *J. Appl. Polym. Sci. 74*, 2488–2498.

40. Motawie, A. M., Sadek, E. M., Awad, M. M. B., & El-Din, A. F. (1998). Coatings form epoxidized (polyurethane-polyester) resin system. *J. Appl. Polym. Sci. 67*, 577–581.

41. Ahmad, S., Ashraf, S. M., Sharmin, E., Zafar, F., & Hasnat, A. (2002). Studies on ambient cured polyurethane modified epoxy coatings synthesized from a sustainable resource. *Prog. Cryst. Growth Charact. Mater. 45*, 83–88.

42. Sounders, K. J. (1988). *Organic Polymer Chemistry*, 2nd edn., Chapman and Hall, USA, pp. 29.

43. Ahmad, S., Ashraf, S. M., Kumar, S., Alam, M., & Hasnat, A. (2005). High performance paints from a sustainable resource. *Indian J. Chem. Tech. 12*, 193–197.

44. Siyanbola, T. O., Sasidhar, K., Anjaneyulu, B., Kumar, K. P., Rao, B. V. S. K., Ramanuj, N., Olaofe, O., Akintayo, E. T., & Raju, K. V. S. N. (2013). Anti-microbial and anti-corrosive poly(ester-amide urethane) siloxane modified ZnO hybrid coatings from *Thevetia peruviana* seed oil. *J. Mater. Sci. 48*, 8215–8287.

45. Ali, G. H., El-Taweel, G. E., & Ali, M. A. (2004). The cytotoxicity and antimicrobial efficiency of Moringa oleifera seeds extracts. *Int. J. Env. Stud. 61*, 699–708.

46. Siyanbola, T. O., James, O. O., Gurunathan, T., Sasidhar, K., Ajanaku, K. O., Ogunniran, K. O., Adekoya, J. A., Olasehinde, G. I., Ajayi, A. A., Olaofe, O., Akintayo, E. T., & Raju, K. V. S. N. (2015). Synthesis and characterization and antimicrobial evaluation of polyesteramide resin from *Moringa oleifera* seed oil (MOSO) for surface coating application. *Canad. J. Pure and Appl. Sci. 9*, 3229–3240.

47. Ahmad, S., Haque, M. M., Ashraf, S. M., & Ahmad, S. (2004). Urethane modified boron filled polyesteramide: a novel anti-microbial polymer from a sustainable resource. *Eur. Polym. J. 40*, 2097–2104.

48. Bharathi, N. P., Khan, N. U., Alam, M., Sheraz, S., & Hashmi, A. A. (2010). Cadmium incorporated oil based bioactive polymers: synthesis, characterization and physico-chemical studies. *J. Inorg. Organomet. Polym. 20*, 833–838.

CHAPTER 5

SUCROSE NON-FERMENTING-1 RELATED KINASE 1 (SNRK1): A KEY PLAYER OF PLANT SIGNAL TRANSDUCTION

BHAVIN S. BHATT,[1] FENISHA D. CHAHWALA,[2] B. SINGH,[3] and ACHUIT K. SINGH[3]

[1]*Shree Ramkrishna Institute of Computer Education and Applied Sciences, Surat, E-mail: bhavin18@gmail.com*

[2]*School of Life Sciences, Central University of Gujarat, Sector 30, Gandhinagar, E-mail: fenisha_chahwala@yahoo.com*

[3]*Crop Improvement Division, ICAR – Indian Institute of Vegetable Research, Varanasi, E-mail: bsinghiivr@gmail.com, achuits@gmail.com*

CONTENTS

ABSTRACT

Protein kinases play important role in signal transduction as well as participate in regulatory functions during different conditions throughout the phyla. These are subfamily of serine/threonine kinases which plays pivotal role as metabolite sensors which constantly checks balance of demand and supply of energy by metabolically active cells. Among all protein kinases, sucrose nonfermenting-1 related kinase-1 (SnRKs) acts as metabolic sensors in plants. These kinases are evolutionary conserved in all eukaryotes from budding yeast (SNF1) to mammal (AMP activated protein kinase – AMPK). Plant SnRKs regulates transcriptional and metabolic programming of cells to promote tolerance against biotic and abiotic stress partially through the general repression of anabolism and an induction of catabolic processes. SnRKs divided in to three subfamilies: SnRK1, SnRK2, and SnRK3. SnRK1 kinases act as a central regulatory component to control metabolism for example carbohydrate, lipid and fatty acid uptake and regulation. It is well established that SnRK2 and SnRK3 cannot fulfill SnRK1 function hence they having similarity in target reorganization. SnRK1 generally functions as heterotrimeric complex having α-catalytic subunit and β,γ-regulatory subunits. Some aspects for the role of SnRK1 kinase that control the fundamental issues have yet remained elusive. In this chapter, we will discuss about the structural, functional as well as regulatory aspects of SnRK1 and its upstream and downstream components along with its role in pathogen defense response.

5.1   INTRODUCTION

In natural environment, plants face different factors such as biotic and abiotic stresses. In response to stress, plants activate different signaling pathways rather than normal singling cascade to regulate homeostasis to overcome the stress condition. There are different types of signaling mechanisms get activated throughout the life of plants. One of the most important consequences of signaling event mechanisms is phosphorylation/dephosphorylation of downstream proteins. It is recognized as regulatory mechanism for certain proteins[1]. These events (phosphorylation/dephosphorylation) regulate the cell signaling cascades of different cells. It regulates many protein activities at different time points of life cycle. Kinases are enzymes that transfer phosphate group form one molecule to other molecule. Eukaryotic protein kinase transfer γ-phosphate via ATP to amino acid of the molecule, which in case of serine/threonine kinase, transfers to terminal hydroxyl (–OH) group of amino acid serine and/or threonine.

This chapter will discuss about some basic groups of protein kinases and its role in living system. It will focus our view through some basic structure and mechanism of SnRK1 in plant system and role of SnRK1 during stress conditions.

5.2 HISTORICAL DEVELOPMENTS AND CLASSIFICATION

The study of protein kinases has a long history over 60 years. It is estimated that 1 to 3% functional eukaryotic genes are responsible to code protein [2]. Budding yeast, *Saccharomyces cerevisiae* has nearly 113 protein kinase genes. Thale Cress, *Arabidopsis thaliana* has about 1000 and human has almost 518 [1]. These kinases have been classified into five major families depending upon sequence similarities. Here, we will discuss plant protein kinases are based on Hanks and Hunter rule [3].

AGC group: AGC group name is given after finding of the protein kinase A (PKA), protein kinase G (PKG) and protein kinase C (PKC). It is basically serine/threonine kinase. It is cyclic nucleotide dependent kinases and calcium dependent kinases family (PKA, PKG, and PKC). This is regulated by secondary messengers like cyclic AMP or lipids. It contains 60 member in the family [4].

CaMK group: This group has calcium/calmodulin dependent kinase and SNF1/AMPK families. Members of these families are activated when there is increase in the concentration of intracellular Calcium (Ca^{++}), which transfer phosphate group from ATP to Ser/Thr kinase to other protein. At the downstream signaling, they are involved in the regulation of transcription factor, resulting in modulation of gene expression [5].

CMGC group: This group is most conserved group almost being present in all types of eukaryotes. This group named after the finding of kinases involved in splicing and metabolic processes, growth kinases, stress responsive kinases and cyclin dependent kinases (CDK). It having nine different family members. It involves families of CDK (Cyclin Dependent Kinases), MAPK (Mitogen Activated Protein Kinase), GSK-3 (Glycogen Synthase Kinase-3) and CKII (Casein Kinase II) [6].

Conventional PTK group: The name is based on protein tyrosine kinase (PTK). These types of kinases exclusively use tyrosine residue rather than Ser/Thr residue for phosphorylation of proteins. This involves Non-membrane spanning protein-tyrosine kinases and membrane spanning protein-tyrosine kinases. There are nine non receptor PTK and thirteen receptor RTK (receptor tyrosine kinase) have been defied till date [7].

Other group: It is a family of protein kinase, which are not involved in above four groups. This group includes RLKs (receptor like kinase), RTKs (receptor tyrosine kinases), CTR1 (constitutive triple response 1) [8], Tsl (thymidylate synthase kinase I) protein kinases.

Apart from the metabolic regulation in healthy plants, these protein kinases are also involved in stress signaling pathways, for example, glycogen synthase kinase (GSK3), mitogen activated protein Kinase (MAPK), calcium dependent protein kinase (CDPK), SNF-1 related protein kinase [9]. There are many kinases which are exclusively present in plants but absent in other eukaryotes. These kinases are therefore responsible for signaling pathways exclusively executed in plants. On the other hand, SNF-1 (yeast sucrose non-fermenting kinase-1), AMPK (mammalian AMP activated protein kinase) and SnRK1 (SNF-1 related protein kinase) related protein kinases are highly conserved serine/threonine kinase ubiquitously present in all eukaryotes.

Alderson et al. first time reported the presence of SnRK1 in rye endosperm from cDNA library in 1991, which showed 48% amino acid sequence identity with AMPK and SNF-1. It encoded 57.7 KDa protein with 570 amino acid residues. SnRK1 was the first plant protein kinase for which molecular and biochemical studies were done together. SnRK1 have conserved catalytic domain showing maximum similarities with mammalian AMPK and yeast SNF-1. They are closely related to CDPK group of kinase family [10]. The SNF-1/AMPK/SnRK are ortholog of each other and act as greatly conserved metabolic sensor throughout eukaryotic protein kinase family [11].

Yeast SNF-1 having a fundamental role when, there is shift from fermentative to oxidative conversion in response to decrease level of glucose. They partly activate repressive gene which are responsible for utilization of alternative carbon sources [12]. In yeast (*Saccharomyces cerevisiae*) SNF1 is basically require for adaptation during glucose starvation. It directs cells to use an alternative carbon source. It controls biosynthesis of carbohydrate as well as autophagy in response to recycle the macromolecules and organelle [13].

Mammalian AMPK act as an energy saving regulator and transcriptional controller during starvation condition [11]. It regulates whole energy metabolism as well as glucose homeostasis via modifying the process like glucose uptake of muscles, production of insulin and its secretion, body lipid management and apatite [14]. In plants, SnRKs regulates metabolism during biotic and abiotic stress. AMPK in mammals implements as an energy saving program via direct enzyme regulation and energy control. They regulate the synthesis of fatty acids and cholesterol synthesis.

SnRK family is classified in to three different families: SnRK1, SnRK2, and SnRK3 [3]. All three are plant specific serine/threonine kinase. SnRK1 is divided into two sub groups: SnRK1a and SnRK1b [15]. They are key regulators of glucose metabolism. Further, SnRK2 is divided into three groups on the basis of its regulation by abscisic acid. It regulates the response of abiotic stress as well as abscisic acid dependent plant development [9]. SnRK3 regulates via WPK4 gene. This gene is regulated by light cytokinins, nutrition deprivations, salt tolerance, salt overly sensitive 2 (SOS2) and sucrose [1]. There is no evidence of idleness between different SnRK families. SnRK2 and SnRK3 do not set off the SnRK1 deletion mutation phenotype. Depending upon some target reorganization similarity, it was suggesting that SnRK2 and SnRK3 arose by duplication of SnRK1 in plants. During plant evolution, they diverged rapidly and involved in regulation of new demand for stress signaling with metabolic signaling.

5.3   STRUCTURE OF SNRK1 COMPLEX

The SnRK1 protein kinases are highly conserved throughout the eukaryotes and showing similarities with SNF1 of yeast and mammalian AMPK. They having α, β and γ heterotrimeric subunits [11].

The α subunit is catalytic subunit and is composed of two parts. One subunit is kinase domain and other is regulatory domain. Kinase domain located at the N-terminal half of the protein. Threonine 175 residue is the conserved domain in the activation loop. Thr175 require phosphorylation by upstream kinase to bestow kinase activity. The kinase domain has canonical fold among 11 sub-domains [12]. They also contain the activation loop, which called as T-loop. Regulatory domain contain ubiquitin associated domain (UBA), which may propose to mediate interaction with ubiquitinated proteins[16]. These kinases also possess KAI (kinase associated I) domain, which is responsible for interaction with regulatory subunit and upstream phosphatase.

The β subunit provide scaffold to keep α and γ subunit together. The γ-subunits are characterized by N-terminus and cystathione-beta-synthase repeats (Batman domains). They can bind with adenosine derivatives. The SnRK1 activation is depends upon phosphorylation of conserved T-loop by upstream kinases. All though there is no evidence of how phosphorylation state is affected by cellular energy level [17]. SnRK1 activity is altered by specific phosphatase.

The γ subunit act as direct regulators to regulate the SNF-1 kinase activity [18]. The γ domain contains four cystathonine synthase domain (CBS). Scott et al. reported the function of γ subunit on carbohydrate metabolism. It required for the formation of heterotrimeric complex with β and α subunit.

Figure 5.1 shows domain subunit structure SnRK1 in *arabidopsis thaliana* (Thale cress). SnRK1 consist of three subunits *viz.* α, β and γ. The α-subunit possess conserved phosphorylation sites, kinase domain, Kinase associated domain (KA1) autoregulation and inhibitory domain (UBA). The β subunit contains starch binding domain and association with the SNF1 complex (ASC) domain. The β subunit interacts with the α subunit at KA1 domain and binds to γ subunit at ASC domain. The γ subunit contains multiple cystathionine β-synthase (CBS) domains.

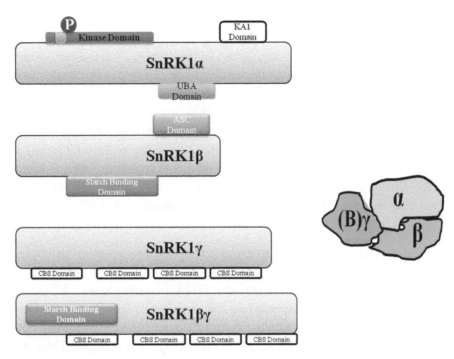

FIGURE 5.1   Domain subunit structure SnRK1 in *arabidopsis thaliana* (Thale cress).

5.4   REGULATION OF SNRK1 KINASE

SnRK1 activates through phosphorylation of a conserved Threonine[175] in the T-loop of catalytic subunit [19]. There are many protein kinases that regulate via phosphorylation, which is apparently require suitable alignment of ATP to regulate its interaction with catalytic lysine [20]. During the studies in total cell extracts of plants, there is no significant difference found in phosphorylation level of activation loop between control and stress condition. These studies suggest that there may be involvement of another phosphorylation domain present to control the activity during stress.

5.5   UPSTREAM KINASE

By sequence comparison, complementation and phosphorylation assays, it is implicated that GIRK1/2 (SnAK1/2) act as upstream kinase in plant.

SnAKs can autophosphorylate and subsequently start phosphorylation cascade downstream to activate SnRK1α *in vivo*. SnAK autophosphorylates at T^{153}/T^{154} residue for their activation and inactivation. On the other hand, as a negative feedback, they are inhibited via phosphorylation of SnRK on T-loop at S^{260}/S^{261}. The occurrence SnAKs as SnRK1 upstream kinase leftover to assess only because of their function in plants is to an overlap between expression of SnAK and phosphorylation of SnRK1in shoot apical meristem. CIPK1 (calcineurin B-like-interacting protein kinase 1) is also one of the proteins that may act as upstream kinase of SnRK1. Likewise, LKB1, there is no requirement of Ca^{+2} for their regulation. Although they are also insensitive to CaMPK specific inhibitor STO-609 [21].

5.6   UPSTREAM PHOSPHATASE

There are also some evidence tells that PP2A-like phophatase, ABI1 and PP2C phosphatase dephosphorylate and inactivate plant SnRK1 in vitro [22]. PP2C are well-known regulators for ABA pathway. They regulate the repression of ABA with SnRK2 and by blocking ABA receptor. There are reports of some other pps, which interact with SnRK1. But the functional significance of their interaction is unknown. PP2C, PP74 interact through SnRK1α2 in vitro. Furthermore, PTP (protein tyrosine phosphatase), KIS1 was also reported to interact with SnRK1α2. This phosphatase was also able to dock CBM domain, which allows binding to starch in vivo as well and also, identified as a component, which is responsible for starch over accumulation for the sex4 mutant.

5.7   ROLE OF SNRK1 IN METABOLIC STRESS

SnRK1, as SNF1 and AMPK plays a role during starvation condition (metabolite stress or darkness) in plants. SnRK1 act as an important mediator molecule during cross talk between stress tolerance and metabolic response of tissue against biotic or abiotic stress. Figure 5.2 summarize multifunctional role of SnRK1 during stress response and metabolic regulation in plants. There is some study on *Physcomitrella*, suggest that SnRK1 mutants impaired their ability to mobilize starch reserve during darkness. SnRK1 also required for expression of different genes that are responsible to regulate low level of sugar. They are also responsible for reallocation of carbon in response to herbivory. In this way, their role in stress response is well established.

Role of SnRk1 on metabolite is notoriously difficult, because no result has been achieved influentially to cater the role of SnRK in metabolism directly. Yet there is report of modulation of glucose and sucrose metabolism by SnRK1. There are some evidences suggest that glucose 6-phosphate inhibit SnRK1 while 5'-AMP and trehalose 6-phosphate (T6P) interact and activate it. 5'AMP modulates the phosphorylatin of SnRK1 by regulating its activity. Thus, the homeostatic level of sucrose is sensed by SnRK1, which ultimately allows cell to regulate sucrose signal transduction.

5.8 SNRK1 REGULATE STORAGE CARBOHYDRATE BIOSYNTHESIS

SnRK1 can be activated through high glucose/sucrose ratio. In addition to starvation, it could be incorrect to tell that SnRK1 only involve in starvation related regulation [4]. In plants, SnRK1 is also required for starch biosynthesis. SnRK1 is also activates ADP-glucose phosphorylase via sucrose dependent redox activation [17]. There are also some evidences that SnRK1 regulate carbon during storage pathway to starch for glycolytic pathway. During several study inhibitions of SnRK1 via antisense mechanism lead to developmental abnormalities in pollen grain as well as total loss of starch accumulation and viability. There is also evidence of expression of SnRK1 concur starch accumulation in rice and Sorghum. SnRK1 also control the genes that control the sucrose synthase and ADP-glucose ribosylase (Figure 5.2).

5.9 ROLE OF SNRK1 IN VIRUS DEFENSE

Geminiviridae is the second largest family of the plant viruses after Potyvirus. They infect economically important crops such as vegetables, pulses and ornamental crops. It is reported that geminivirus order many signaling cascades during infection to propagate the copy number of it. In response to defense, plant activates many defense related pathway to overcome and

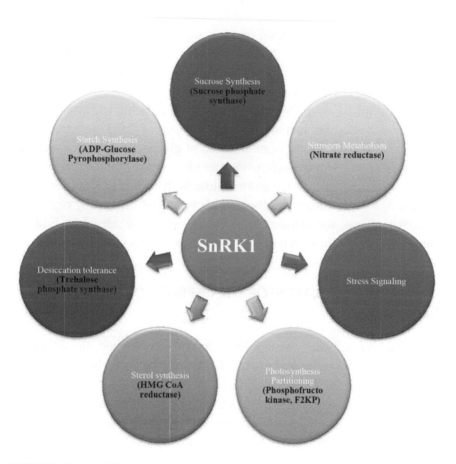

FIGURE 5.2   Multifunctional role of SnRK.

control the homeostasis of plants. Geminivirus interfere with this plant defense system by interacting with SAMDC1 and ADK cycle to suppress the plant methyl cycle. Hao et al. reported that AL2 and L2 protein of geminivirus enhanced the susceptibility and increase the viral infectivity. They showed that AL2 and L2 inactivate the SNF1 kinase. Where increase in the level of SNF1 kinase enhance the resistance to the plant. This study indicate that amendments arbitrated by SNF1 kinase in the system may enhance the dieses tolerance in plants [23]. Remorins are proteins exclusively present in the plant, first time reported as plasma membrane protein [24]. It is positive regulator of Geminivirus by interacting with SnRK1[25].

Geminivirus rep interacting kinase (GIRK1) a phylogenetically related to SnRK1 kinase shows interaction with SnRK1. GIRKs unambiguously bind with SnRK1 catalytic subunit and phosphorylate threonine residue [21]. This study shows that interaction of SnRK1 with GIRK may function as a signaling cascade to provide metabolic requirement during infection.

5.10 CRITICAL DISCUSSION

In plants, SnRK1 kinase function as heterotrimeric compound, in which α subunit as catalytic subunit and β, γ as regulatory subunit. This SnRK1 is controlled by phoshorylation mechanism. Alteration of certain regulatory processes during starvation condition is regulated by SnRK1 and direct a new way for various anabolic and catabolic pathway during developmental and stress tolerance in plants.

5.11 SUMMARY

Over the 60 years' study, the full mechanism, activity and control of SnRK1 protein kinases is yet to be elucidated. There are many upstream and downstream molecules which controls complex mechanism of SnRK1 activation. Because SnRK1 complex itself shares very interactive network, it will be trustworthy to say its pivotal role in essential metabolic mechanisms and cell's vital functions. They involved in normal growth as well as stress responsive conditions fundamental life processes in plants.

KEYWORDS

- AMPK
- protein kinase
- serine/threonine kinase
- signal transduction
- SnRK-1

REFERENCES

1. Halford, N. G., & Hey, S. J., (2009). "SNF1-related protein kinases (SnRKs) act within an intricate network that links metabolic and stress signalling in plants," *Biochem. J., 419*, 247–259.
2. Stone, J. M., & Walker, J. C., (1995). "Plant protein kinase families and signal transduction," *Plant Physiol., 108*, 451–457.
3. Hrabak, E. M., Chan, C. W. M., Gribskov, M., Harper, J. F., Choi, J. H., Halford, N., Kudla, J., Luan, S., Nimmo, H. G., Sussman, M. R., Thomas, M., Walker-Simmons, K., Zhu, J.-K., & Harmon, A. C., (2003). "The Arabidopsis CDPK-SnRK superfamily of protein kinases," *Plant Physiol., 132*, 666–680.
4. Rolland, F., Thevelein, J. M., Sheen, J., & Baena-gonza, E., (2007). "LETTERS: A central integrator of transcription networks in plant stress and energy signalling," *Nature, 448*, 938–942. .
5. Haribabu, B., Hook, S. S., Selbert, M. A., Goldstein, E. G., Tomhave, E. D., Edelman, A. M., Snyderman, R., & Means, A. R., (1995). "Human calcium-calmodulin dependent protein kinase I: cDNA cloning, domain structure and activation by phosphorylation at threonine-177 by calcium-calmodulin dependent protein kinase I kinase," *EMBO J., 14*(15), 3679–3686.
6. Varjosalo, M., Keskitalo, S., VanDrogen, A., Nurkkala, H., Vichalkovski, A., Aebersold, R., & Gstaiger, M., (2013). "The Protein Interaction Landscape of the Human CMGC Kinase Group," *Cell Rep., 3*(4), 1306–1320.
7. Schenk, P. W., & Snaar-Jagalska, B. E., (1999). "Signal perception and transduction: The role of protein kinases," *Biochim. Biophys. Acta - Mol. Cell Res., 1449*(1), 1–24.
8. Mayerhofer, H., Panneerselvam, S., & Mueller-Dieckmann, J., (2012). "Protein kinase domain of CTR1 from arabidopsis thaliana promotes ethylene receptor cross talk," *J. Mol. Biol., 415*(4), 768–779.
9. Kulik, A., Wawer, I., Krzywińska, E., Bucholc, M., & Dobrowolska, G., (2011). "SnRK2 Protein Kinases—Key Regulators of Plant Response to Abiotic Stresses," *Omi. A J. Integr. Biol., 15*(12), 859–872.
10. Halford, N. G., Hey, S., Jhurreea, D., Laurie, S., McKibbin, R. S., Paul, M., & Zhang, Y., (2003). "Metabolic signalling and carbon partitioning: Role of SNF1-related (SnRK1) protein kinase," *J. Exp. Bot., 54*(382), 467–475.
11. Hardie, D. G., (2007). "AMP-activated/SNF1 protein kinases: conserved guardians of cellular energy," *Nat. Rev. Mol. Cell Biol., 8*, 774–785.
12. Hedbacker, K., & Carlson, M., (2008). "SNF1/AMPK pathways in yeast.," *Front. Biosci., 13*, 2408–2420.
13. Celenza, J. L., & Carlson, M., (1984). "Structure and expression of the SNF1 gene of Saccharomyces cerevisiae," *Mol. Cell. Biol., 4*(19), 54–60.
14. Hardie, D. G., Ross, F. A. & Hawley, S. A. (2012). "AMP-activated protein kinase: A target for drugs both ancient and modern," *Chem. Biol., 19*(10), 1222–1236.
15. Alderson, A., Sabelli, P. A., Dickinson, J. R., Cole, D., Richardson, M., Kreis, M., Shewry, P. R., & Halford, N. G., (1991). "Complementation of SNF1, a mutation affecting global regulation of carbon metabolism in yeast, by a plant protein kinase cDNA," *Proc. Natl. Acad. Sci. U.S.A., 88*(19), 8602–8605.

16. Nietzsche, M., Schießl, I., & Börnke, F., (2014). "The complex becomes more complex: protein-protein interactions of SnRK1 with DUF581 family proteins provide a framework for cell- and stimulus type-specific SnRK1 signaling in plants," *Front. Plant Sci.*, *5*, p. 54.

17. Tiessen, A., Prescha, K., Branscheid, A., Palacios, N., McKibbin, R., Halford, N. G., & Geigenberger, P., (2003). "Evidence that SNF1-related kinase and hexokinase are involved in separate sugar-signalling pathways modulating post-translational redox activation of ADP-glucose pyrophosphorylase in potato tubers," *Plant J.*, *35*, 490–500.

18. López-Paz, C., Vilela, B., Riera, M., Pagès, M., & Lumbreras, V., (2009). "Maize AKINβγ dimerizes through the KIS/CBM domain and assembles into SnRK1 complexes," *FEBS Lett.*, *583*(12), 1887–1894.

19. Tsai, A. Y.-L., & Gazzarrini, S., (2014). "Trehalose-6-phosphate and SnRK1 kinases in plant development and signaling: the emerging picture," *Front. Plant Sci.*, *5*, 119.

20. Sheen, J., (2014). "Master regulators in plant glucose signaling networks," *J. Plant Biol.*, *57*, 67–79.

21. Shen, W., Reyes, M. I., & Hanley-Bowdoin, L., (2009). "Arabidopsis protein kinases GRIK1 and GRIK2 specifically activate SnRK1 by phosphorylating its activation loop," *Plant Physiol.*, *150*, 996–1005.

22. Balderas-Hernández, V. E., Alvarado-Rodríguez, M., & Fraire-Velázquez, S., (2013). "Conserved versatile master regulators in signalling pathways in response to stress in plants," *AoB Plants*, *5*, p. plt033.

23. Hao, L., Wang, H., Sunter, G., & Bisaro, D. M., (2003). "Geminivirus AL2 and L2 proteins interact with and inactivate SNF1 kinase," *Plant Cell*, *15*(4), 1034–1048.

24. Farmer, E. E., Pearce, G., & Ryan, C. A. (1989). "In vitro phosphorylation of plant plasma membrane proteins in response to the proteinase inhibitor inducing factor," *Proc. Natl. Acad. Sci. U.S.A.*, *86*, 1539–1542.

25. Wirthmueller, L., Maqbool, A., & Banfield, M. J., (2013). "On the front line: structural insights into plant-pathogen interactions," *Nat. Rev. Microbiol.*, *11*, 761–776.

CHAPTER 6

B-VITAMINS IN RELATION TO SUSTAINABLE CROP PRODUCTIVITY IN CROP PLANTS

ANOOP KUMAR,[1] DILDAR HUSAIN,[2] JAGJEET SINGH,[1] SATYA PAL VERMA,[1] and ZAFAR ABBAS[1]

[1]PG Department of Botany G.F. College (M.J.P. Ruhilkhand University) Shahjahanpur – 242001, India

[2]Department of Botany, Maulana Azad Institute of Arts, Science and Humanities, Mahmudabad, Sitapur – 262003, U.P., India, E-mail: zafarabbas1255@yahoo.com

CONTENTS

ABSTRACT

This chapter deals with B-vitamins, which is involved in various physiological processes of plants, viz. participation in nutrient uptake, seed germination, vegetative growth, crop productivity in the improvement of vegetables, cereals, leguminous, aromatic oil crop, and particularly sugarcane crop. However, our understanding of the mode of B-vitamin action in plant is lagging far behind the level of understanding of the phytohormone action. The role of vitamins in some areas of plant physiology including vernalization, photoperiodism, phytohormone action, biological nitrogen fixation, and enzymatic action has still not been studied completely so far. The possibility of commercial exploitation of vitamins in agricultural domain particularly vitamin B_6 (pyridoxine) in sugarcane crop improvement has also been discussed with soil applied rice bran and seed-sett pyridoxine soaking with improved cane yield and quality.

6.1   INTRODUCTION

The problem of vitamins started with man from him. All the progress accomplished in this domain contributes to a better understanding of the problem in general and ultimately to human well being. It can no longer be doubted that vitamins are essential factors for the growth of plants. They are regulators for metabolism and serve to establish chemical reactions between the various organs of a complex plant. The need of vitamins is due to a loss of the ability to synthesize them. When this condition prevails the plant reacts like a heterotrophic animal. The reactions of plant and animal cells are similar, for the fundamental functions of vitamins are the same in the two kingdoms. The only aspects that differ are the morphological expressions to which the vitamin deficiency gives rise.

Funk in 1912 was the first to isolate an amine from rice husk and polishing that alleviated the symptoms of the disease "beriberi." He also proposed the generic term "vitamine" for it [1, 2] dropped the terminal "e" of "vitamine" because many of the compounds of this group were not amines. The term "vitamin" coined by him was accepted by later vitaminologists [3] classified all vitamins into two groups: (i) the fat soluble, and (ii) the water-soluble vitamin. The fat soluble vitamin included A, D, E, and K, while water soluble vitamin covered B and C. Morton [4] defined vitamin as: (a) organic compounds; (b) components of natural food but distinct from carbohydrates, fats or proteins; (c) present in normal food in extremely small concentration; (d) essential for normal health and growth; (e) causing specific deficiency symptoms when absent or not properly absorbed from the diet; and (f) not synthesized by the host. Therefore, they must be obtained exclusively from the diet, (distinction between vitamin and hormone). It is interesting to note that among the diverse living beings, only plants are capable of synthesizing vitamins. Animals depend directly or indirectly upon plants for their requirements of these essential dietary components. Ironically, the physiological roles of vitamins in plants are not as clearly understood as in animals. However, synthesis, distribution, translocation and function of B-vitamins were investigated to some extent in plants during the first half of the present century after aseptic cultivation of excised plant parts became possible. These studies indicated to the most B-vitamins are synthesized in leaves and were subsequently translocated to the site of action via the phloem. These have been proved to be indispensable for growth and differentiation of excised, particularly roots. These facts led to the establishment of B-vitamins as plant hormones and encouraged agricultural scientists to test them for augmenting the performance of various crops. Bonner and Bonner [5], and Aberg [6] presented excellent reviews, dealing with synthesis, distribution and translocation of these vitamins in plants and with their role in vitro-cultivation of excised organs including stem cutting. However, these reviews did not cover literature pertaining to the role and utility of the B-vitamins in crop improvement. With this brief background our review of literature presents status of most important and relevant research work done on pre-sowing, soaking and foliarly applied B-vitamins with soil applied rice bran (a natural source of pyridoxine) on different crop plants giving emphasis on sugarcane crop has been categorized as given in the following sections.

6.2 ROLE OF THE B-VITAMINS

6.2.1 THIAMIN

Thiamin was first isolated in pure form by Jansen and Donath in 1926 and its empirical formula ($C_{12}H_{18}N_4OSCl_2$) was proposed by Williams in 1936 [7]. Thiamin has two-ring system, a pyrimidine and a thiazole, which are united by a methylene bridge. It is found in the form of thiaminepyrophosphate (TPP) which functions as a co-enzyme in several enzymatic reactions in which an aldehyde groups are transferred from donor to an acceptor.

Thiamin

6.2.2 RIBOFLAVIN

Osborne and Mendel in 1913 first recognized this vitamin in milk, which promotes growth. In 1933, riboflavin was isolated in pure form by Ellinger and Koschara; and Kuhn, Gyorgy and Wagner-Jaur-egg. The structure of riboflavin was established by Kuhn and Karrer in 1935 [7]. Riboflavin has the empirical formula $C_{17}H_{20}N_4O_6$. It is a component of two closely related co-enzyme, flavin mononucleotide (FMN) and flavin adenine dinucleotide (FAD). They functions as tightly bound prosthetic groups of a class of dehydrogenases known as flavoproteins or flavin dehydrogenases. The isoalloxazine ring of the flavin nucleotides serves as transient carrier of a pair of hydrogen atoms removed from the substrate molecule in the reactions catalyzed by these enzyme [1].

Vitamin B₂ (riboflavin)

6.2.3 PANTOTHENIC ACID

In 1933, R.J. Williams found that naturally occurring compound of unknown chemical composition, stimulates the growth of yeast. Later, in 1938, its chemical nature was worked out by him. Kuhn and Wieland in 1940 synthesized the pantothenic acid. It has the empirical formula $C_9H_{17}O_5N$. Pantothenic acid is a component of co-enzyme A, which is a transient carrier of acyl groups in enzymatic reactions involved in fatty acid oxidation, fatty acid synthesis, pyruvate oxidation and biological acetylations. The chemical mechanism by which co-enzyme A carries acyl groups was established by F. Lynen in 1951. He isolated an "active" form of acetate from yeast and showed it to consist of a thioester of acetic acid with thiol or sulfhydryl group of co-enzyme A [1].

Pantothenic acid

6.2.4   NIACIN (NICOTINIC ACID)

In 1912–14 Funk and Suzuki isolated nicotinic acid from yeast and rice bran, but failed to recognize its vitamin character. Elvehjem and Woolley in 1937, identified it as the nutritional factor preventing black-tongue in dogs and pellagra in man. Nicotinamide is the amide form of nicotinic acid. In 1937, Fouts et al. [7] reported the first successful treatment of human pellagra with nicotinamide. Nicotinic acid and nicotinamide has empirical formula $C_6H_5O_2N$ and $C_6H_6ON_2$, respectively [7]. Nicotinamide is component of two related co-enzymes, nicotinamide adenine dinucleotide (NAD) and nicotinamide adenine dinucleotide phosphate (NADP). These two co-enzymes are also referred to as pyridine co-enzymes or pyridine nucleotides, since nicotinamide is a derivative of pyridine. The pyridine nucleotides function as the co-enzymes of a large number of oxidoreductases, collectively called as pyridine-linked dehydrogenases. They act as electron acceptors during the enzymatic removal of hydrogen atoms from specific substrate molecules. One hydrogen atom from the substrate is transferred as a hydride ion to the nicotinamide portion of the oxidized forms of the coenzymes (NAD and NADP) to yield the reduced coenzyme ($NADH^+$ and $NADPH^+$, respectively) [1].

Niacin

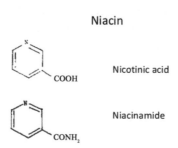

Nicotinic acid

Niacinamide

6.2.5   PYRIDOXINE

First in 1932 Ohdake isolated a compound of the formula $C_8H_{11}O_3N$. HCl from rice polishing, but failed to recognize its vitamin character. Gyorgy was the first to define vitamin B_6 as distinct entity in 1934. Later on isolation of the pure crystalline vitamin B_6 was done independently by various research workers namely, Lepkovsky and coworkers in 1939 [7]. Vitamin

B$_6$ groups includes three closely related compounds viz. pyridoxine, pyridoxal, and pyridoxamine. The active form of vitamin B$_6$ is pyridoxal phosphate, which also occurs in its amino form pyridoxamine phosphate. Pyridoxal phosphate serves as the tightly bound prosthetic group of a number of enzymes catalyzing reactions of amino acids. The best common of these are transminations reaction in which an amino group of α-amino acid is reversely transferred to α-carbon atom of an α-keto acids [1].

Pyridoxine

6.2.6 BIOTIN

In 1927, Boas found that certain foods contain an organic substance which protects against egg-white toxicity. In 1931, Gyorgy recognized the necessity of an egg-white injury factor (vit. H) for man. In 1935, Kogl was the first to isolate it in pure form and in 1942 its chemical structure was given by Duvigneaud [7]. Biotin has empirical formula ($C_{10}H_{16}O_3N_2S$). It contains fused imidazole and thiopene rings. Biotin is the active component of biocytin. Biotin is transient carrier of carbondioxide during the action of certain carboxylating enzymes, e.g., Propionyl CoA carboxylase and acetyl-CoA carboxylase [1].

Biotin

6.2.7 FOLIC ACID

Pfiffner in 1943 reported the isolation of a compound, which was active for *Lactobacillus casei* and which promoted growth and hemoglobin formation in chick. This factor was designated "vitamin B_6" because of its relation to the chick nutrition. Later on in 1945, Pfiffner announced that this active compound is present in yeast and yeast extracts [8]. Folic acid has three major compounds; glutamic acid, P-amino benzoic acid and Pteridine derivatives. Folic acid does not act as co-enzyme itself, but it is enzymatically reduced in the tissues to tetrahydrofolic acid (FH_4), which is the active co-enzyme form of this vitamin. FH_4 functions as an intermediate carriers of 1-carbon groups in a number of complex enzymatic reactions in which -CH_3, -CH_2-, -CH=, -CHO, -CH=NH groups are transferred from one molecule to another [1].

HOOC-CE₂-CH₂-CHNH

glutamic acid   P-aminobenzoic acid   Pteridine

6.2.8   VITAMIN B₁₂ (CYANOCOBALAMINE)

It is discovered by George Minot and William Murphy in 1926 and isolated in crystalline form in 1948 by E.L. Smith and by E. Rickes and K. Folkers. Its empirical formula is $C_{63}H_{90}O_{14}N_{14}.PCO$. It is the precursor of coenzyme B_{12}. Vitamin B_{12} is unique among all the vitamins in that it contains not only a complex organic molecule but also an essential trace element, cobalt. Vitamin B_{12} is called cyanocobalamine because it contains a cyano group attached to the cobalt. Enzymes requiring co-enzyme B_{12} have a common denominator the ability to carryout the shift of a hydrogen atom from one carbon atom to an adjacent one, in exchange for an alkyl, carboxyl, hydroxyl, or amino group [1].

Vitamin B₁₂

6.3   SOIL APPLICATION: RICE BRAN

As the aim of all agro-physiological research is to increase economic yield of crop plants. One way to achieve this objective is to devise techniques for the most efficient-utilization of the input including growth regulators, irrigation fertilizer, and of the vitamins tested, pyridoxine was noted to have the best effect on root growth. Since then these researches have been diversified to investigate the practical implication of various physiological processes in cereals, legumes and oil crops enabling them to utilize the costly input more efficiently and thus pave the way for higher productivity and improved quality of the produce [10–12]. For sugarcane, this technique could be of added practical significance as soil-applied rice bran (a natural source of pyridoxine) besides giving pyridoxine shot/prime for growth of root would not only result in higher water and nutrient uptake but also conserved more water at formative stage (shoot emergence + tillering) which probably helped plants to tiller profusely [13, 14]. It had become abundantly clear in the early forties that besides plant hormones, e.g., auxins, gibberellins, etc. growth promoting substances included another set of heterogeneous organic compounds (vitamins) with varying modes of action. These compounds are required in small amount only, and in a more or less unchanged form participate at various stages in the labyrinth called metabolism. However, knowledge of the vitamin requirement of plants had to wait the development of organ and tissue culture techniques because, although whole plants are capable of synthesizing them, their organs cannot do so separately. Thus, within the intact plant, by virtue of their translocation from the site of synthesis to that of action in plants [5, 15]. Nevertheless, detailed consideration of the effect of these compounds on intact plants remained unduly neglected, under these dismal circumstances the conclusions drawn by Ref. [6] in this review are no doubt remarkable. He opined that "normal green plants can certainly grow well without any external vitamin application, but it is "naturally not excluded that certain plants or certain developmental stages may contain slightly sub-optimal vitamin amounts and therefore, may be stimulated by further addition." For more than two decades of the role of B- vitamins in plants has attracted sporadic attention. These studies have indicated that various physiological processes, e.g., nutrient uptake, seed germination, photosynthesis, as well as chlorophyll and protein synthesis, depend more or less on the availability of the B-vitamins. However, in-depth investigation is required so as to ascertain their exact modes of action in plants. The effect of rice bran (a natural source of pyridoxine) has been worked out to a very little extent so far [13, 14, 16–25].

6.4   B-VITAMINS AND PLANT PHYSIOLOGY

B-vitamins constitute a heterogeneous group of varied organic compounds that are designated for simplification as B_1 (thiamine), B_2 (riboflavin), nicotinic acid (niacin or pellegra preventive, i.e., PP factor) pantothenic acid, B_6 (pyridoxine), biotin, folic acid and B_{12} (cyanocobalamine). These vitamins act as coenzymes whose functions in microorganisms and animals are well understood and are included in various standard textbooks of biochemistry [1]. Earlier, it was believed that in higher plants also vitamins operate by playing similar to their established roles in other organisms [5]. However, this view has been elaborated to a great extent in the light of research conducted during the last three to four decades. It has now been fairly established that in plants, B-vitamins act not only as coenzymes and phytohormones [5, 6] but also play a regulatory role in various physiological processes [9–12, 26] observed that thiamine, riboflavin and nicotinamide promoted respiratory activity in one-leaf seedlings of wheat, oats, and barley. Iijima [27, 28] also noted that foliar spray of thiamine increased respiration in kidney beans, potatoes and sweet potatoes. The respiratory quotient was found to be unity indicating that thiamine promoted the catabolism of carbohydrate. Moreover, leaf applied thiamine was also reported to enhance sugar and starch content and C/N ratio in stems.

Some members of B vitamins are reported to be involved in chlorophyll synthesis. For example, Ref. [29] found that spray of 50 mg/L vitamin B2 checked the decomposition of chlorophyll (caused by foliar spray of 20 mg/L gibberellic acid). The treatment also increased the chlorophyll content in sugarbeet. Similarly, exogenous application of pyridoxine in solution culture promoted chlorophyll content in wheat and pea seedling raised from pyridoxine deficient seeds [30]. Besides [31], working with thiamine deficient albino mutant tomato plants, made a systematic study with regard to the role of thiamine in chlorophyll synthesis. They applied 2 mg thiamine with 10ml water twice a week to the leaves. The treated plants showed normal chlorophyll formation: but their progeny remained albino. Interestingly, when the albino plants were provided with and aminolevulenic acid and prophobilinogen (chlorophyll precursors), the plants did not assimilate these compounds into chlorophyll, suggesting that thiamine deficiency impaired chlorophyll synthesis before the formation of pyrrole rings. Further, supply of thiamine precursore, *viz.* thiazole and pyrimidine, individually or in combination showed that mutant plants responded only to pyrimidine, indicating impairment of its synthesis. On the other hand, vitamim B_{12} has been noted to stimulate the development of chloroplast [32]. However, working with seedlings of early and late maturing varieties of groundnut opined, on the basis of close relationship between contents of B-vitamins and chlorophyll, that all the B group vitamins may be related to chlorophyll synthesis [33]. However, this cause and effect relationship needs to be substantiated with direct experimental proof.

There are reports showing participation of B-vitamins in photosynthesis. Foliar spray thiamine solution was noted to enhance photosynthesis in kidney beans and cabbage [34]. Contrary to this, riboflavin application promoted photorespiration and inhibited photosynthesis by 50 to 70% [35]. However, pyridoxal phosphate proved to be an effective stimulator of photosynthetic electron transport in isolated chloroplast [36]. Similarly, in Ref. [37] it is observed in vivo stimulation of $^{14}CO_2$ fixation in chloroplasts isolated at two vegetative and two reproductive stages from cluster bean plants administered foliar spray of thiamine [200 mg/L), riboflavin [50 mg/L) and folic acid [100 mg/L) at 15 and 30 days after sowing. Moreover, isolated chloroplasts from non-treated cluster bean plants also showed enhanced $^{14}CO_2$ fixation when they were incubated in vitro with the above B-vitamins at concentration ranging from 0.1 to 20 mg/L. However, there was a marked difference between in vivo and in vitro $^{14}CO_2$ fixation as far as vitamin doses were concerned. Thus, whereas isolated chloroplasts exposed to $^{14}CO_2$ in vitro responded most when vitamin concentration was below 5 mg/L, similarly obtained chloroplasts showed maximum fixation of $^{14}CO_2$ when the concentration of vitamins administered in vivo was as high as 50 to 200 mg/L. Of the vitamins tested, niacin proved the most effective and folic acid, the least. The authors presumed that these vitamins were indirectly involved in $^{14}CO_2$ fixation via increasing endogenous level of certain phytohormones.

The other investigations revealed the participation of the vitamins in amino acid and protein synthesis. Williams et al. [38] established that biotin stimulated aspartic acid synthesis. Similarly riboflavin mediated the incorporation of ^{14}C into serine, tryptophan and tyrosine and reversed the chloramphenicol inhibited protein synthesis partially in green (*Phaseolus radiatus* L.) seedlings [39, 40]. Subsequently, Ref. [41] suggested that B-vitamins enhanced protein synthesis either at transcriptional or translational level.

Regulation of enzyme activity by B-vitamins is well established [1]. Present researches have thrown light on their influence on some new enzyme systems [9–12]. Foliar spray of 50 mg/L thiamine, riboflavin, and nocotinic acid restored the activities of peroxidase and carbonic anhydrase inhibited by 2,4-dichlorophenoxy acetic acid in oats and beans. However, riboflavin was the most effective in restoring enzyme activity while nicotinic acid proved least effective [42]. Inclusion of biotin, niacin and pyridoxine in nutrient solution promoted the activity of succinic dehydrogenase in root and shoot of 4d old seedlings of *Phaseolus radiatus* [40]. Moreover, [43] observed an increase in α-amylase activity in green gram seedlings as a result of treatment with a low dose of riboflavin and high dose of thiamine. They opined that these vitamins probably mimicked the role of gibberellic acid.

The response of stomata to B-vitamins has recently bean studied in cluster bean by Ref. [44]. Or the vitamins tested, pantothenic acid increased the stomatal index and frequency at early vegetative stage. On the other hand, pyridoxine and, thiamine were the most effective for stomatal index and frequency, respectively at the last stage of reproductive growth. Riboflavin affected the opening of stomata to the maximum extent at early hours of the day during entire period of plant growth. However, a peak in stomatal opening was observed with all vitamins especially from 9.00 h to 21.00 h. The concluded that these vitamins presumably enhanced endogenous levels of phytohormones, particularly cytokinins, leading to increased cell division and differentiation as well as starch hydrolysis. The former action seemed to be reflected in increased stomatal index and frequency and the latter, in the opening of stomata. Schneider et al. [9] reported that pyridox-ine is a cofactor for many enzymatic reactions, especially those involved in amino acid metabolism.

6.5   B-VITAMINS AND NUTRIENT UPTAKE

Absorption of nutrients against their concentration gradient in plants is closely associated with various physiological processes. This act, com-monly known as "accumulation." is primarily accomplished by certain carrier proteins, running across the cell membrane, at the expense of meta-bolic energy [45]. Since B-vitamins regulate various processes of metabo-lism, including release of energy via respiration, synthesis of proteins etc. [26–28, 41, 46], these vitamins may, therefore, help in the absorption of nutrients from the soil solution. However, critical studies on this aspect are meager.

Kudrev and Pandey [47] noted that spraying with thiamine or pyridox-ine accelerated uptake of nitrogen in wheat plants from seedling stage to ear emergence but did not alter the general trend. Similarly, pyridoxine treatment to seeds for 1–3 h also increased the nitrogen and phosphorus content in 2d old seedlings of cotton. However, this increase was more in the presence of potassium dihydrogen orthophosphate than that of super-phosphate [48]. Dimitrova Russeva and Lilova [49] studied the effect of application of thiamine, pyridoxine and nicotinic acid on the uptake of nitrogen and phosphorus by *Mentha piperita* in nutrient solution and in soil. The uptake of nutrients was increased by the application of these vitamins, phosphorus responding particularly to single application of nic-otinic acid and double application of others. Pandey [50] performing a sand culture experiment, noted greater uptake of nitrogen in wheat as a result of application of vitamin B, separately or in combination with indol 3-acetic acid. Gopala Rao and Raghava Reddy [51] studied the compara-tive efficacy of B-vitamins in sodium, potassium, calcium and phosphorus uptake in one-week-old Vigna radiata seedlings. Thiamine and biotin were ineffective in phosphorus uptake, but riboflavin, pyridoxine and panto-thenic acid facilitated the uptake of all the nutrients studied. However, pyridoxine, nicotinic acid and pantothenic acid particularly exhibited more influence on potassium and phosphorus uptake than on that of the other two nutrients. In an extensive study for two years, [52] reported that soaking lentil seeds for 12 h and mung bean seeds of 4 h in 0.3% pyridoxine solution enhanced leaf nitrogen, phosphorus and potassium content at vegetative, flowering and fruiting stages of the two crops. Moreover, foliar spray of 0.2% pyridoxine solution at flowering stage increased the content of these nutrients in leaves (estimated at fruiting stage) of both crops. He assumed that treatment with vitamins, particularly pyridoxine might have facilitated the uptake of these nutrients by increasing the permeability of the root cell membrane. Alternatively, they might have behaved like coenzymes of certain carrier proteins, which are considered to be responsible for transporting the nutrients across the membrane. These assumptions, however, need experimental verification and present new research dimensions in vitaminology. Houshmandfar et al. [10] in order to study the effect of seed pyridoxine-priming and different nitrogen levels on corn (Zea mays L. var. Sc 704), a field experiment was conducted during 2007 at corn cultivation season in Saveh region of Iran. The experiment was laid out in a split-plot design with tree nitrogen levels included total application of 80 (control), 130 and 180 kg N ha^{-1} as urea in the main plots and tree seed pyridoxine-priming treatments included 0 (water-soaked control), 100 and 200 mg/L in subplots in three replications. Pre-sowing seed treatments with pyridoxine were applied for eight hours duration. Leaf area index (LAI) and total dry matter (TDM) were calculated from 30th to 105th day after sowing (DAS). Both pyridoxine and nitrogen treatments showed significant positive effects on yield and yield components such as grain yield, number of grains per ear, number of grains per row, and protein content of grain. Pyridoxine concentration of 200 mg/L with a total application of 180 kg N/ha produced maximum value of growth and yield parameters. The experiment was set up to investigate that how seed pyridoxine-priming treatment and different nitrogen levels affected growth and yield characteristics of corn plant. Pyridoxine-priming and nitrogen fertilizer improved growth parameter consist of LAI and TDM, yield and yield components such as grain yield, protein content of grain, and number of grains per ear. The higher growth and yield characteristics in the treated plants could be resulted in an advanced nutrient uptake related to nutrient bioavailability. An increase in N application could enhance nitrogen bioavailability, which helped in nitrogen uptake. It is also could be viewed as a function of better root growth as pyridoxine acts as a root growth factor. An increase in root growth also could enhance nutrient uptake, which helped in the proliferation of the leaf. Improvement of leaf proliferation is clear from increased leaf surface shown by LAI resulting into enhanced TDM. Higher nutrient uptake and translocation of nitrogen in plant also could subsequently lead to enhanced amino acid and protein synthesis. Similar high nutrient uptake in *Mentha piperita, Brassica juncea* and *Triticum aestivum* treated with pyridoxine have also been reported by other workers. Pyridoxine concentration of 200 mg/L with a total application of 180 kg N/ha produced maximum value of grain yield and protein content. In conclusion, these results have practical implications in that pre-sowing seed treatment with pyridoxine solution could enhance the availability of nitrogen to the crop, and could ensure the production of more nutritious corn grain yield. Seed pyridoxine-priming is a simple economical way to improve the quality and quantity of corn yield.

6.6   B-VITAMINS AND SEED GERMINATION

The seed is the originator of the next generation and, therefore, contains adequate quantities of reserve food, minerals and growth regulators to sustains life until the emerging seedling becomes self sufficient. However, impaired partitioning of photosynthates and other substances during seed formation results in low germinability of seeds. Generally, it has been observed that low content of growth regulators in seeds severely reduces germination, which is improved by soaking the seeds with these substances. Iijima [27] achieved augmented seed germination in kidney bean, maize and radish by soaking their seed for 24 h in 0.01 ppm thiamine solution. However, old seeds of these crops showed much higher% germination in response to thiamine treatment. Aizikovick [53] also noted that treating rice seeds with vitamin B_1 and B_2 resulted in high rate of seed germination under field conditions. Orcharov and Kulieva [48] observed a slight improvement in the germination in cotton as a result of soaking its seeds for 1–3 h in 0.1% pyridoxine solution. On the other hand, [54] found that niacin at a concentration of 250 ppm had no effect on seed germination. Serebryakova [55] treated seeds of *Rosa cinnamonea* with solutions of nicotinic acid or vitamin B_1. The emergence of seedlings was noted to be 43 and 53% more in 0.01% nicotinic acid and 0.02% vitamin B_1 solution, respectively. In addition, [56] applied seven different B-vitamins to two cultivars of Capsicum through seed soaking and secured high seed germination rates coupled with improved seedling growth. Soaking of rose seeds in nicotinic acid, vitamin B_1 and B_2 promoted germination by 43, 51, and 30%, respectively, over the control, Triticale exhibited remarkable improvement in seed germination under field conditions on receiving exogenous supply of 0.2% pyridoxine solution through seeds [57].

The above studies indicate that seed treatment with B-vitamins trigger some unknown physiological processes which subsequently influence germination in various crops. It is well established that commencement of seed germination results in the enzymatic breakdown of reserve food material (carbohydrates, fats, and proteins). Therefore, it is of paramount importance to study the response of various hydrolytic enzymes, particularly amylases, lipases and proteases to the application of vitamins. Neelmathi et al. [58] observed early induction of axillary shoots in sugarcane has been obtained by using MS medium supplemented with riboflavin 10 mg/L along with growth regulators BAP [0.25 mg/L), NAA [0.5 mg/L), and kinetin [1.07 mg/L). Lowest level of riboflavin (mg/L) with the combination of 2.0% sucrose induced higher frequency of multiplication.

6.7   B-VITAMINS AND VEGETATIVE GROWTH

Bonner and Greene [59] observed the vitamin requirement of slow and fast growing plants under green house conditions. The slow growing species of plants, e.g., Aleurites, Bougainvillea, Arbutus, Eucalyptus, and Comellia, responded to thiamine supplied with Hoagland's solution in sand culture. After two months, it was recorded that shoot length became double compared to the respective control. Similarly, root system of the treated plants also exhibited luxuriant growth. But fast growing species of annual plants, like pea, radish, and tomato, showed no response to the vitamin treatment. They also noted considerable amount of vitamin B_1 in the applied organic manure that might have exerted beneficial effect on plant development. In another study, addition of 0.1 mg vitamin B_1 with nutrient solution favored dry matter accumulation in plants cultivated in sand or soil. Moreover, carrot seedlings, receiving this vitamin on alternate days for a year, showed higher growth rate than the control. Similarly Cosmos, Poa, and Brassica (with low endogenous level of vitamin B_1), responded positively to the vitamin treatment, white tomato and pea, possessing high content of the vitamin in their leaves, proved to be non-responsive. Analysis of responsive and non-responsive varieties of pea leaves also revealed that the former varieties possessed only half of the vitamin content noted in those of the latter, suggesting vitamin B_1 level in leaves to provide a reliable criterion B_1 for deciding whether or not to treat the plants with this vitamin for the improvement of their performance [59]. Hitchcock and Zimmerman [60] observed a beneficial effect of thiamine on aster plants. Weekly treatment with this vitamin produced taller plants with higher fresh weight than control plants receiving only tap water.

Murneek [61] conducted two parallel experiments on tomato dill, Rudbeckia, cosmos and ornamental peppers in pots under greenhouse conditions. The plants were grown on poor and rich soils, as judged by the quantity of organic matter present. Of these plants, cosmos proved to be highly responsive to the application of vitamin B_1 and tomato was totally non-responsive. In the parallel experiment, in place of vitamin B_1 the same species were supplied with leaf mold that contained appreciable quantity of vitamin B_1. The composition of soil of each pot in both experiments was in the ratio of 1/3 loam: 1/3 sand 1/3 mold. Supply of vitamin B_1 exerted a conspicuous beneficial (though variable) effect on these plants. Roots were noted to be particularly influenced resulting in beneficial effect on the entire plant body. Spreading of 1/2 to 3/4 inch (1.27 to 1.90 cm) leaf mold on the soil surface proved good, and sometimes yielded better results than those with vitamin B_1, as leaf mold might have also provided other organic stimulants. He further concluded that in agricultural practices success with vitamin B_1 and other growth regulators would seems to depend largely upon the types of soil and fertilizer used. Mariat [62] applied vitamin B_1 at the rate of 2.6 mg/L with symbiotic-jillied and sugared Knop's solution and compared the performance of treated cattleya orchid seedling with those receiving no vitamin. He observed that 42% of the total seedlings receiving the vitamin produced leaves in three months, while only 3% of untreated seedlings could do so.

Brusca and Hass [63] found 0.01 and 0.02 g/plant of vitamin B_6 and 0.02 g/plant vitamin B_{12} added to the nutrient solution stimulated the growth of citrus plants. They observed that application of vitamin B_1 and B_2 enhanced height, leaf number, fresh and dry weight of pea, broad bean, beet and wheat in pot culture. The effect of both vitamins was most pronounced on beet and the least, on pea. For example, these vitamins at 0.01 mg/L increased the leaf number by 25% in beet seedlings.

Mel' Tser [64] reported interesting results in field trials on two spring wheat cultivars (Krasnozernaya and Remo) receiving 0.01% nicotinic acid solution through seed treatment. The vitamin stimulated the growth of Krasnozernaya at the early stage and that of Remo at the late stage. In general the number of florets/spikelet was increased in both cultivars. Orcharov and Kulieva [48] noted two to three fold increase in leaf area of cotton seedlings receiving 0.01% pyridoxine solution for 1.3 h as pre-sowing treatment.

Vergnano [65] applied vitamin B_1 and B_6 (at the rate of 0.01) mg/L with nutrient solution) to Colutca arborescens, Hedera helix, and Rosa cuttings in sand culture. These vitamins induced rooting in the latter two plants. Treated plants of Rosa also produced more buds and leaves with broader leaf blades that the controls. Similarly, enhanced growth of root, stem, leaf, umbel and fruit in Ammi visnaga was obtained as a result of seed treatment with 50 mg/L thiamine [66]. Contrary to the above findings, the authors of Refs. [67–72] did not observe any growth stimulating effect of vitamin B_1, B_2, and nicotinic acid on various horticultural crops, including tomato, lettuce, cosmos, mustard, radish, cauliflower, musk melon, pepper, rutabagas, snap beans, and summer squash. Likely, it seems that the varieties of crops tested by them contained sufficient amounts of endogenous vitamins, and therefore, did not respond to the exogenous supply of the vitamin. Sajjad and Samlullah [12]. In one field experiment conducted according to simple randomized block design on mustard (*Brassica juncea* L.) Var. Varuna to establish the most suitable concentration for soaking of mustard seed in thiamine hydrochloride solution of concentration that is 0.01, 0.02, 0.03, 0.06, 0.09, and 0.12% on the performance of mustard. The parameter such as plant dry weight, leaf area, leaf area index (LAI), crop growth rate (CGR), relative growth rates (RGR), and net assimilation rate (NAR) were calculated from 40 to 120 days sowing (DAS), Yield and its components such as number of pods per plant, 1000 seed weight, yield; biological yield, harvest index and oil yield were measured at harvest. In general soaking of seeds in 0.03% of thiamine hydrochloride solution was found to be the best in comparison to any other concentration of the treatment for most of the parameter studied. In growth parameters, leaf area, plant dry weight showed significant results at the early stages of sampling (that is, 40–80 DAS). CGR also followed the same trend; however, RGR and NAR did not follow the same at any stage of the sampling. Yield parameter like the pod number per plant, biological yield and seed yield as well as oil yield registered significant result. Thus it was concluded that soaking of mustard seeds in 0.03% thiamine hydrochloride ensured better growth and yield. Judicious application of fertilizers can improve the productivity of the crop. However, application of large quantities of inorganic fertilizers has disadvantages too. On one hand, it increases burden on our already exhausted foreign reserves for importing these fertilizers and on the other hand it cause pollution hazards too. As B-vitamins are known to be involved in various physiological activities of the plants, it can be concluded that growth and yielding ability of mustard could be improved by pre-sowing seed treatment with economical dilutions of aqueous thiamine solution and fertilizer requirement of mustard variety could be cut down.

6.8   B-VITAMINS AND CROP PRODUCTIVITY

Review of literature in earlier sections clearly points out a pleiotropic role of B-vitamins, establishing that treatment with these substances modifies various physiological processes. As crop performance at harvest is the manifestation of these processes, it is not beyond expectation to ameliorate crop productivity by administering B-vitamins to the seeds or to the seedlings at the beginning. The findings discussed below bear testimony to this hypothesis. Kjelvick [73] studied the effect of soaking vegetable seeds for 24 h in nicotinic acid on their yield performance. He observed that yield of radish and outdoor cucumber was enhanced by this treatment.

Alzikovick [53] also reported 2% increase in rice yield as a result of pre-sowing seed treatment with vitamin B_1 and B_2. Similarly, soaking of pea seeds in 0.01% vitamin B_1 solution enhanced seed yield by 15% [74]. Dimitrova Russeva and Lilova [49] observed the effect of application of thiamine, pyridoxine and nicotinic acid on *Mentha piperita*. Nicotinic acid enhanced the essential oil yield, as did two applications of thiamine, whereas pyridoxine reduced it.

In a field trial, [75] investigated the effect of soaking wheat seeds for 12 h in 0.005. 0.05, and 0.5% nicotinamide solution. Grain yield was enhanced from 16.7 q/ha in the control to 21.9, 18.8, and 14.5 q/ha by the application of 0.0005 and 0.05% nicotinic acid and 0.5% of nicotionamide treatment, respectively, while it was reduced to 14.3 q/ha due to 0.5% nicotinic acid application. Further, the grain so produced possessed 89.2, 100.2, 96.3, 87.9, and 89.3 mg protein g dry grain, respectively, as compared with 82.7 mg in the control It indicated that only treatments with nicotinic acid improved the grain quality.

Polyanskaya and Kuvadov [76] reported an increase in the number of both plant bolls, boll weight and cotton yield as a result of soaking cotton seeds in nicotinic acid before sowing. In another study, seed treatment with nicotinic acid resulted in early maturation of two cultivars of Capsicum accompanied by high yield, but other B-vitamins proved ineffective in this regard [56, 77] studied the effect of thiamine and nicotinic acid on tomato under field conditions. Plants grown from vitamin treated seeds produced 23.1 to 30.2% higher yield than the control plants.

Kulieva et al. [78] investigated the response of melon and watermelon to vitamin treatments under laboratory and field conditions. Seeds of these plants received various concentrations ranging from 0.01 to 0.0001% of thiamine, nicotinic acid, pyridoxine or cyanocobalamine. Number and weight of fruit at 45 and 90 days was noted to be increased by thiamine (0.001%), cyanocobalamine (0.0001%) or nicotinic acid (0.0001%). Seed treatment with 50 mg/L thiamine solution also promoted fruit growth and increased the concentration of total hormones and their main components, khallin and visnagin in *Ammi visnaga* [66]. Similarly, vitamin B_1 enhanced ascorbic acid formation by 50% and B_2 and nicotinic acid, by 25% in young rose plants, which were, raised form the seeds treated with vitamin B_1, B_2 or nicotinic acid.

6.9 FOLIAR APPLICATION OF B-VITAMINS

Compared to seed treatment, less attention has been paid to the study of spray of B-vitamins on standing crops. However, in most of the cases, foliar application of the vitamins improved growth performance, which was subsequently manifested in enhanced economic yield of various crops. The available literature is reviewed below.

Iijima [79] observed a beneficial effect of spraying thiamine solution on the leaves of sweet potatoes. Foliar application of 1 ppm vitamin solution (at the rate of 2 mL/plant at intervals of 10 d) increased the number and length of roots accompanied by high percentage of large storage roots. The fresh weight of stems and leaves was also increased by the treatment. Moreover, spraying was more effective in the first half than in the second half of the growing season. Cajlahjan [80] noted that vitamin B_1 applied to plants by vacuum infiltration or spray promoted flowering in perilla and maize, but nicotinic acid proved to be ineffective. Besides, combination of thiamin spray with acidic chemicals or with urea resulted in high yield responses of various garden crops; but in the presence of alkaline chemicals, the vitamin spray curtailed their yield [81]. Boukin [82] reported growth-stimulating effect of nicotinic acid or thiamine spray on beans, tobacco tomato, mulberry and carrot. Application of 0.001% solution of these vitamins promoted leaf growth in general and enhanced root yield more than 50% in carrot. Galachalova et al. [83] observed that spray of thiamine enhanced grain yield of a cultivar of wheat containing low content of seed thiamine.

Popova et al [84] sprayed the pistils of *Capsicum annum* just after pollination with thiamine, riboflavin, pyridoxine and nicotinic acid in various combinations. The vitamins increased fruit set and number of seeds fruit, shortened the growing period and increased plant height in F_1 generation. All the vitamins, except riboflavin, caused earliness in the F_1. Arsen'eva [85] investigated the effect of physiologically active substances, including B-vitamins (thiamine, nicotinic acid, and pyridoxine) and vitamin combination and completion of flowering, flower size and color and shoot growth in lilac, hydrangea and spiraea. The spray of vitamins was given during reproductive organ differentiation and floral bud development. Among these plants lilac gave the best response for parameters studied to 100 mg pyridoxine spray, while a mixture of thiamine, nicotinic acid and vitamin at the concentration of 100 mg/L proved optimum for spiraea.

Et-kholy and Saleh [86] sprayed *Matricaria chamomilla* with 10 mL solution each of thiamine and vitamin C [25, 50, 100, and 150 ppm) at 30 and 45 after transplantation. Of these, 150 ppm of either thiamine or vitamin C significantly increased the number of flowers and head production/plant by 47.22 and 56.60%, respectively, in first growing season over the control; but essential oil and chamazulene percentage in treated plants remained unaltered. However, the total average yield of these economically important ingredients was increased by the application of `50 ppm of either of these vitamins. It had become abundantly clear in the early forties that besides plant hormones, e.g., auxins, gibberellins, etc., growth promoting substances included another set of heterogeneous organic compounds (vitamins) with varying modes of action. These compounds are required in small amounts only, and, in a more or less unchanged form, participate at various stages in the labyrinth called metabolism. However, knowledge of the vitamin requirement of plants had to wait the development of organ and tissue culture techniques because, although shole plants are capable of synthesizing them, their organs cannot do so separately. Thus, with the intact plant, by virtue of their translocation from the site of synthesis to that of action in plants. B-vitamins conform to the definition of hormones in plants [5, 15]. Nevertheless, detailed consideration of the effect of these compounds on intact plants remained unduly neglected. Under these dismal circumstances the conclusions drawn by [6] in his review are no doubt remarkable. He opined that "normal green plants can certainly grow well without any external vitamin application, but it is naturally not excluded that certain plants or certain developmental stages may contain slightly suboptimal vitamin amounts and, therefore, may be stimulated by further additions." For more than two decades, study of the role of B-vitamins in plants has attracted sporadic attention. These studies have indicated that various physiological processes, e.g. nutrient uptake, seed germination, respiration, photosynthesis as well as chlorophyll and protein synthesis, depend more or less on the availability of the B-vitamins. However, in depth investigations are required so as to ascertain their exact modes of action in plants. Kodendara-maiah and Gopala Rao [44] suggested that B-vitamins participate in plant growth and development indirectly by enhancing the endogenous levels of various growth factors such as cytokinins and gibberellins. However, this assumption needs verification by monitoring the change, if any in the concentration of various endogenous growth regulators in response to the application of the B-vitamins. The risk could be simpli-fied by using labeled vitamins to probe their involvement in various processes. The involvement of B-vitamins in the expression of genes particularly during seed germination, is also a definite possibility that may be investigated by separating and identifying proteins (gene prod-ucts) by gel electrophoresis after treating seeds or other plant organs with the vitamins. A comparison of the protein pattern of the treated parts with that of the control could be expected. Zanjan and Asli [11] made objective of this study to evaluate the effects of seed Pyridoxine-priming duration on germination and early seeding growth character-istics of two wheat genotypes included inbred lines of PBW-154 and PBW-343. The experiments were carried out with completely random-ized design (CRD) with five seed priming treatments in three replica-tions. The seed priming treatments included three Pyridoxine-priming duration treatments consist of 6, 12, and 24 h were compared with the unsoaked seed control and hydro-priming with distilled water for 12 h. The Pyridoxine concentration of 200 mg/L prepared in distilled water was used as Pyridoxine-priming media. Seed Pyridoxine-priming treatments improved seed germination and early seeding growth traits included germination percentage, coleoptiles, and radicle length, seed-ing dry matter accumulation, mean germination time (MGT), germina-tion index (GI), vigor index (VI), and time to 50% germination (T_{50}) of both genotypes. Seed pyridoxine-priming duration of 12 h produced maximum value for most of the germination and early seedling growth characteristics of wheat inbred lines of PBW-154 and PBW-343. These results have practical implications in the pre-sowing seed treatment with pyroxene solution could enhance the seed germination and early seedling growth characteristics of wheat plant. It has been established that pyridoxine is required for root development, which can positively influence the early seedling growth. Seed soaking application of graded aqueous pyridoxine solutions increased the dry matter accumulation of mung bean. Pyridoxine requirement for optimum performance of mustard cultivars of PK-8203 and Varuna were 0.05% and 0.0125%, respectively. Seed pyridoxine-priming duration of 12 h produced maximum value for most of the germination and early seedling growth characteristics of wheat inbred lines of PBW-154 and PBW-343. In conclusion, these results have practical implications in that pre-sowing seed treatment with pyridoxine solution could enhance the seed germination and early seedling growth characteristics of crops. Seed pyridoxine-priming is a simple economical way to improve the seedling establishment of wheat plant. Seed priming is an important factor influenced germination and early seedling growth of annual crops. The effects of seed priming on crops are dependent on the complex interaction of factors such as priming substance, plant genotype, and priming duration. Germination and seedling establishment are influential stages, which affected both quality and quantity of crop yields. We have investigated that how seed pyridoxine–priming duration influenced seed germination and early seedling growth characteristics of two wheat genotypes. Seed pyridoxine-priming treatments improved seed germination and early seedling growth traits of both genotypes. The increment in seed germination and early seedling growth due to seed priming treatment is also in conformity with the findings of others. Priming of wheat seed using polyethylene glycol or potassium salts (K_2HPO_4 or KNO_3) resulted in accelerated seed germination. A significant enhancement in seed germination and seedling growth characteristics of wheat was through the hydro-priming of seeds for 24 h. The three early phases in seed germination are [1] imbibition, [2] lag phase, and [3] protrusion of the radicle through the testa. Generally, priming affects the lag phase and causes early DNA replication increased RNA and protein synthesis and greater ATP availability. Furthermore, pyridoxine is a cofactor for many enzymatic reactions, especially those involved in amino acid metabolism.

6.10   B-VITAMINS AND SUGARCANE CROP

Sugarcane is one of the world's economically most important cultivated crop [87]. It is the chief source of centrifugal sugar in the world and it contributed nearly 60% world sugar production [88]. It is one of the most important cash crops of tropics and sub-tropics in India. It plays a pivotal role in both agricultural and industrial economy of our country. India is one of the largest producers of sugar and is in neck-to-neck race with Brazil for first position. The Indian share in sugar production is about 13% of the world and 41% of Asia [89]. In India, sugarcane is grown under different agro-climatic conditions and occupies about 22% area (4.4 m ha) of the GCA with an average productivity of 68.2 t/ha [90]. The area under sugarcane cultivation in Uttar Pradesh (U.P.) alone is about 48% of India. But from the productivity point of view it is lagging behind. It has two main reasons one is the agro-climatic conditions, which is not up to the desired level and second one is the lack of cane production technology management. As far as planting technology is concerned in Punjab, Haryana, and Western U.P., the majority of farmers mostly prefer late planting i.e. sugarcane cultivation after wheat crop harvesting. Specially, in Western U.P. about 60% area is covered under this process which adversely affect the productivity of sugarcane due to poor tillers formation on account of early monsoon rains and moreover the emergence of plenty of weeds. It had been established that 32 to 35% deterioration in yield is due to late planting adaptation against autumn and spring planting. Sundra [91] reported that summer planted crop gives very low yield, particularly due to poor germination and tillering because of extremely high temperature coupled with moisture stress. The cane productivity declines by 30 to 50% if planting is delayed upto end of April or early May [92, 93]. It is also well known that the autumn (October) planted cane yields 20% more than the spring (February–March) planted cane, and matures earlier but it needs 30% more time. In fact, the first Green Revolution, though mostly unsung, took place in sugarcane in 1920s, when the new inter-specific hybrids (*Saccharum officinarum* x *S. spontaneum*) starting with cultivar 'Co205' – yielded 2½ times more than the traditional canes (varieties belonging to *S. barberi* and *S. sinense*, mostly used for sugar making). This success of new hybrids may be gauged from the fact that within 10 years number of sugar mills in India rose from 20 in 1920 to 110 in 1933–34. The architect of this unsung revolution, Dr. T.S. Venkataraman, the then Director of Sugarcane Breeding Centre, Coimbatore, was knighted in 1942 by the then imperial government. Today, the cultivated sugarcane is no longer *S. officinarum* but a *Saccharum* hybrid complex involving different species of *saccharum* and allied genera. The process of producing cane hybrids using *S. officinarum*, as female parent is called '*Nobilization of Sugarcane.*' The Coimbatore-breed hybrid canes (popularly known as 'co-canes') supported the sugar industries of different countries including those of the United States of America, South Africa, and many others. In India, before the advent of these hybrids more than 80% sugar was produced in the states of Indo-Gangetic plains. The development of hybrids gradually increased the cane cultivation in the tropical states. Sugarcane is considered a tropical plant and grows well in that environment. Thus, tropical states with longer growing period of sugarcane always have more cane yield than the subtropical states, where the growing period for sugarcane is shorter due to climatic extremes (very high temperature during the formative phase and very low temperature during the maturity phase). The natural disparity may be minimized through appropriate agronomic interventions. Hence, the choice of a suitable cultivar, season and adoption of viable crop management strategies play a decisive role in augmenting sugarcane production.

The ultimate economic product of sugarcane is not the seed but the sugar, which is stored in the stalk and cannot be seen by naked eye. Sugarcane maturity, therefore, cannot be judged by the considerations applicable to grain crops. Accumulation of sugar in the stalk begins soon after completion of elongation phase when glucose produced during photosynthesis is not utilized mainly for further growth but stored largely in the stalk as sucrose. When concentration of this sucrose exceeds 16% and juice purity increase over 85% the cane is said to be mature. The prime concern of the plant in the early stages is growth as opposed to sugar storage. As the plant advances in age, the rapid growth processes tend to slacken as water and nitrogen reserves are depleted. As the growth declines, less of the sugar produced each day is diverted in building new tissues and more is stored as sucrose. Ripening is the process of sugar accumulation. It is the culmination or perfection of maturity. It is a stage of growth when the sugar content is at its best in the stalks. Agronomically, it is based upon the appearance of internodes no longer subtended by green leaves and a parallel accumulation of sucrose in each successive internode towards a common higher value. The two process run parallel and are complementary to each other depicting the metabolic transformations of biological molecules associated with sugar accumulation [94].

Considerable work with various aspects on the metabolism of ripening and maturity in sugarcane has been done by different workers [95–110] but the effect of vitamins also having catalytic and regulatory functions on metabolic aberrations during ripening and maturity of sugarcane crop seems to be the neglected aspect.

The function of different vitamins in animals can be easily studied, by merely removing a particular vitamin from the diet. This study is, however, difficult in plants, since they synthesize vitamins themselves but the concentration level may not be enough to carry out regulatory and other biochemical processes at optimum level, therefore, externally applied vitamins to plants increase growth and yield potential in different crop plants [5, 6, 13, 14, 21–25, 111, 139]. Moreover, [59] maintained that a plant would respond to exogenous supply of the vitamins only if its endogenous vitamin levels were low. The hypothesis has been confirmed in the authors laboratory were highly encouraging for peppermint crop [21, 22]. With the new high yielding different genotypes of sugarcane available for research, the present author decided to test these from the point of view of growth characters and correlations between biochemical alterations under the influence of soil applied rice bran (a natural source of pyridoxine) along with pyridoxine soaked sett on ripening and maturity of sugarcane cultivars, since, the growth and storage are reciprocally related presumably because of competition for the available photosynthate [112]. It may not be out of place to mention here that work done at Shahjahanpur, during the last two decades or so on these important aspects of the applied physiology of cereals, oil crops, medicinal, and sugar crops under the supervision of Professor Dr. Abbas and others has yielded valuable results [5, 6, 13, 14, 20–25, 111–128, 144, 145].

A field experiment was conducted during [2011–2012] to study the effect of different doses of soil-applied rice bran (0, 10, 20, 40 and 60 kg/ha), T_1, T_2, T_3, T_4, T_5, respectively, along with pre-sowing 0.03% pyridoxine soaked setts (T_6) with ten sugarcane varieties namely CoS 8436 (V_1), CoS 95255 (V_2), CoSe 96268 (V_3), CoSe 98231(V_4), CoSe 01235 (V_5), CoS 01424 (V_6), CoSe 95422 (V_7), UP 097 (V_8), CoSe 92423 (V_9) and CoS 96275 (V_{10}) on growth (Cane length, fresh and dry weight/plant), leaf biochemical components (acid and neutral invertases) during various growth stages (pre-monsoon, post-monsoon, early ripening, late ripening, and maturity) of growth and development, including cane yield and juice quality (Brix, pol %, sucrose content, purity) at harvest. The experiment was based on a factorial randomized design, each treatment replicated thrice. Thus, different doses of soil-applied rice bran together with basal fertilizer dose were supplied in soil before placing the pyrogallol sterilized setts. A uniform basal dose of 150 kg N, 60 kg P, and 80 kg K/ha was applied to each bed. Nitrogen was given in [1/3 + 1/3 + 1/3] split doses as top dress ing. Urea, Monocalcium superphosphate and muriate of potash were used as perspective sources of N P K. The growth characteristics leaf biochemical analyses at different stages and cane yield as well as juice quality at harvest were studied. The first five sugarcane varieties (CoS 8436, CoS 95255, CoS 96268, CoSe 98231 and CoSe 01235) harvested one month before (at 330 days) due to early maturation as compared to the remaining five varieties.

It was observed that increasing levels of soil-applied rice bran as well as 0.03% pyridoxine sett-soaking significantly increased cane length, fresh and dry weight gradually at every stage of development till harvest in sugarcane as compared to control [139], Tables 6.1–6.15. Such results are well expected as the beneficial effects of rice bran in other crops with various aspects have also been noted by [16–19] and in sugarcane, recently, by the author himself [13, 14, 139]. Similarly, the better effect of soil-applied rice bran may be attributed besides supplying water soluble natural pyridoxine also conserved more water at formative stage (shoot emergence + tillering) which probably helped plants to tiller profusely and abundant nutrient availability [13, 14]. Similarly, pyridoxine having positive impact on physiological and biochemical pool of the crop plants corroborates [9–12]. The soil-applied rice bran @ 60 kg/ha before planting proved toxic for cane yield and proved less effective. The next best treatment was soil-applied 40 kg/ha rice bran (T_4). The above observed positive impact may be associated with higher nitrate reductase activity [111]. Presumably, the vitamin is involved in synthesis of nitrate reductase or alters the permeability of cell membrane or enhance the activity of carrier proteins by acting as coenzymes, thus favoring nutrient uptake. The conclusions drawn by [6] in his review are no doubt relevant. He opined that "normal green plants can certainly grow well without any external vitamin application, but it is naturally not excluded that certain plants or certain developmental stages may contain slightly sub-optimal vitamin amounts and therefore, may be stimulated by further additions. In addition, probably pyridoxine participates in plant growth and development indirectly by enhancing the endogenous levels of various growth factors [37]. Thus, it has been established that pyridoxine participate in diverse physiological processes and seem to be pleiotropic in its action [12].

Sugarcane variety CoSe 92423 (V_9) produced highest cane yield of 106.73 t/ha followed by UP 097 (V_8) with 99.05 t/ha (Table 6.26) due to highest positive association with cane length, fresh and dry weight at every stage of growth and development by the growth parameters associative impact of the treatments till harvest, Tables 6.1–6.15. It is quite understandable as species of a genus, and even varieties of a species, differ under the same environmental conditions, in their utilization of inputs [129, 130]. The variety CoS 8436 (V_1) gave lowest cane yield due to inferior values for cane length, fresh and dry weight at all stages till harvest, Tables 6.1–6.15.

Foliar biochemical components may be regarded as mini imitation of the concentration levels present in the stem tissues. Hence, the foliar biochemical analyses for various components might become a biochemical parameter indicating their status at different growth stages and their associations with sucrose, ripening, maturity and cane yield etc. would be highly valuable. From the results of foliar biochemical analyses, it

TABLE 6.1 Effect of Soil-Applied Rice Bran And Pre-Sowing Pyridoxine Soaking of Setts on Length/Cane (cm) at Pre-Monsoon (120 days) in Sugarcane (*Sachharum officinarum* L.)

Varieties	Soil-applied rice bran (kg/ha)					0.03% Pyridoxine (T_6) Soaking	Mean
	0 (T_1)	10 (T_2)	20 (T_3)	40 (T_4)	60 (T_5)		
CoS8436 (V_1)	126.60	136.30	145.30	154.30	141.60	156.40	143.42
CoS 95255(V_2)	136.60	138.60	142.30	169.30	166.30	171.30	154.07
CoS 96268 (V_3)	140.60	152.00	156.00	162.30	165.60	165.30	156.97
CoSe 98231 (V_4)	141.00	155.60	172.60	184.60	170.60	189.90	169.05
CoSe 01235(V_5)	134.30	158.00	178.00	191.00	179.00	195.00	172.55
CoS 01424 (V_6)	117.00	129.00	144.00	183.00	171.00	188.80	155.47
CoSe 95422 (V_7)	150.30	168.60	173.30	187.00	163.00	189.90	172.02
UP 097 (V_8)	120.00	133.00	149.00	185.60	168.60	188.80	157.50
CoSc 92423 (V_9)	153.00	161.00	189.60	199.60	182.00	199.80	180.83
CoS 96275 (V_{10})	118.00	135.30	153.60	156.00	143.60	165.00	145.25
Mean	133.74	146.74	160.37	177.27	165.13	181.02	

	CD at 5%	F-value
Treatments	5.14	*
Varieties	6.63	*
Treatments × Varieties	16.25	*

Mean of three replicates.

* – Significant.

can be concluded that It was interesting to note that the first five varieties (CoS 8436, CoS 95255, CoS 96268, CoSe 98231, and CoSe 01235) matured and therefore, harvested one month earlier to final harvesting due to early maturation signals in all the respective foliar (acid and

TABLE 6.2   Effect of Soil-Applied Rice Bran and Pre-Sowing Pyridoxine Soaking of Setts on Length/Cane (cm) at Post-Monsoon (210 days) in Sugarcane (*Sachharum officinarum* L.)

Varieties	Soil-applied rice bran (kg/ha)					0.03% Pyridoxine (T_6) Soaking	Mean
	0 (T_1)	10 (T_2)	20 (T_3)	40 (T_4)	60 (T_5)		
CoS8436 (V_1)	170.60	181.00	193.00	203.30	191.60	205.60	190.85
CoS 95255 (V_2)	206.60	216.60	268.60	308.30	296.00	309.40	267.58
CoS 96268 (V_3)	206.60	215.60	241.60	293.30	259.30	295.50	251.98
CoSe 98231 (V_4)	213.30	221.00	241.60	251.60	247.00	256.50	238.50
CoSe 01235 (V_5)	218.30	217.00	253.30	271.60	258.60	275.00	248.97
CoS 01424 (V_6)	208.30	243.30	265.00	280.60	255.30	285.00	256.25
CoSe 95422 (V_7)	226.60	239.30	261.00	275.30	261.00	285.00	258.03
UP 097 (V_8)	251.30	264.30	291.30	305.60	295.00	307.30	285.80
CoSe 92423 (V_9)	248.30	267.30	291.00	311.00	306.00	315.10	289.78
CoS 96275 (V_{10})	200.30	221.30	247.00	275.00	270.30	280.50	249.07
Mean	215.02	228.67	255.34	277.56	264.01	281.49	

	CD at 5%	F-value
Treatments	9.66	*
Varieties	12.47	*
Treatments × Varieties	30.56	NS

Mean of three replicates.

* = Significant, NS = Non-significant.

neutral invertases) analysis Tables 6.16–6.25. And cane traits as well as in juice quality [13, 14], Tables 6.26–6.31. The 0.03 % pyridoxine sett-soaking (T_6) as well as 40 kg/ha soil- applied rice bran (T_4), a natural source of pyridoxine, may be regarded as a promoter to the metabolic

TABLE 6.3   Effect of Soil Applied Rice Bran and Pre-Sowing Pyridoxine Soaking of Setts on Length/Cane (cm) at Early Ripening (270 days) in Sugarcane (*Sachharum officinarum* L.)

Varieties	Soil-applied rice bran (kg/ha)					0.03% Pyridox-ine (T$_6$) Soaking	Mean
	0 (T$_1$)	10 (T$_2$)	20 (T$_3$)	40 (T$_4$)	60 (T$_5$)		
CoS8436 (V$_1$)	205.00	250.00	245.00	255.50	254.50	259.90	244.98
CoS 95255(V$_2$)	310.00	325.00	329.00	345.00	340.00	358.00	334.50
CoS 96268 (V$_3$)	301.00	310.00	325.00	346.00	330.00	350.00	327.00
CoSe 98231 (V$_4$)	260.00	270.00	275.00	281.00	271.00	285.00	273.67
CoSe 01235(V$_5$)	250.50	260.00	265.00	269.90	260.50	270.50	262.73
CoS 01424 (V$_6$)	270.00	280.00	289.50	290.00	279.90	295.50	284.15
CoSe 95422 (V$_7$)	245.50	269.90	275.50	281.00	265.00	298.90	272.63
UP 097 (V$_8$)	265.50	275.50	285.50	289.50	280.00	310.50	284.42
CoSe 92423 (V$_9$)	285.50	291.50	295.50	309.90	301.50	330.30	302.37
CoS 96275 (V$_{10}$)	250.50	268.80	275.00	285.00	280.00	285.30	274.10
Mean	264.35	280.07	286.00	295.28	286.24	304.39	

	CD at 5%	F-value
Treatments	5.77	*
Varieties	7.45	*
Treatments × Varieties	18.25	NS

Mean of three replicates.

* = Significant, NS = Non-significant.

labyrinth providing a strong organic impact on metabolic centers for accelerating ripening and maturity, quality as well as cane yield and juice quality in sugarcane crop. Mishra and Gupta [131]; Thangavelu and Chiranjivi Rao [132]; Alexander [140]; Das and Prabhu [141]; Sachdeva et

TABLE 6.4   Effect of Soil-Applied Rice Bran and Pre-Sowing Pyridoxine Soaking of Setts on Length/Cane (cm) at Late Ripening (330 days) in Sugarcane (*Sachharum officinarum* L.)

Varieties	Soil-applied rice bran (kg/ha)					0.03% Pyridoxine (T₆) Soaking	Mean
	0 (T₁)	10 (T₂)	20 (T₃)	40 (T₄)	60 (T₅)		
CoS8436 (V₁)	225.00	280.00	275.00	280.00	280.00	285.00	270.83
CoS 95255(V₂)	340.00	345.00	340.00	369.00	360.00	375.00	354.83
CoS 96268 (V₃)	321.00	330.00	346.00	375.00	350.00	380.00	350.33
CoSe 98231 (V₄)	295.00	299.00	299.50	301.15	298.00	310.00	300.44
CoSe 01235(V₅)	280.00	285.00	288.00	290.00	285.50	295.00	287.25
CoS 01424 (V₆)	315.50	320.00	355.00	360.00	350.00	365.00	344.25
CoSe 95422 (V₇)	280.00	285.00	285.00	290.00	280.50	300.50	286.83
UP 097 (V₈)	295.00	305.50	350.00	359.50	360.00	340.50	335.08
CoSe 92423 (V₉)	320.00	331.50	345.50	360.00	340.00	380.00	346.17
CoS 96275 (V₁₀)	280.00	285.00	289.10	293.50	290.00	298.00	289.27
Mean	295.15	306.60	317.31	327.82	319.40	332.90	

	CD at 5%	F-value
Treatments	9.18	*
Varieties	11.85	*
Treatments × Varieties	29.04	NS

Mean of three replicates.

* = Significant, NS = Non-significant.

al. [142]; Gopala Rao and Negi Reddy [143] also reported that in the early part of season had higher amount of reducing sugars and at optimum maturity the reducing sugar was minimum. Similarly, [13, 14, 139]

TABLE 6.5   Effect of Soil-Applied Rice Bran and Pre-Sowing Pyridoxine Soaking of Setts on Length/Cane (cm) at Maturity (360 days) in Sugarcane (*Sachharum officinarum* L.)

Varieties	Soil-applied rice bran (kg/ha)					0.03% Pyridox-ine (T_6) Soaking	Mean
	0 (T_1)	10 (T_2)	20 (T_3)	40 (T_4)	60 (T_5)		
CoS8436 (V_1)	-	-	-	-	-	-	-
CoS 95255(V_2)	-	-	-	-	-	-	-
CoS 96268 (V_3)	-	-	-	-	-	-	-
CoSe 98231 (V_4)	-	-	-	-	-	-	-
CoSe 01235(V_5)	-	-	-	-	-	-	-
CoS 01424 (V_6)	332.00	344.60	367.00	375.00	365.00	381.20	360.80
CoSe 95422 (V_7)	301.60	309.30	323.30	339.00	337.00	351.00	326.87
UP 097 (V_8)	328.30	355.00	360.00	391.00	381.00	398.50	368.97
CoSe 92423 (V_9)	341.00	351.60	386.60	413.30	370.00	419.60	380.35
CoS 96275 (V_{10})	301.60	313.30	331.30	355.60	342.60	365.50	334.98
Mean	160.45	167.38	176.82	187.39	179.56	191.58	

	CD at 5%	F-value
Treatments	6.29	*
Varieties	8.12	*
Treatments × Varieties	19.89	*

Mean of three replicates.

* = Significant.

noted a decrease in the total sugars and starch content during ripening and maturity under the influence of soil-applied rice bran and pyridoxine sett-soaking, indicated that starch, like sugars is an alternate reservoir

TABLE 6.6   Effect of Soil-Applied Rice Bran and pre-Sowing Pyridoxine Soaking of Setts on Fresh Weight/Cane (gm) at Pre-Monsoon (120 days) in Sugarcane (*Sachharum officinarum* L.)

Varieties	Soil-applied rice bran (kg/ha)					0.03% Pyridox-ine (T_6) Soaking	Mean
	0 (T_1)	10 (T_2)	20 (T_3)	40 (T_4)	60 (T_5)		
CoS8436 (V_1)	14.00	19.10	21.20	30.10	29.30	35.00	24.78
CoS 95255(V_2)	23.10	24.00	29.90	59.80	38.10	60.00	39.15
CoS 96268 (V_3)	13.80	16.50	59.40	63.00	38.20	69.90	43.47
CoSe 98231 (V_4)	26.30	34.30	50.00	61.00	55.60	70.00	49.53
CoSe 01235(V_5)	26.40	31.50	32.50	97.00	82.80	99.80	61.67
CoS 01424 (V_6)	15.90	18.70	19.30	36.10	38.60	45.50	29.02
CoSe 95422 (V_7)	29.10	31.00	59.10	84.60	38.60	95.50	56.32
UP 097 (V_8)	16.80	21.80	44.50	66.00	44.00	68.00	43.52
CoSe 92423 (V_9)	20.60	30.80	37.10	65.60	44.10	76.90	45.85
CoS 96275 (V_{10})	15.30	18.40	26.10	31.00	36.30	41.90	28.17
Mean	20.13	24.61	37.91	59.42	44.56	66.25	

	CD at 5%	F-value
Treatments	1.60	*
Varieties	2.06	*
Treatments × Varieties	5.06	*

Mean of three replicates.

* = Significant.

of carbohydrate for energy and these may be treated as a (biochemical traits) signal for attaining maturity.

In addition, there was an increase in acid invertase activity upto early ripening (270 days) stage and a sharp drop in the last final two stages under the influence of 0.03% pyridoxine sett-soaking (T_6) followed by 40 kg/ha soil-applied rice bran (T_4), (Tables 6.16–6.25) showing its positive association with maturation and an important trait for assessing ripening and maturity in sugarcane crop. Further, there was a significant gradual increase in neutral invertase activity not only with the progress in age but also due to impacts received from 0.03%

TABLE 6.7   Effect of Soil-Applied Rice Bran and Pre-Sowing Pyridoxine Soaking of Setts on Fresh Weight/Cane (gm) at Post-Monsoon (210 days) in Sugarcane (*Sachharum officinarum* L.)

Varieties	Soil-applied rice bran (kg/ha)					0.03% Pyridox-ine (T₆) Soaking	Mean
	0 (T₁)	10 (T₂)	20 (T₃)	40 (T₄)	60 (T₅)		
CoS8436 (V₁)	175.10	184.00	211.00	301.60	272.50	315.00	243.20
CoS 95255(V₂)	255.30	290.00	367.60	483.60	410.60	490.50	382.93
CoS 96268 (V₃)	237.50	273.30	368.30	424.30	342.00	449.30	349.12
CoSe 98231 (V₄)	201.50	259.00	330.10	361.00	350.60	390.50	315.45
CoSe 01235(V₅)	232.30	277.00	357.60	431.60	400.60	490.50	364.93
CoS 01424 (V₆)	237.50	346.00	375.30	403.30	376.80	480.50	369.90
CoSe 95422 (V₇)	212.30	299.60	341.60	382.30	222.80	405.60	310.70
UP 097 (V₈)	279.00	285.00	393.30	499.10	444.30	499.50	400.03
CoSe 92423 (V₉)	355.30	365.00	439.10	555.60	397.80	580.90	448.95
CoS 96275 (V₁₀)	227.90	299.30	307.30	355.30	351.10	391.60	322.08
Mean	241.37	287.82	349.12	419.77	356.91	449.39	

	CD at 5%	F-value
Treatments	8.83	*
Varieties	11.40	*
Treatments × Varieties	27.93	*

Mean of three replicates.

* – Significant.

pyridoxine sett-soaking (T_6) as well as 40 kg/ha soil-applied rice bran (T_4) (Tables 6.21–6.25) showing its higher level closely associated with ripening and maturity and readiness of the crop for harvest (Table 6.26).

TABLE 6.8   Effect of Soil-Applied Rice Bran and Pre-Sowing Pyridoxine Soaking of Setts on Fresh Weight/Cane (gm) at Early Ripening (270 days) in Sugarcane (*Sachharum officinarum* L.)

Varieties	Soil-applied rice bran (kg/ha)					0.03% Pyridoxine (T_6) Soaking	Mean
	0 (T_1)	10 (T_2)	20 (T_3)	40 (T_4)	60 (T_5)		
CoS8436 (V_1)	192.70	217.60	286.00	351.60	305.80	365.60	286.55
CoS 95255 (V_2)	307.50	311.60	456.00	421.00	360.50	455.20	385.30
CoS 96268 (V_3)	303.30	373.30	392.50	404.80	385.80	410.70	378.40
CoSe 98231 (V_4)	265.10	320.00	363.60	490.00	359.00	495.00	382.12
CoSe 01235 (V_5)	288.30	372.60	462.60	496.60	489.00	525.00	439.02
CoS 01424 (V_6)	448.60	463.60	519.30	672.30	654.30	699.00	576.18
CoSe 95422 (V_7)	345.10	387.00	403.00	412.50	453.30	501.10	417.00
UP 097 (V_8)	328.30	394.30	458.30	582.60	543.30	599.10	484.32
CoSe 92423 (V_9)	412.10	445.00	529.00	663.30	631.60	715.20	566.03
CoS 96275 (V_{10})	430.00	495.00	499.30	562.60	622.50	690.50	549.98
Mean	332.10	378.00	436.96	505.73	480.51	545.64	

	CD at 5%	F-value
Treatments	8.65	*
Varieties	11.17	*
Treatments × Varieties	27.36	*

Mean of three replicates.

* = Significant.

It is also well known that acid invertase frequently exhibit a relationship with cell expansion [133, 143], in any developing organelle suggesting that hydrolysis of sucrose to hexoses may provide the substrate necessary for growth. On the other hand, the spatially separated neutral invertase functions towards controlling the hexose pool size in the mature cells [134, 135, 143].

Regarding juice quality parameters, it may be observed that, brix, pol (%), sucrose (%), and purity (%), (Tables 6.27–6.31) were negatively associated with the cane yield (Table 6.26) with respect to CoSe 92423 (V_9) the highest yielder (recorded lowest values for these traits) and CoSe 95422

TABLE 6.9　Effect of Soil-Applied Rice Bran and pre-Sowing Pyridoxine Soaking of Setts on Fresh Weight/Cane (gm) at Late Ripening (330 days) in Sugarcane (*Sachharum officinarum* L.)

Varieties	Soil-applied rice bran (kg/ha)					0.03% Pyridox- ine (T_6) Soaking	Mean
	0 (T_1)	10 (T_2)	20 (T_3)	40 (T_4)	60 (T_5)		
CoS 8436 (V_1)	285.50	310.00	315.00	451.50	355.00	495.00	368.67
CoS 95255 (V_2)	375.00	430.00	590.00	610.00	510.00	601.00	519.33
CoS 96268 (V_3)	360.00	385.50	485.00	499.50	480.50	570.00	463.42
CoSe 98231 (V_4)	301.00	356.00	495.00	510.00	401.00	620.00	447.17
CoSe 01235 (V_5)	355.00	410.50	490.00	501.30	390.50	515.50	443.80
CoS 01424 (V_6)	472.00	498.00	499.30	595.60	565.00	675.00	550.82
CoSe 95422 (V_7)	385.00	409.10	468.00	595.00	485.00	660.50	500.43
UP 097 (V_8)	510.00	550.00	580.00	695.00	670.50	720.00	620.92
CoSe 92423 (V_9)	495.00	560.00	630.00	780.00	799.00	801.00	677.50
CoS 96275 (V_{10})	455.00	499.50	510.00	601.50	531.30	679.00	546.05
Mean	399.35	440.86	506.23	583.94	518.78	633.70	

	CD at 5%	F-value
Treatments	9.99	*
Varieties	12.90	*
Treatments × Varieties	31.61	*

Mean of three replicates.

* = Significant.

(V_7) and CoS 96275 (V_{10}) as well as CoS8436 (V_1) recorded relatively higher values (Table 6.26). Also, between stalk weight and juice sucrose (%) [13, 14, 139] has also been noted by [136, 137].

It may be concluded that all the quality characters (Tables 6.27–6.31) starts almost significantly favoring from early ripening (270 days) and continued to do so till February/March (the harvest) irrespective of the varieties and their maturity groups under the influence of 0.03% pyridox- ine sett-soaking (T_6) and 40 kg/ha soil-applied rice bran (T_4). Since the temperature in the plains of Northern India start declining from October

TABLE 6.10   Effect of Soil-Applied Rice Bran and Pre-Sowing Pyridoxine Soaking of Setts on Fresh Weight/Cane (gm) at Maturity (360 days) in Sugarcane (*Sachharum officinarum* L.)

Varieties	Soil-applied rice bran (kg/ha)					0.03% Pyri-doxine (T_6) Soaking	Mean
	0 (T_1)	10 (T_2)	20 (T_3)	40 (T_4)	60 (T_5)		
CoS 8436 (V_1)	-	-	-	-	-	-	-
CoS 95255 (V_2)	-	-	-	-	-	-	-
CoS 96268 (V_3)	-	-	-	-	-	-	-
CoSe 98231 (V_4)	-	-	-	-	-	-	-
CoSe 01235 (V_5)	-	-	-	-	-	-	-
CoS 01424 (V_6)	517.00	559.60	595.60	709.30	671.30	780.80	638.93
CoSe 95422 (V_7)	411.60	516.00	586.30	706.60	595.00	785.50	600.17
UP 097 (V_8)	640.80	644.00	667.00	828.30	838.30	910.00	754.73
CoSe 92423 (V_9)	590.00	660.00	779.00	933.30	912.30	980.00	809.10
CoS 96275 (V_{10})	465.10	590.60	676.00	736.60	644.10	810.00	653.73
Mean	262.45	297.02	330.39	391.41	366.10	426.63	

	CD at 5%	F-value
Treatments	11.28	*
Varieties	14.56	*
Treatments × Varieties	35.67	*

Mean of three replicates.

* = Significant.

onwards and by this time the growth of the cane is almost completed and glucose start converted into sucrose. The rate of sugar accumulation within the cane stalk is much faster in early maturing varieties as compared to late maturing varieties [13, 14, 138, 139].

TABLE 6.11   Effect of Soil-Applied Rice Bran and Pre-Sowing Pyridoxine Soaking of Setts on Dry Weight/Cane (gm) at Pre-Monsoon (120 days) in Sugarcane (*Sachharum officinarum* L.)

Varieties	Soil-applied rice bran (kg/ha)					0.03% Pyridoxine (T_6) Soaking	Mean
	0 (T_1)	10 (T_2)	20 (T_3)	40 (T_4)	60 (T_5)		
CoS 8436 (V_1)	2.616	3.713	4.213	6.216	6.110	6.300	4.861
CoS 95255 (V_2)	4.610	4.710	5.210	11.910	4.313	11.990	7.124
CoS 96268 (V_3)	2.910	3.310	10.310	13.416	11.010	13.500	9.076
CoSe 98231 (V_4)	5.413	6.910	10.116	12.410	11.013	12.950	9.802
CoSe 01235 (V_5)	4.416	6.616	7.310	19.410	15.410	19.480	12.107
CoS 01424 (V_6)	3.113	3.910	4.600	6.713	7.713	8.900	5.825
CoSe 95422 (V_7)	5.416	5.513	9.613	13.713	6.510	14.960	9.288
UP 097 (V_8)	3.513	4.116	9.110	16.213	8.513	16.560	9.671
CoSe 92423 (V_9)	4.013	6.313	7.616	13.310	9.016	16.990	9.543
CoS 96275 (V_{10})	2.716	3.713	6.316	8.916	7.513	10.300	6.579
Mean	3.874	4.882	7.441	12.223	8.712	13.193	

	CD at 5%	F-value
Treatments	1.40	*
Varieties	1.85	*
Treatments × Varieties	4.55	*

Mean of three replicates.

* = Significant.

TABLE 6.12   Effect of Soil-Applied Rice Bran and Pre-Sowing Pyridoxine Soaking of Setts on Dry Weight/Cane (gm) at Post-Monsoon (210 days) in Sugarcane (*Sachharum officinarum* L.)

Varieties	Soil-applied rice bran (kg/ha)					0.03% Pyridoxine (T_6) Soaking	Mean
	0 (T_1)	10 (T_2)	20 (T_3)	40 (T_4)	60 (T_5)		
CoS 8436 (V_1)	34.62	37.12	42.60	60.51	54.62	63.50	48.83
CoS 95255 (V_2)	51.12	57.61	73.31	91.31	75.00	97.00	74.23
CoS 96268 (V_3)	46.81	54.31	74.01	85.01	68.51	91.30	69.99
CoSe 98231 (V_4)	40.22	51.71	66.12	74.12	70.01	76.50	63.11

TABLE 6.12   *(Continued)*

CoSe 01235 (V_5)	46.62	55.51	72.31	87.91	81.12	89.90	72.23
CoS 01424 (V_6)	47.31	68.62	74.61	81.12	75.11	89.30	72.68
CoSe 95422 (V_7)	43.12	60.01	68.51	75.81	64.71	83.50	65.94
UP 097 (V_8)	54.81	57.01	78.31	99.11	83.80	109.90	80.49
CoSe 92423 (V_9)	67.01	71.01	89.31	112.00	79.80	115.00	89.02
CoS 96275 (V_{10})	45.62	61.01	66.62	83.81	70.51	89.00	69.43
Mean	47.72	57.39	70.57	85.07	72.32	90.49	

	CD at 5%	F-value
Treatments	4.06	*
Varieties	5.24	*
Treatments × Varieties	12.84	NS

Mean of three replicates.

* = Significant.

TABLE 6.13   Effect of Soil-Applied Rice Bran and Pre-Sowing Pyridoxine Soaking of Setts on Dry Weight/Cane (gm) at Early Ripening (270 days) in Sugarcane (*Sachharum officinarum* L.)

Varieties	Soil-applied rice bran (kg/ha)					0.03% Pyridox-ine (T_6) Soaking	Mean
	0 (T_1)	10 (T_2)	20 (T_3)	40 (T_4)	60 (T_5)		
CoS8436 (V_1)	38.616	44.013	56.110	70.310	64.113	73.000	57.694
CoS 95255(V_2)	81.313	98.313	124.116	161.616	150.616	169.900	130.979
CoS 96268 (V_3)	75.616	94.116	115.010	140.501	117.716	165.500	118.077
CoSe 98231 (V_4)	53.116	63.813	72.616	81.116	73.110	89.900	72.279
CoSe 01235 (V_5)	59.010	73.616	94.313	120.116	101.500	143.000	98.593
CoS 01424 (V_6)	82.616	87.000	118.616	122.813	103.913	139.570	109.088
CoSe 95422 (V_7)	106.116	115.813	122.116	143.313	132.313	150.000	128.279
UP 097 (V_8)	106.116	135.116	170.510	176.010	169.116	186.600	157.245

TABLE 6.13   *(Continued)*

CoSe 92423 (V$_9$)	121.116	130.813	165.010	191.813	185.010	201.500	165.877
CoS 96275 (V$_{10}$)	108.300	117.010	133.010	165.510	145.110	189.900	143.140
Mean	83.194	95.962	117.143	137.312	124.252	150.887	

	CD at 5%	F-value
Treatments	3.61	*
Varieties	4.67	*
Treatments × Varieties	11.44	*

Mean of three replicates.

* = Significant.

TABLE 6.14   Effect of Soil-Applied Rice Bran and Pre-Sowing Pyridoxine Soaking of Setts on Dry Weight/Cane (gm) at Late Ripening (330 days) in Sugarcane (*Sachharum officinarum* L.)

Varieties	Soil-applied rice bran (kg/ha)					0.03% Pyridoxine (T$_6$) Soaking	Mean
	0 (T$_1$)	10 (T$_2$)	20 (T$_3$)	40 (T$_4$)	60 (T$_5$)		
CoS 8436 (V$_1$)	46.60	53.50	61.30	76.30	67.10	79.10	63.98
CoS 95255 (V$_2$)	83.45	99.95	132.30	167.00	153.10	171.30	134.52
CoS 96268 (V$_3$)	76.70	96.30	118.01	144.30	121.90	160.90	119.69
CoSe 98231 (V$_4$)	55.50	66.90	75.30	84.30	78.30	91.30	75.27
CoSe 01235 (V$_5$)	61.30	75.80	98.80	131.30	105.10	142.10	102.40
CoS 01424 (V$_6$)	83.70	92.30	126.60	129.80	108.60	146.90	114.65
CoSe 95422 (V$_7$)	108.10	119.30	131.00	146.70	135.80	156.60	132.92
UP 097 (V$_8$)	107.75	131.10	169.30	199.90	188.70	205.60	167.06
CoSe 92423 (V$_9$)	123.70	138.90	171.90	199.95	173.70	215.30	170.58
CoS 96275 (V$_{10}$)	87.10	119.90	136.60	169.30	148.90	196.60	143.07
Mean	83.39	99.40	122.11	144.89	128.12	156.57	

	CD at 5%	F-value
Treatments	5.57	*
Varieties	7.19	*
Treatments × Varieties	17.61	*

Mean of three replicates.

* = Significant.

TABLE 6.15 Effect of Soil-Applied Rice Bran and Pre-Sowing Pyridoxine Soaking of Setts on Cane Dry Weight/Cane (gm) at Maturity (360 days) in Sugarcane (*Sachharum officinarum* L.)

Varieties	Soil-applied rice bran (kg/ha)					0.03% Pyridox-ine (T_6) Soaking	Mean
	0 (T_1)	10 (T_2)	20 (T_3)	40 (T_4)	60 (T_5)		
CoS 8436 (V_1)	-	-	-	-	-	-	-
CoS 95255 (V_2)	-	-	-	-	-	-	-
CoS 96268 (V_3)	-	-	-	-	-	-	-
CoSe 98231 (V_4)	-	-	-	-	-	-	-
CoSe 01235 (V_5)	-	-	-	-	-	-	-
CoS 01424 (V_6)	103.30	115.00	139.60	148.00	134.10	165.20	134.20
CoSe 95422 (V_7)	109.00	121.30	135.30	166.60	137.10	165.20	139.08
UP 097 (V_8)	114.60	143.50	199.10	103.60	189.60	215.50	160.98
CoSe 92423 (V_9)	150.10	139.00	192.00	200.60	178.30	260.55	186.76
CoS 96275 (V_{10})	106.00	119.95	136.60	188.10	149.60	199.90	150.03
Mean	58.30	63.88	80.26	80.69	78.87	100.64	

	CD at 5%	F-value
Treatments	7.33	*
Varieties	9.47	*
Treatments × Varieties	23.20	*

Mean of three replicates.

* = Significant.

TABLE 6.16 Effect of Soil-Applied Rice Bran and Pre-Sowing Pyridoxine Soaking of Setts on Leaf Acid Invertase at Pre-Monsoon (120 days) in Sugarcane (*Sachharum officinarum* L.) (units/mg protein)**

Varieties	Soil-applied rice bran (kg/ha)					0.03% Pyridox-ine (T_6) Soaking	Mean
	0 (T_1)	10 (T_2)	20 (T_3)	40 (T_4)	60 (T_5)		
CoS 8436 (V_1)	0.30	0.33	0.34	0.36	0.38	0.40	0.35
CoS 95255 (V_2)	0.42	0.44	0.43	0.46	0.49	0.49	0.46
CoS 96268 (V_3)	0.36	0.38	0.39	0.40	0.40	0.42	0.39
CoSe 98231 (V_4)	0.40	0.42	0.44	0.43	0.46	0.48	0.44

TABLE 6.16   *(Continued)*

CoSe 01235 (V$_5$)	0.35	0.38	0.39	0.40	0.44	0.45	0.40
CoS 01424 (V$_6$)	0.46	0.47	0.49	0.52	0.52	0.55	0.50
CoSe 95422 (V$_7$)	0.48	0.50	0.53	0.55	0.56	0.58	0.53
UP 097 (V$_8$)	0.45	0.49	0.51	0.54	0.58	0.59	0.53
CoSe 92423 (V$_9$)	0.49	0.51	0.53	0.53	0.54	0.56	0.53
CoS 96275 (V$_{10}$)	0.42	0.44	0.46	0.48	0.50	0.49	0.47
Mean	0.41	0.44	0.45	0.47	0.49	0.50	

	CD at 5%	F-value
Treatments	0.0067	*
Varieties	0.0086	*
Treatments × Varieties	0.0213	*

Mean of three replicates.

* = Significant **= where unit stands for the amount of enzyme required to invert 1 u mole of sucrose in 1 hour at 37 C.

TABLE 6.17   Effect of Soil-Applied Rice Bran and Pre-Sowing Pyridoxine Soaking of Setts on Leaf acid Invertase at Post-Monsoon (210 days) in Sugarcane (*Sachharum officinarum* L.) (units/mg protein)**

Varieties	Soil-applied rice bran (kg/ha)					0.03% Pyridoxine (T$_6$) Soaking	Mean
	0 (T$_1$)	10 (T$_2$)	20 (T$_3$)	40 (T$_4$)	60 (T$_5$)		
CoS 8436 (V$_1$)	0.42	0.44	0.48	0.50	0.52	0.52	0.48
CoS 95255 (V$_2$)	0.49	0.50	0.53	0.54	0.53	0.54	0.52
CoS 96268 (V$_3$)	0.45	0.48	0.49	0.50	0.52	0.55	0.50
CoSe 98231 (V$_4$)	0.49	0.52	0.51	0.54	0.59	0.60	0.54
CoSe 01235 (V$_5$)	0.49	0.53	0.55	0.58	0.61	0.64	0.57
CoS 01424 (V$_6$)	0.52	0.54	0.56	0.59	0.60	0.65	0.58
CoSe 95422 (V$_7$)	0.55	0.58	0.60	0.62	0.65	0.66	0.61
UP 097 (V$_8$)	0.54	0.60	0.62	0.63	0.68	0.70	0.63
CoSe 92423 (V$_9$)	0.56	0.65	0.69	0.70	0.72	0.79	0.69
CoS 96275 (V$_{10}$)	0.50	0.54	0.56	0.59	0.62	0.66	0.58
Mean	0.50	0.54	0.56	0.58	0.60	0.63	

TABLE 6.17   *(Continued)*

	CD at 5%	F-value
Treatments	0.0103	*
Varieties	0.0133	*
Treatments × Varieties	0.0327	*

Mean of three replicates.

* = Significant.

**= where unit stands for the amount of enzyme required to invert 1 u mole of sucrose in 1 hour at 37 C.

TABLE 6.18   Effect of Soil-Applied Rice Bran and Pre-Sowing Pyridoxine Soaking of Setts on Leaf Acid Invertase at Early Ripening (270 days) in Sugarcane (*Sachharum officinarum* L.) (units/mg protein)**

Varieties	Soil-applied rice bran (kg/ha)					0.03% Pyri-doxine (T_6) Soaking	Mean
	0 (T_1)	10 (T_2)	20 (T_3)	40 (T_4)	60 (T_5)		
CoS 8436 (V_1)	0.21	0.25	0.28	0.30	0.29	0.32	0.28
CoS 95255 (V_2)	0.20	0.34	0.36	0.40	0.41	0.43	0.36
CoS 96268 (V_3)	0.28	0.31	0.36	0.39	0.40	0.44	0.36
CoSe 98231 (V_4)	0.30	0.35	0.38	0.38	0.40	0.42	0.37
CoSe 01235 (V_5)	0.29	0.30	0.33	0.36	0.38	0.40	0.34
CoS 01424 (V_6)	0.32	0.33	0.36	0.39	0.39	0.41	0.37
CoSe 95422 (V_7)	0.29	0.31	0.33	0.36	0.37	0.64	0.38
UP 097 (V_8)	0.35	0.38	0.40	0.41	0.39	0.42	0.39
CoSe 92423 (V_9)	0.34	0.38	0.39	0.42	0.44	0.46	0.41
CoS 96275 (V_{10})	0.32	0.35	0.36	0.38	0.36	0.38	0.36
Mean	0.29	0.33	0.36	0.38	0.38	0.43	

	CD at 5%	F-value
Treatments	0.0056	*
Varieties	0.0072	*
Treatments × Varieties	0.0178	*

Mean of three replicates.

* = Significant.

**= where unit stands for the amount of enzyme required to invert 1 u mole of sucrose in 1 hour at 37 C.

TABLE 6.19 Effect of Soil-Applied Rice Bran and Pre-Sowing Pyridoxine Soaking of Setts on Leaf Acid Invertase at Late Ripening (330 days) in Sugarcane (*Sachharum officinarum* L.) (units/mg protein)**

Varieties	Soil-applied rice bran (kg/ha)					0.03% Pyridoxine (T$_6$) Soaking	Mean
	0 (T$_1$)	10 (T$_2$)	20 (T$_3$)	40 (T$_4$)	60 (T$_5$)		
CoS 8436 (V$_1$)	0.06	0.06	0.05	0.03	0.03	0.02	0.04
CoS 95255 (V$_2$)	0.09	0.06	0.04	0.02	0.02	0.02	0.04
CoS 96268 (V$_3$)	0.09	0.06	0.03	0.02	0.02	0.01	0.04
CoSe 98231 (V$_4$)	0.08	0.04	0.03	0.03	0.01	0.01	0.03
CoSe 01235 (V$_5$)	0.06	0.03	0.02	0.03	0.02	0.02	0.03
CoS 01424 (V$_6$)	0.11	0.10	0.09	0.06	0.04	0.04	0.07
CoSe 95422 (V$_7$)	0.09	0.06	0.07	0.05	0.04	0.03	0.06
UP 097 (V$_8$)	0.10	0.08	0.08	0.06	0.05	0.04	0.07
CoSe 92423 (V$_9$)	0.10	0.08	0.05	0.04	0.02	0.01	0.05
CoS 96275 (V$_{10}$)	0.08	0.06	0.04	0.04	0.03	0.02	0.05
Mean	0.09	0.06	0.05	0.04	0.03	0.02	

	CD at 5%	F-value
Treatments	0.0012	*
Varieties	0.0015	*
Treatments × Varieties	0.0038	*

Mean of three replicates.

* = Significant.

**= where unit stands for the amount of enzyme required to invert 1 u mole of sucrose in 1 hour at 37 C.

TABLE 6.20 Effect of Soil-Applied Rice Bran and Pre-Sowing Pyridoxine Soaking of Setts on Leaf Acid Invertase At Maturity (360 days) in Sugarcane (*Sachharum officinarum* L.)

Varieties	Soil-applied rice bran (kg/ha)					0.03% Pyridoxine (T$_6$) Soaking	Mean
	0 (T$_1$)	10 (T$_2$)	20 (T$_3$)	40 (T$_4$)	60 (T$_5$)		
CoS 8436 (V$_1$)	-	-	-	-	-	-	-
CoS 95255 (V$_2$)	-	-	-	-	-	-	-

TABLE 6.20   *(Continued)*

CoS 96268 (V_3)	-	-	-	-	-	-	-
CoSe 98231 (V_4)	-	-	-	-	-	-	-
CoSe 01235 (V_5)	-	-	-	-	-	-	-
CoS 01424 (V_6)	0.050	0.040	0.030	0.02.	0.020	0.010	0.030
CoSe 95422 (V_7)	0.080	0.060	0.040	0.020	0.010	0.010	0.037
UP 097 (V_8)	0.060	0.060	0.040	0.030	0.020	0.020	0.038
CoSe 92423 (V_9)	0.040	0.040	0.030	0.020	0.010	0.010	0.025
CoS 96275 (V_{10})	0.060	0.060	0.020	0.020	0.010	0.010	0.030
Mean	0.029	0.026	0.016	0.010	0.007	0.006	

	CD at 5%	F-value
Treatments	0.0011	*
Varieties	0.0015	*
Treatments × Varieties	0.0037	*

Mean of three replicates.

* = Significant.

**= where unit stands for the amount of enzyme required to invert 1 u mole of sucrose.

TABLE 6.21   Effect of Soil-Applied Rice Bran and Pre-Sowing Pyridoxine Soaking of Setts on Leaf Neutral Invertase at Pre-Monsoon (120 days) in Sugarcane (*Sachharum officinarum* L.) (units/mg protein)**

Varieties	Soil-applied rice bran (kg/ha)					0.03% Pyridox-ine (T_6) Soaking	Mean
	0 (T_1)	10 (T_2)	20 (T_3)	40 (T_4)	60 (T_5)		
CoS 8436 (V_1)	0.190	0.160	0.140	0.100	0.100	0.080	0.128
CoS 95255 (V_2)	0.140	0.120	0.130	0.100	0.090	0.080	0.110
CoS 96268 (V_3)	0.170	0.140	0.120	0.100	0.090	0.060	0.113
CoSe 98231 (V_4)	0.150	0.150	0.140	0.120	0.100	0.080	0.123
CoSe 01235 (V_5)	0.120	0.100	0.100	0.080	0.080	0.060	0.090
CoS 01424 (V_6)	0.060	0.040	0.040	0.030	0.030	0.020	0.037
CoSe 95422 (V_7)	0.100	0.080	0.060	0.050	0.040	0.030	0.060
UP 097 (V_8)	0.120	0.100	0.060	0.040	0.020	0.062	0.067
CoSe 92423 (V_9)	0.150	0.120	0.110	0.090	0.080	0.060	0.102
CoS 96275 (V_{10})	0.140	0.120	0.100	0.080	0.060	0.030	0.088
Mean	0.134	0.113	0.100	0.079	0.069	0.056	

TABLE 6.21 *(Continued)*

	CD at 5%	F-value
Treatments	0.0046	*
Varieties	0.0060	*
Treatments × Varieties	0.0148	*

Mean of three replicates.

* = Significant.

**= where unit stands for the amount of enzyme required to invert 1 u mole of sucrose in 1 hour at 37 C.

TABLE 6.22 Effect of Soil-Applied Rice Bran and Pre-Sowing Pyridoxine Soaking of Setts on Leaf Neutral Invertase at Post-Monsoon (210 days) in sugArcane (*Sachharum officinarum* L.) (units/mg protein)**

Varieties	Soil-applied rice bran (kg/ha)					0.03% Pyridox- ine (T₆) Soaking	Mean
	0 (T₁)	10 (T₂)	20 (T₃)	40 (T₄)	60 (T₅)		
CoS 8436 (V₁)	0.25	0.22	0.20	0.19	0.16	0.14	0.19
CoS 95255 (V₂)	0.14	0.12	0.10	0.08	0.08	0.06	0.10
CoS 96268 (V₃)	0.27	0.24	0.20	0.16	0.12	0.11	0.18
CoSe 98231 (V₄)	0.25	0.25	0.24	0.22	0.16	0.13	0.21
CoSe 01235 (V₅)	0.22	0.20	0.20	0.18	0.15	0.16	0.19
CoS 01424 (V₆)	0.16	0.12	0.11	0.09	0.06	0.06	0.10
CoSe 95422 (V₇)	0.24	0.22	0.20	0.21	0.18	0.16	0.20
UP 097 (V₈)	0.30	0.24	0.21	0.19	0.14	0.12	0.20
CoSe 92423 (V₉)	0.35	0.30	0.26	0.22	0.20	0.19	0.25
CoS 96275 (V₁₀)	0.34	0.31	0.25	0.21	0.17	0.11	0.23
Mean	0.25	0.22	0.20	0.18	0.14	0.12	

	CD at 5%	F-value
Treatments	0.0074	*
Varieties	0.0096	*
Treatments × Varieties	0.0236	*

Mean of three replicates.

* = Significant.

**= where unit stands for the amount of enzyme required to invert 1 u mole of sucrose in 1 hour at 37 C.

TABLE 6.23   Effect of Soil-Applied Rice Bran and Pre-Sowing Pyridoxine Soaking of Setts on Leaf Neutral Invertase at Early Ripening (270 days) in Sugarcane (*Sachharum officinarum* L.) (units/mg protein)**

Varieties	Soil-applied rice bran (kg/ha)					0.03% Pyridox-ine (T_6) Soaking	Mean
	0 (T_1)	10 (T_2)	20 (T_3)	40 (T_4)	60 (T_5)		
CoS 8436 (V_1)	0.29	0.32	0.35	0.39	0.41	0.43	0.37
CoS 95255 (V_2)	0.21	0.26	0.29	0.32	0.34	0.36	0.30
CoS 96268 (V_3)	0.35	0.39	0.41	0.45	0.46	0.48	0.42
CoSe 98231 (V_4)	0.28	0.30	0.34	0.36	0.36	0.39	0.34
CoSe 01235(V_5)	0.26	0.30	0.34	0.37	0.39	0.40	0.34
CoS 01424 (V_6)	0.22	0.29	0.31	0.38	0.42	0.44	0.34
CoSe 95422 (V_7)	0.30	0.36	0.41	0.44	0.48	0.52	0.42
UP 097 (V_8)	0.35	0.40	0.42	0.46	0.52	0.56	0.45
CoSe 92423 (V_9)	0.44	0.46	0.50	0.55	0.59	0.65	0.53
CoS 96275 (V_{10})	0.40	0.42	0.46	0.49	0.52	0.55	0.47
Mean	0.31	0.35	0.38	0.42	0.45	0.48	
	CD at 5%		F-value				
Treatments	0.0077		*				
Varieties	0.0100		*				
Treatments × Varieties	0.0245		*				

Mean of three replicates.

* = Significant.

**= where unit stands for the amount of enzyme required to invert 1 u mole of sucrose in 1 hour at 37 C.

TABLE 6.24   Effect of Soil-Applied Rice Bran and Pre-Sowing Pyridoxine Soaking of Setts on Leaf Neutral Invertase at Late Ripening (330 days) in Sugarcane (*Sachharum officinarum* L.) (units/mg protein)**

Varieties	Soil-applied rice bran (kg/ha)					0.03% Pyridoxine (T_6) Soak-ing	Mean
	0 (T_1)	10 (T_2)	20 (T_3)	40 (T_4)	60 (T_5)		
CoS 8436 (V_1)	0.32	0.34	0.38	0.40	0.44	0.45	0.39
CoS 95255 (V_2)	0.26	0.29	0.32	0.34	0.38	0.40	0.33

TABLE 6.24 *(Continued)*

CoS 96268 (V$_3$)	0.39	0.43	0.46	0.49	0.52	0.56	0.48
CoSe 98231 (V$_4$)	0.33	0.38	0.39	0.42	0.44	0.45	0.40
CoSe 01235 (V$_5$)	0.34	0.34	0.36	0.40	0.46	0.48	0.40
CoS 01424 (V$_6$)	0.31	0.35	0.39	0.42	0.44	0.49	0.40
CoSe 95422 (V$_7$)	0.35	0.40	0.44	0.49	0.53	0.58	0.47
UP 097 (V$_8$)	0.40	0.45	0.49	0.54	0.56	0.60	0.51
CoSe 92423 (V$_9$)	0.46	0.50	0.52	0.60	0.68	0.72	0.58
CoS 96275 (V$_{10}$)	0.41	0.44	0.46	0.49	0.55	0.65	0.50
Mean	0.36	0.39	0.42	0.46	0.50	0.54	

	CD at 5%	F-value
Treatments	0.0080	*
Varieties	0.0103	*
Treatments × Varieties	0.0254	*

Mean of three replicates.

* = Significant.

**= where unit stands for the amount of enzyme required to invert 1 u mole of sucrose in 1 hour at 37 C.

TABLE 6.25   Effect of Soil-Applied Rice Bran and Pre-Sowing Pyridoxine Soaking of Setts on Leaf Neutral Invertase at Maturity (360 days) in Sugarcane (*Sachharum officinarum* L.) (units/mg protein)**

Varieties	Soil-applied rice bran (kg/ha)					0.03% Pyridox-ine (T$_6$) Soaking	Mean
	0 (T$_1$)	10 (T$_2$)	20 (T$_3$)	40 (T$_4$)	60 (T$_5$)		
CoS 8436 (V$_1$)	-	-	-	-	-	-	-
CoS 95255 (V$_2$)	-	-	-	-	-	-	-
CoS 96268 (V$_3$)	-	-	-	-	-	-	-
CoSe 98231 (V$_4$)	-	-	-	-	-	-	-
CoSe 01235 (V$_5$)	-	-	-	-	-	-	-
CoS 01424 (V$_6$)	0.42	0.44	0.49	0.50	0.53	0.55	0.49
CoSe 95422 (V$_7$)	0.45	0.44	0.50	0.54	0.59	0.63	0.53
UP 097 (V$_8$)	0.49	0.51	0.55	0.59	0.62	0.65	0.57
CoSe 92423 (V$_9$)	0.52	0.60	0.64	0.69	0.70	0.75	0.65
CoS 96275 (V$_{10}$)	0.48	0.50	0.53	0.59	0.65	0.69	0.57
Mean	0.24	0.25	0.27	0.29	0.31	0.33	

TABLE 6.25   *(Continued)*

	CD at 5%	F-value
Treatments	0.010	*
Varieties	0.013	*
Treatments × Varieties	0.032	*

Mean of three replicates.

* = Significant.

**= where unit stands for the amount of enzyme required to invert 1 u mole of sucrose in 1 hour at 37 C.

TABLE 6.26   Effect of Soil-Applied Rice Bran and Pre-Sowing Pyridoxine Soaking of Setts on Cane Yield (t/ha) at Harvest in Sugarcane (*Sachharum officinarum* L.)

Varieties	Soil-applied rice bran (kg/ha)					0.03% Pyridox-ine (T_6) Soaking	Mean
	0 (T_1)	10 (T_2)	20 (T_3)	40 (T_4)	60 (T_5)		
CoS 8436 (V_1)	50.00	53.30	61.60	72.30	71.00	78.50	64.45
CoS 95255 (V_2)	73.30	83.30	103.30	104.60	101.00	105.50	95.17
CoS 96268 (V_3)	70.60	84.30	91.00	96.60	94.60	99.80	89.48
CoSe 98231 (V_4)	56.30	66.60	83.60	103.30	95.60	105.50	85.15
CoSe 01235 (V_5)	52.60	60.30	78.00	96.60	93.60	99.90	80.17
CoS 01424 (V_6)	72.00	83.60	95.30	101.30	99.00	105.50	92.78
CoSe 95422 (V_7)	58.00	68.00	75.30	101.00	98.30	110.00	85.10
UP 097 (V_8)	83.60	92.30	92.00	111.60	99.80	115.00	99.05
CoSe 92423 (V_9)	87.60	96.30	106.30	120.60	106.50	123.10	106.73
CoS 96275 (V_{10})	60.30	76.00	80.50	103.60	103.00	110.10	88.92
Mean	66.43	76.40	86.69	101.15	96.24	105.29	

	CD at 5%	F-value
Treatments	3.81	*
Varieties	4.92	*
Treatments × Varieties	12.06	*

Mean of three replicates.

* = Significant.

TABLE 6.27 Effect of Soil-Applied Rice Bran and Pre-Sowing Pyridoxine Soaking of Setts on Brix Obtained (%) in Sugarcane (*Sachharum officinarum* L.)

Varieties	Soil-applied rice bran (kg/ha)					0.03% Pyridoxine (T_6) Soaking	Mean
	0 (T_1)	10 (T_2)	20 (T_3)	40 (T_4)	60 (T_5)		
CoS 8436 (V_1)	20.63	20.47	19.84.	19.62	19.00	19.85	19.49
CoS 95255 (V_2)	20.03	19.52	19.58	19.50	20.41	20.41	20.11
CoS 96268 (V_3)	19.42	19.73	20.42	21.26	19.98	21.30	20.85
CoSe 98231 (V_4)	18.00	18.30	18.40	19.40	19.20	19.60	19.40
CoSe 01235 (V_5)	19.50	19.70	19.65	20.00	19.90	20.30	20.07
CoS 01424 (V_6)	19.56	19.99	19.53	19.90	19.58	19.95	19.81
CoSe 95422 (V_7)	21.27	21.48	19.10	21.15	20.16	21.30	20.87
UP 097 (V_8)	19.74	19.65	19.90	19.95	19.92	20.30	20.06
CoSe 92423 (V_9)	18.09	19.55	19.43	19.93	19.90	20.10	19.98
CoS 96275 (V_{10})	20.30	20.80	19.89	19.68	21.08	21.40	20.72
Mean	19.65	19.92	19.54	20.04	19.91	20.45	
		CD at 5%		F-value			
Treatments		0.48		*			
Varieties		0.62		*			
Treatments × Varieties		1.53		*			

Mean of three replicates.

* = Significant.

TABLE 6.28 Effect of Soil-Applied Rice Bran and Pre-Sowing Pyridoxine Soaking of Setts on Brix Corrected in Sugarcane (*Sachharum officinarum* L.)

Varieties	Soil-applied rice bran (kg/ha)					0.03% Pyridoxine (T_6) Soaking	Mean
	0 (T_1)	10 (T_2)	20 (T_3)	40 (T_4)	60 (T_5)		
CoS 8436 (V_1)	20.72	20.66	20.07	20.81	19.19	19.95	20.23
CoS 95255 (V_2)	20.22	19.71	19.77	19.69	20.60	20.65	20.11
CoS 96268 (V_3)	20.61	19.92	20.61	21.45	20.17	21.55	20.72
CoSe 98231 (V_4)	18.19	18.49	18.59	19.59	19.39	20.05	19.05
CoSe 01235 (V_5)	19.69	19.89	19.84	20.19	20.09	20.50	20.03
CoS 01424 (V_6)	19.82	20.25	19.79	20.16	19.84	20.10	19.99

TABLE 6.28 *(Continued)*

CoSe 95422 (V$_7$)	21.46	21.67	19.29	21.34	20.35	21.65	20.96
UP 097 (V$_8$)	19.93	19.84	20.09	20.14	20.11	20.65	20.13
CoSe 92423 (V$_9$)	18.18	19.74	19.62	20.12	20.09	20.45	19.70
CoS 96275 (V$_{10}$)	20.49	20.99	20.09	19.87	21.27	21.95	20.78
Mean	19.93	20.12	19.78	20.34	20.11	20.75	

	CD at 5%	F-value
Treatments	0.34	*
Varieties	0.44	*
Treatments × Varieties	1.09	*

Mean of three replicates.

* = Significant.

TABLE 6.29 Effect of Soil-Applied Rice Bran and Pre-Sowing Pyridoxine Soaking of Setts on Pol (%) in Sugarcane (*Sachharum officinarum* L.)

Varieties	Soil-applied rice bran (kg/ha)					0.03% Pyridoxine (T$_6$) Soaking	Mean
	0 (T$_1$)	10 (T$_2$)	20 (T$_3$)	40 (T$_4$)	60 (T$_5$)		
CoS 8436 (V$_1$)	75.20	75.00	71.20	70.10	69.20	70.30	71.83
CoS 95255 (V$_2$)	72.21	69.30	69.40	69.30	73.80	71.30	70.89
CoS 96268 (V$_3$)	69.70	71.00	74.40	76.60	73.00	77.60	73.72
CoSe 98231 (V$_4$)	63.00	63.20	63.60	68.80	68.10	69.10	65.97
CoSe 01235 (V$_5$)	68.50	69.20	68.90	72.00	71.90	73.10	70.60
CoS 01424 (V$_6$)	71.40	74.00	71.30	73.60	71.50	74.30	72.68
CoSe 95422 (V$_7$)	78.10	78.50	69.40	76.00	74.30	76.70	75.50
UP 097 (V$_8$)	70.00	69.40	71.90	72.00	72.00	73.10	71.40
CoSe 92423 (V$_9$)	63.20	69.30	69.10	70.60	70.50	71.30	69.00
CoS 96275 (V$_{10}$)	74.60	76.80	72.90	72.20	75.40	76.30	74.70
Mean	70.59	71.57	70.21	72.12	71.97	73.31	

	CD at 5%	F-value
Treatments	1.16	*
Varieties	1.50	*
Treatments × Varieties	3.69	*

Mean of three replicates.

* = Significant.

TABLE 6.30   Effect of Soil-Applied Rice Bran and Pre-Sowing Pyridoxine Soaking of Setts on sucrose (%) in Sugarcane (*Sachharum officinarum* L.)

Varieties	Soil-applied rice bran (kg/ha)					0.03% Pyridox-ine (T$_6$) Soaking	Mean
	0 (T$_1$)	10 (T$_2$)	20 (T$_3$)	40 (T$_4$)	60 (T$_5$)		
CoS 8436 (V$_1$)	18.05	18.00	17.15	16.90	16.73	16.95	17.30
CoS 95255 (V$_2$)	17.39	16.71	16.74	16.71	17.73	16.99	17.05
CoS 96268 (V$_3$)	16.81	17.12	17.88	18.37	17.58	17.85	17.60
CoSe 98231 (V$_4$)	15.29	15.30	15.40	16.59	16.46	16.70	15.96
CoSe 01235 (V$_5$)	16.52	16.69	16.62	17.34	17.32	17.30	16.97
CoS 01424 (V$_6$)	17.22	17.82	17.19	17.73	17.24	17.85	17.51
CoSe 95422 (V$_7$)	18.72	18.79	16.78	18.23	17.89	17.85	18.04
UP 097 (V$_8$)	16.88	16.74	17.10	17.34	17.34	17.60	17.17
CoSe 92423 (V$_9$)	15.34	16.71	16.66	17.01	16.98	17.30	16.67
CoS 96275 (V$_{10}$)	17.93	18.41	17.56	17.40	18.08	17.90	17.88
Mean	17.02	17.23	16.91	17.36	17.34	17.43	

	CD at 5%	F-value
Treatments	0.30	*
Varieties	0.39	*
Treatments × Varieties	0.96	*

Mean of three replicates.

* = Significant.

TABLE 6.31   Effect of Soil-Applied Rice Bran and Pre-Sowing Pyridoxine Soaking of Setts on Purity in Sugarcane (*Sachharum officinarum* L.)

Varieties	Soil-applied rice bran (kg/ha)					0.03% Pyridox-ine (T$_6$) Soaking	Mean
	0 (T$_1$)	10 (T$_2$)	20 (T$_3$)	40 (T$_4$)	60 (T$_5$)		
CoS 8436 (V$_1$)	87.11	87.12	85.45	84.21	87.18	86.80	86.31
CoS 95255 (V$_2$)	86.56	84.77	84.67	84.86	86.06	85.90	85.47
CoS 96268 (V$_3$)	81.56	85.84	86.75	85.64	87.15	86.60	85.59
CoSe 98231 (V$_4$)	84.05	82.74	82.84	84.68	84.89	84.90	84.02
CoSe 01235 (V$_5$)	83.90	83.91	83.77	85.88	86.21	86.30	85.00
CoS 01424 (V$_6$)	86.88	88.00	86.76	87.95	86.90	87.30	87.30

TABLE 6.31　*(Continued)*

CoSe 95422 (V_7)	87.23	86.70	86.98	85.43	87.91	86.77	86.84
UP 097 (V_8)	84.70	84.37	85.71	86.09	86.22	86.55	85.61
CoSe 92423 (V_9)	84.37	84.65	84.51	84.54	84.51	84.80	84.56
CoS 96275 (V_{10})	87.50	87.70	87.45	87.56	85.00	87.10	87.05
Mean	85.39	85.58	85.49	85.68	86.20	86.30	

	CD at 5%	F-value
Treatments	0.79	NS
Varieties	1.02	*
Treatments × Varieties	2.50	NS

Mean of three replicates.

* = Significant, NS = Non-significant.

ACKNOWLEDGMENT

The authors are thankful to the Principal, G.F. College, Shahjhanpur – 242001, India for providing necessary facilities together with CRC Press, New Jersey, U.S.A. for kind favors and moral supports.

KEYWORDS

- **crop productivity**
- **growth**
- **pyridoxine**
- **sugarcane**

REFERENCES

1. Lehninger, R. A. (1978). 'Principles of Biochemistry' (2nd ed.), Worth Publishers, Inc., New York, pp. 335–353.
2. Drumond, J. C. (1920). Researches on the fat soluble accessory substances. III Technique for carrying out folding tests for vitamin A (fat soluble A). *Biochem. J., 14*, 660–664.
3. Mc Collum, E. V., & Davis, M. (1915). The nature of the dietary deficiencies of rice. *J. Biol. Chem., 23*, 181–230.

4. Morton, R. A. (1971). The vitamin concept. *Vitam. Horm., 32,* 155–165.
5. Bonner, J., & Bonner, H. (1948). The B-vitamins as Plant Hormones. *Vitam. Horm., 6,* 225–275.
6. Aberg, B. (1961). Vitamins as growth factors in higher plants. In: Encyclopedia of Plant Physiology, W. Ruhland, (ed.). Springer-Verlag: *Berlin, 14,* 418–449.
7. Rosenberg, H. R. (1942). 'Chemistry and Physiology of the Vitamins'. Inter-Science Publishers, Inc. New York.
8. Harris, R. S., & Thimann, K. V. (1948). Vitamins and Hormones, 6, Academic Press Inc. Publishers, New York.
9. Schneider, G., H. Kack, & Y. Lindqvist, (2005). The manifold of vitamin B6 dependent enzymes. *H. Chen, L. Xiong., Plant J., 44*(3), 396–408.
10. Houshmandfar, Alireza Asli, & Davood Eradatmand. (2011). Response of Corn (Zea Mays L.) to Seed Pyridoxine-priming and Different Levels of Nitrogen., *Adv. Env. Biol., 5*(1), 53–57.
11. Zanjan, M. G., & Asli, D. E. (2012). A Study of seed germination and early seeding growth of wheat genotypes affected by different seed pyridoxine-priming duration., *Ann. Biol. Res., 3*(12), 5687–5691.
12. Sajjad, A., & Samlullah, E. (2015). Enhancement of growth and yield of Mustard (*Brassica juncea* L.) Var. Varuna by Thiamine Hydrochloride (Vitamin-B$_1$) Application, *J. Func. Env. Biol., 5*(1), 24–30.
13. Kumar, A., Shukla, & S. K., Abbas, Z. (2009a). Effect of soil-applied rice bran and pyridoxine soaking on sugarcane yield. *Sugarcane International (The Journal of Cane Agriculture), 27*(5), 210–211.
14. Kumar, A., Kumar, A, Kumar, P., & Abbas, Z. (2009b). Effect of soil-applied rice bran on germination percentage and cane yield of ten sugarcane (*Saccharum officinarum* L.) cultivars. *Life Science Bulletin, 6* (3), 161–164.
15. Burstrom, H. (1949). The chemical regulation of growth–General survey. In: *Encyclopedia of Plant Physiology*, W. Ruhlanad, (ed.) Springer-Verlag, Berlin, (1961). 14, 324–329.
16. Hirano, M., Sugiyama, M., Hatakeyama, Y., Kuroda, E, & Murata, T. (1998). Effect of the application of rice bran on the carbohydrate metabolism in leaves and stems of rice variety. Iwate Univ., Morioka (Japan). *J. Crop Sci., 67*(2), 208–215.
17. Hwang, Young-Hee, Jang, Young-Su, Kim, Moo-Kcy, Lee, & Hoi–Seon. (2002). Effect of rice bran extract and rice bran oil on Bifidobacterium and lactobacillus. Inst. Agri. Sci. Tech., College of Agriculture, Chondbuk National University, *Chondbuk (Print), 45*(2), 77–80.
18. Lowilai, P., Kabata, K., Okamoto, C., & Kikuchi, M. (1994). Effect of rice bran and wheat bran on fermentation quality and chemical composition of water hyacinth silage. Kyushu Tokai University, Choyo, Kumamoto (Japan). Faculty of Agriculture. *J. Japanese Soc. Grassland Science, 40,* 271–277.
.19. Jhune, C. S., & Kim, G. (2000). Effects of rice bran added at spawn making on the cultivation of Oyster mushroom, Pleurotus spp., PO–National Institute of Agricultural Science and Technology, *The Korean Journal of Mycology, 28* (1), 1–5.
20. Kishor, B., & Abbas, Z (2003). Effect of pyridoxine on the growth, yield, oil content and amino–nitrogen content of *Mentha piperita* L. Proc. Nat. Symp. on "Plant

Biology and Biodivesity in changing Environment, " held on Dec 29–31, (2003). Jamia Hamdard University, New Delhi, p. 58.

21. Kishor, B. (2006). Effect of pyridoxine and nitrogen on growth, yield, essential oil and Biochemical components of *Mentha piperita* L., under salt stess. PhD thesis, M. J. P. Rohilkhand University, Bareilly.

22. Kishor, B., Kanaujia, S. N., & Abbas, Z. (2006). Effect of vitamin B6 (Pyridoxine) on the growth, yield, oil content and biochemical components of peppermint *Mentha piperita. L. Nat. Jour. Life Sci., 3*(3), 215–220.

23. Kumar A., & Abbas Z. (2014a). Effect of pyridoxine and gibberellins spray on growth yield quality of Sugarcane *Saccharum officinarum* L. An International Journal of Biological and Chemical Research (JBCR), *31*, 349–354.

24. Kumar A., & Abbas Z. (2014b). Effect of soil applied rice bran and pyridoxine soaking on leaf starch and amylase activity in Sugarcane (*Saccharum officinarum* L.) cultivars. Proc. Nat. Sem. on 'Technological advances and their impact on environment held on April 2014 at Poorima College of Engineering, Jaipur, 302–322, p. 60.

25. Kumar A., & Abbas Z. (2014c). Studies on leaf total free amino acid content and invertases as a result of soil applied rice bran and seed soaked pyridoxine in ten Sugarcane (*Saccharum officinarum* L.) cultivars. Proc. Nat. Sem. on 'Technological advances and their impact on environment held on April 2014 at Poorima College of Engineering, Jaipur, 302–322, p. 61.

26. De Capite, (1949). Thiamine, riboflavin and nicotinamide in the ketabolic processes. *Ann. Fac. Agri. Univ., Perugia, 6*, 59–68.

27. Iijima, T. (1952). On the Physiology and utilization of vitamin B1 in garden crops. I. The effect of vitamin B1 on the germination of kidney beans. *J. Hort. Ass., Japan,* (1953). *21*, 117–122 (Cited from Hort. Abstr. 23, (3074).

28. Iijima, T. (1956a). On the physiology and utilization of vitamin B1 in garden crops vi. The effect of thiamine application on respiration of some horticultural crops. *J. Hort. Ass., Japan, 25*, 11–16.

29. Artimonov, V. I. (1966). The synthesis and the destruction of chlorophyll in plants induced gibberellin and vitamin B2. *Soviet Pl. Physiol., 13*, 379–383.

30. Zavenyagina, T. N., Bukin, & Yu. V. Study of experimental B6 a vitaminosis of pea and wheat seedlings. *Soviet Plant Physiol., 16*, 253–260. 128, 482–487.

31. Langridge, J., & Brock, R. D. (1961). A thiamine requiring mutant of tomato. *Aust. J. Biol. Sci., 14*, 66–69.

32. Easley, L. W. (1969). Modification of protein synthesis by vitamin B12 in the marine algal flagellate *Neachloris. J. Protozool., 16*, 286–287.

33. Gopala Rao, P., & Sastry, K. S. (1972). B group vitamins during seedling growth of late and early varieties of groundnut (*Arachis hypogaea* L.). *J. Indian Bot. Soc., 51*, 155–161.

34. Iijima, T. (1957a). On the physiology and utilization of vitamin B1 in garden crops viii. The effects of vitamin B1 on the photoynthesis of some horticultural plants. *J. Hort. Ass., Japan, 26*, 247–250.

35. Tregunna, B, (1966). Flavin mononucleotide control of glycolic acid oxidase and photorespiration in corn leaves. *Science, 151*, 1239–1241.

36. Black, C. C., & Sanpietro, A. (1987). Vitamin B6 activity in photosynthetic reactions. *Arch Biochem. Biophys.*, Abbas, Z., & Kumar, S. Also, see, Response of son-

alika wheat and triticale varieties to timings of nitrogen application coinciding with irrigation schedule. *Geobios, 14*, 27–30.

37. Kodendaramaiah, J., & Gopala Rao, P. (1984). Photosynthesis by isolated chloroplasts of *Cyamopsis tetragonoloba* (L.) Taub., as influenced by B-vitamins. *Indian J. Plant Physiol., 27*, 166–171.

38. Williams, R. J., Eakin, R. E. Beerstecher, E., & Shive, E. (1950). The Biochemistry of B-vitamins. Reinhold Publishing Corporation, New York.

39. Gopala Rao, P. (1973). Influence of riboflavin on growth, respiration chlorophyll and protein contents in green gram. *Curr. Sci., 42*, 580–581.

40. Gopala Rao, P. Nagi Reddy, A., & Rajkumar, N. (1974). Activation of succinic dehydrogenase activity on the shoot, respiration and protein synthesis of seedlings of *Phaseolus radiatus* by B vitamins. *Ibid. 43*, 796–797.

40. Gopala Rao, P. Nagi Reddy, A., & Rajkumar, N. (1974). Activation of succinic dehydrogenase activity on the shoot, respiration and protein synthesis of seedlings of *Phaseolus radiatus* by B vitamins. *Ibid. 43*, 796–797.

41. Gopala Rao, P., & Kodendaramaiah, J. (1982). Interaction of kinetin with B group vitamins on the seedling growth of green gram (*Phaseolus radiatus* L.). *Proc. Indian Acad. Sci., (Plant Sci.), 91*, 183–188.

42. Sruoginite, A. V, & Shpokene, A. P. (1968). Effect of group B-vitamins on the activity of carbonic anhydrase, catalase and respiration intensity in plants treated with the herbicide 2, 4 *D. Tr. Acad. Nauk. Litov. S.S.R.B., 3*, 161–171.

43. Gopala Rao, P., & Sudarsanam, G. (1984). Carbohydrate changes induced by temperature and vitamins in green gram (*Vigna radiata* L. Wilezek) seedlings. *Proc. Indian Acad. Sci., (Plant Sci.), 93*, 111–117.

44. Kodendaramaiah, J., & Gopala Rao, P. (1985). Influence of B-vitamins on stomatal index, frequency and diurnal rhythms in stomatal opening in *Cyamopsis tetragonoloba (L.) Taub. J. Biol. Res., 50*, 68–73.

45. Sutcliffe, J. F. (1962). Mineral Salts Absorption in Plants. Pergamon Press, New York.

46. Iijima, T. (1956b). On the physiology and utilization of vitamin B1 in garden crops vii. The effect of foliage thiamine sprays on the chemical composition of taste of some garden crops. *J. Hort. Ass., Japan, 25*, 194–198.

47. Kudrev, I, & Pandey, S. (1965). Raising yield of swamp damaged wheat by vit. B1 spraying. *Centr. Acad. Bulg. Sci., 18*, 555–557.

48. Orcharov, K. E., & Kulieva L. (1968). Effects of vitamin B6 and pp on germination of seeds. *Khlopkovodstvo. 18* (3), 41–42. *(Cited from Field Crop Abstr. 21*(4), 2739).

49. Dimitrova Russeva, E., & Lilova, T. (1969), Growth of *Mentha piporita* and synthesis of essential oil as affected by thiamin (vit. B1). phyridoxine (Vit. B6) and nicotinic acid (vit. pp.). *Rasten. Nauki., 6*, 73–83.

50. Pandey, S. (1979). Effect of concentration of solution and of plant treatment with indolyl-3 acetic acid and vitamin B1 on nitrogen abrorption and yield of wheat. *Raslenickdni Nauki, 16*, 39–46.

51. Gopala Rao, P., & Raghava Reddy, B. N. (1985). Uptake of major elements as influenced by B vitamins in green gram. *Geobios, 12*, 70–73.

52. Ansari, S. A. (1986). Physiomorphological response of *Lens culinaris* L. Medic. and *Vigna radiata* L. Wilezek to pyridoxine application PhD Thesis, Aligarh Muslim University, Aligarh.

53. Aizikovick, L. E. (1967). Application of physiologically active substances to rice under O2 deficiency. *Trudy Denepropetr Selpkhez. Inst., 9*, 53–56.

54. Mullick, P., & Chatterji, U. N. (1974). Niacin inhibition of root growth in lettuce seedlings. *Curr. Sci. 40*, 40–41.

55. Serebryakova, N. V. (1971). The effect of vitamins on seed germination, and growth and development of *Rosa cinnamomea. Sbornik Nauchnylah Rabot vasesoyuzno go Nauchno Issledovaateskogo Instituta Lekarstvennykh Rastenu No. 4*, 56–84. (*Cited from Hort. Abstr. 43*, 2990, 1973).

56. Sinkovicks, M. (1974). Studies on increasing the effectiveness of vitamins of B-complex for melons. *Zoldegtermesztes 4*, 125–133 (*Cited from Hort. Abstr. 40*(3), 1557. 1974).

57. Ashfaque, N., Afridi, M. M. R. K., & Ansari, S. A. (1983). Effect of pyridoxine treatment of grain on yield of triticale. Annual Conference of the Society for Advancement of Botany, Hissar, May 28–29, p. 1.

58. Neelmathi, D., Ramya, P., & Mallika, S. (2005). Effect of riboflavin on initiation and multiplication of shoot cultures of sugarcane. *Sugar Tech. 7*, 83.

59. Bonner, J., & Greene, J. (1939). Vitamin-B1 and the growth of green plants. *Bot. Gaz., 100*, 226–237.

60. Hitchcock, A. E., & Zimmerman, P. W. (1941). Further tests of vitamin B1 on established plants and on cuttings. *Centra, Boyce Thompson Inst., 12*, 143–155.

61. Murneek, A. E. (1941). Vit, B1 vs Organic matter for plant growth. *J. Proc. Amer. Hort Sci. 38*, 715–717.

62. Mariat, F. (1944). Favorable influence of vit. B1 on the germination of *Cottleya* orchids. *Rev. Hort. Paris, 116*, 68–69.

63. Brusca, J. N., & Hass, A. R. C. (1957). Organic chemical on citrus. *Calif. Agric., 11*, 4.

64. Mel' Tser, & R. F. (1967). Reaction of different morphophysiological types of spring wheat to seed treatment with nicotinic acid, *Nauch. Dokl. Vyssh. Shk. (biol. Nauki) No. 12* (48), 94–98.

65. Vergnano, O (1959). Vitamin B1 and B6 on the rooting of certain cuttings *Nuovo G. Bot. Ital. 66*, 1–13.

66. Reda, F., Fadi, M Abdel-Alla, R. S., & El-Mourse A. (1977). Physiological studies on *Ammi visnaga* L. V. The effect of thiamine and ascorbic acid on growth and chromone yield. Egyptian *J. Physio. Sci. 18*, 19–27.

67. Arnon, D. I. (1940). Vit. B1 in relation to growth of green plants. *Science, 92*, 264–266.

68. Hamner C. L. (1940). Effect of vitamin B1 upon the development of some flowering plants. *Bot. Gaz, 102*, 156.

69. Minnum, E. C. (1941a). Effects of vitamins on growth of radish and cauliflower. *Bot. Gaz., 103*, 397–400.

70. Minnum, E. C. (1941b). Effect of vitamin B1 on the yield of several vegetable crop plants. *J. Proc. Amer. Soc. Hort. Sci., 38*, 475–476.

71. Templeman, W. G., & Pollard, N. (1941). The effect of vitamin B1 and nicotinic acid upon growth and yield of spring oats and tomatoes in sand culture. *Ann. Bot. 5*, 133–147.

72. Minarik, C. E. (1942). Effect of vit. B1 on growth of rice plants. *Plant Physiol., 17*, 141–142.

73. Kjelvick, S. (1965). Soaking vegetable seeds in nocotinic acid. *Gratner. Yrest., 55*, 1156.

74. Gutmanis, K. (1967). The effect of vitamins on the yield and chemical composition of garden peas. *Izv. Akad. Nauklart. S.S.S.R. No. 5*, pp. 105–108 (cited from Hort. Abstr. *38*, 7776, 1968).

75. Genkel, K. P. (1970). Effect of pre-sowing treatment of wheat seeds, with vitamin PP or sodium fluoride on protein change in seeds. *Fiziol. Rast. 17*, 605–609. *(cited from Field Crop Abstr. 24*, 1689).

76. Polyanskaya, L. A., & Kuvadov, M. (1974). Development of cotton as affected by treatment with nicotinic acid. *Khlopkovodstvo, 12*, 31–32.

77. Radzevicius, A., & Bluzmanas, P. (1975). The effect of thiamin and nocotinic acid on some physiological processes in tomatoes. Naunchyne ir. Trudy vysshikh uchebnykhrzavedenu Lit. S. S. R. Biologiva, *14*, 70–74. (Cited from Hort. Abstr. *46*, p. 802, IARI, 1976, 9414).

78. Kulieva, L. K. Azizova, A., Abdullaeva, M, & Movlamova, M. (1976). The reaction of melons to additional treatment with vitamins. *Biol, Zhivotnykhi Rastenii Turkmenestana, 3*, 85–91 (Cited from Hort. Abstr. *48*, 8127, 1978).

79. Iijima, T. (1955). On the physiology and utilization vitamin B1 in garden crops. 111. The effects of foliage thiamine sprays on growth and yield of sweet potatoes. *J. Hort. Ass. Japan, 24*, 228–236.

80. Cajlahjan, M. H. (1956). The effects of vitamins on the growth and developments of plants. *Dokhydy Acad. Nauk. S. S. S. R., 111*, 8974–8977.

81. Iijima, T. (1956b). On the physiology and utilization of vitamin B1 in garden crops vii. The effect of foliage thiamine sprays on the chemical composition of taste of some garden crops. *J. Hort. Ass., Japan, 25*, 194–198.

82. Boukin, V. N. (1958). Notes on the study on vitamins in plants. *Qual. Plant. Maveg., 3/4*, 374–380.

83. Galachalova, Z. N., Kungurtseva, V. V., Marusina, T. M., & Makhoskina, G. A. (1967). Cause of physiological deficiencies in cereal seeds in Western Siberia. In: *Physiology of Cold Resistance and Field Emergence in Siberia*, F. E. Reimers, (ed.). Nauka Moscow, pp. 49–57.

84. Popova, D., Kamenova, V., & Mullick, P. (1974). Studied on the effects of vitamins during pollination on the F1 generation of *Capsicum annum, Genetika, 7*, 31–35.

85. Arsen'eva. G. S. (1977). Effect of physiologically active substances on the growth and flowering of shrubs. *Nauch. Trudy. Akad. komunun khoz., No. 151*; 15–23.

86. Et-kholy, S. A., & Saleh, M. M. (1980). Effect of thiamine and ascorbic acid on the yield, essential oil and chamzulene formation in *Matricaria chammomila*. In: *Shams Univ. Fac. Agric. Res. O*, 1409.

87. Rehim, S., & Espig, G. (1991). The cultivated plants of the tropics and subtorpics. *Priese*. Gmbh, Berlin, Germany. pp. 66–69.

88. Gupta, R., Kumar, R., & Tripathi, S. K. (2004). Study on agroclimatic condition and productivity pattern of sugarcane in India. *Sugar Tech., 6* (3), 141–149.

89. Chinnasamy, R. T. C., & Jayanthi, C. (2004). Bio-organic nutrient management in sugarcane production, *Agric. Rev., 25* (3), 201–210.

90. Patel, V. S., Bafna, A. M., Raj, V. C., Colambe, B. N., & Patel, D. D. (2008) Effect of different levels and source of organics on sugarcane. (var. CoLK 8001). *Indian Sugar,* 65–70.

91. Sundra, B. (1998). 'Sugarcane cultivation' pp. 27.

92. Verma, R. S. (2004). 'Sugarcane Production Technology in India'. pp. 114–132.

93. Singh, J. P., Pal, S., Ram, S., & Lal, K. (2008). Sugarcane production technology under late spring planting condition for western tract of Uttar Pradesh, At a Glance. *Coop. Sugar, 40*(4), 57–67.

94. Mohan Naidu, K., & Arulraj, S. (1987). 'Sugar Technologies', Platinum jubilee (ed.), Sugarcane Breeding Institute (SBI), Coimbatore.

95. Agarwal, M., Sehtiya, H. L., Dendsay, J. P. S., & Agarwal, M. (1998). Starch hydrolysis activity from internodes of sugarcane. *Sugarcane, 5*, 16–17.

96. Alexander, A. G., Samuels, G., Spain, G. L., & Montalvo-Zapata, R., (1972). Physiology of sugarcane under water stress; invertase, ATP-ase and amylase behavior in plants experiencing water deficiency night flooding and continuous flooding Journal of Agriculture of the university of *Puerto-Rico, 56* (2), 115–133.

97. Ali, S. A., Srivastava, R. P., & Lal, K., (1996). Singh, G. P. Relative performance of some agro-chemicals on growth yield and juice quality of sugarcane. *Indian sugar, 46* (4), 263–266.

98. Dhamankar, V. S., Joshi, S. S., Sawant, R. A., Tawar, P. N., & Hapase, D. G. (1993). Invertase activity pattern during the growth period of early, mid, late and late maturing sugarcane varieties. Sugarcane (United Kingdom), (no.) p. 2–5 (Record 18 of 153–AGRIS 1999–2003/09).

99. Mohan Naidu, K., & Arulraj, S. (1987). *'Sugar Technologies'*, Platinum jubilee (ed.), Sugarcane Breeding Institute (SBI), Coimbatore.

100. Hasan M. F., Sikder, M. A., Miah, M. A. S., & Rehman, M. H. (2002). Effect of enzyme activities on sucrose accumulation in different stages of sugarcane (*Saccharum officinarum* L.). *Pakistan Sugar Journal, 17*(5), 2–9.

101. Kumar, A., & Pande, H. P. (1991). Effects of late application of growth regulators on sugars and hydrolytic enzymes of sugarcane stubble buds and subsequent effects on ratoon regrowth. *Coop. Sugar, 22*(7), 449–453.

102. Kumar, A., Pande, H. P., & Kumar, A. (1989). Effect of growth regulators on foliar enzymes and sugars, relationships with the juice quality of sugarcane. *Bhartiya Sugar, 15*(2), 47–54.

103. Li- Yuquian, Xie–Jiusheng, Tan–Zhongwen, Li–YQ, Xie–J. S., & Tan, Z. W. (1995). Studies on the correlations among metabolisms of foliar carbohydrate and nitrogen and cane yield quality in sugarcane. *Scientia Agricultura Sinica, 28* (4), 46–53.

104. Mohandas, S. Naidu, V., & Naidu, M. (1983). Catalytic effect of calcium on invertase activity in sugarcane. *Tropical Agriculture, 60*(2), 148.

105. Robertson, M. J., Muchow, R. C., Wood, A. W., & Campbell, J. A. (1996). Accumulation of reducing sugars by sugarcane, effects of crop age, nitrogen supply and cultivar. *Field Crops Research, 49*(1), 39–50.

106. Sooknohthan, K. Rao, & M. S. S. (1969). Effect of growth regulating substances on juice quality and yield of sugarcane. *Indian Sugar, 19*(1), 19–36, 39–40.

107. Shukla, S. P., & Srivastava, M. K. (1990). Regulation of enzymic activity of sucrose synthctase and invertase by Indole-3-acetic acid in sugarcane. *Bhartiya Sugar, 16*(1), 153–163.

108. Sugiharto, B., Sumadi, Hutasoit, G. F., Jusuf, M., & Mathis, P. (1995). Effects of nitrogen nutrition on activities of phosphoenolpyruvate carboxylase and sucrose phosphate synthetase and accumulation of sucrose in sugarcane leaf. In: Photosynthesis, from light to biosphere, v. 5, *Proc. xth Int. Photosynth. Congress, Montpellier*, France, 20–25.

109. Veith, R., & Komor, E. (1993). Regulation of growth, sucrose storage and ion content in sugarcane cells measured with suspension cells in continuous culture grown under nitrogen, phosphorus or carbon limitation. *J. Plant Physiol. 142*(4), 414–424.

110. Yang, T. T., & Hseih, T. S. (1977). A study on ripening of sugarcane plants III. Decrease of reducing sugar in cane stalks during ripening. Report of the Taiwan Sugar Research Institute, Taiwan, No. *8*, 45–54.

111. Samiullah, Ansari, S. A., Afridi, M. M. R. K., & Akhtar, M. (1985). Pyridoxine application enhances nitrate reductase activity and productivity in Vigna radiata. *Experientia, 41*, 1412–1414.

112. Hatch, M. D., Sacher, J. A., & Glasziou, K. T. (1963). The sugar accumulation cycle in sugarcane. I. Studies on the enzymes of the cycle. *Plant Physiol., 38*, 338–343.

113. Abbas, Z., Samiullah, Afridi, M. M. R. K., & Inam, A. (1980). Effect of different levels of nitrogen on the fodder yield of five varieties of rainfed sorghum. *Comp. Physiol. Ecol., 5*, 143–145.

114. Abbas, Z., & Kumar, S. (1987). Response of sonalika wheat and triticale varieties to timings of nitrogen application coinciding with irrigation schedule. *Geobios, 14*, 27–30.

115. Kumar, S., & Abbas, Z. (1992). Studies on different sources and methods of zinc application on wheat and triticale. *Geobios, 19*, 156–159.

116. Hasan, A., Abbas, & Z. (2007a). Effect of different concentrations of some leaf-applied agro-chemicals on growth, yield and quality in mustard (*Brassica juncea* L.). Paper presented at Indo-Hungarian Workshop, hcld on 24 March, 2007 at I. V. R. I., Bareilly, sponsored by DST, New Delhi.

117. Hasan, A., Abbas, & Z. (2007b). Nutritional aspects of some selected crop plants. Inc., *"Plant Physiology–Current Trends,"* P. C. Trivedi, (ed.) Pointer Publishers, Jaipur, pp. 153–203.

118. Hasan, A., & Abbas, Z. (2008). Studies on the effects of Agrochemicals, boron and Sulphur on growth and quality of mustard (*Brassica juncea* L.). In: *"Plant Physiology in Agriculture and Forestry,"* P. C. Trivedi, (ed.) Aawishkar Publishers, Jaipur. pp. 16–56.

119. Hasan, A., Kumar, A, Kumar, P., & Abbas, Z. (2007). Effect of nitrogen levels on growth, herb yield and essential oil content of *Ocimum basilicum* var. garbratum (sweet basil). Indian *J. Trop. Biodiv., 15*(2), 140–143.

120. Hasan, A., Kumar, A., & Abbas, Z (2009). Effect of different sources of sulphur and boron on yield characteristics and oil yield in *Brassica juncea L. Life Science Bulletin, 6*(2), 233–240.

121. Kumar, S., Ahmad, A., & Masood, A. (2007). Salinity induced % seed germination and seedling growth of *Brassica napus* cv. Agami. Indian *J. Trop. Biodiv., 15*, 90–93.

122. Kumar, S., Ahmad, A., & Masood, A. (2008). Salinity induced % seed germination and seedling growth, oil yield of sunflower *(Helianthus annuus* L.). cv. PAC-3776. *Research on Crops, 9*, 274–277.

123. Kumar, A, Verma, N Kumar, A, & Abbas, Z. (2010). Crop growth period and population density variations in relation to leaf nitrogen, Floral Initiation and Pod yield of Groundnut (*Arachis hypogaea* L.), *Life Science Bulletin, 7*(2), 193–195.

124. Sharma, A. (2007). Ethno-botanical studies on the Tharu tribe of Udham Singh Nagar, Uttaranchal, PhD Thesis, M. J. P. Rohilkhand University, Bareilly.

125. Kumar, S. (2008). Salinity induced% seed germination and seedling growth, oil yield of sunflower (*Helianthus annuus* L.). PhD thesis, M. J. P. Rohilkhand University, Bareilly.

126. Kanaujia, S. N. Effect of different sources of soil and leaf applied zinc and sulphur nutrients on growth and oil.

127. Verma, N. (2011). Studies on balanced nutrition integrated with organic manures in groundnut (*Arachis hypogaea* L.) PhD Thesis, M. J. P. Rohilkhand University, Bareilly.

128. Sajjad, A., Kumar, A., Kishor, B., & Abbas, Z. (2011). Morphophysiological performance of seedling of three varieties of chick pea (*Cicer arietinum L.*). *Life Science Bulletin, 81*, 79–81.

129. Millikan, C. R. (1961). Plant varieties and species in relation to the occurrence of deficiencies and excess of certain nutrient elements. *J. Aust. Inst. Agric. Sci., 26*, 220.

130. Evans, H. J., & Sorger, G. J. (1966). Role of mineral elements with emphasis on the univalent cations. *Ann. Rev. Plant Physiol., 17*, 47–76.

131. Mishra, A., & Gupta, A. P., (1988). Correlation between sucrose v/s non sucrose (inorganic) constituents in cane juice. *Proc. Ann. Conv. Sug. Tech. Assoc., India, 51*, Ag 157–Ag161.

132. Thangavelu, S., & Chiranjivi Rao, K. (2006). Assessment of reducing sugars in top and bottom portions of sugarcane genetic stocks and its associations with other quality characters. Indian Sugar, May, 29–34.

133. Morris, D. A., & Arthur, E. D., (1984). An association between acid invertase activity and cell growth during leaf expansion in *Phaseolus vulgaris L. J. Exp. Bot., 35*, 1369–1379.

134. Gayler, K. R., & Glasziou, K. T., (1972). Physiological function of acid and neutral invertases in growth and sugar storage in sugarcane. *Physiol. Plant., 27*, 25–31.

135. Willenbrink, J., (1982). Storage of sugars in higher plants. In: F. A. Loewus and W. Tanner (Eds.), *Encyclopedia of Plant Physiology*, New series, 13A, Springer-Verlag, Berlin, pp. 684–699.

136. Rai, P. K., Kumar, R., & Madan, V. K. (1998). Patterns of plant growth and development in Saccharum genotypes, Indian Sugar, December 695–704.

137. Sharma, M. L. Singh, G. P., & Katiyar, R. B., (1989). Testing of maturity and association among its parameters in sugarcane. *Indian Sugar, August*, 291–297.

138. Taneja, A. D., Punia, M. S., Chaudhary, B. S., & Sharma, A. P., (1986). Effect of crop age on the quality of early, mid, and late maturing varieties of sugarcane. *Indian Sugar, July*, 155–160.

139. Kumar, A. (2009). Effect of soil applied rice bran and foliarly applied its extract on carbohydrate and protein metabolism during ripening and maturity in *Saccharum officinarum* L. PhD thesis. M. J. P. Rohilkhand University, Bareilly.

140. Alexander, A. G. (1973). "Sugarcane Physiology," Elsevier Scientific Publishing Company, New York, pp. 231–257.

141. Das G., & Prabhu, K. A. (1990). Study of some hydrolytic enzymes in different varieties of sugarcane during maturity period. Indian Sugar, December, 653–657.

142. Sachdeva, M. Mann, A. P. S., & Batta, S. K. (2003). Sucrose metabolism and expression of key enzyme activities in low and high sucrose storing sugarcane genotypes. *Sugar Tech., 5*(4), 265–271.

143. Gopala Rao, P., & Negi Reddy, A. (1976). [14]C incorporation into amino acids in chloramphenicol inhibited growth and its reversal by riboflavin in green gram (*Phaseolus radiatus* L.) *Z. Pflazenphysiol, 80*, 279–282.

144. Kumar, A., Sajjad, A., Kumar, K., Verma, N., & Abbas, Z. (2011). Effect of different transplanting dates on herb and oil yield content and uptake of plant nutrients in *Ocimum basilcum* L. (Sweet Basil), *Indian Journal of Functional and Environmental Botany, 1*(2) 119–121.

145. Shukla, S. K. (2010). Effect of some agro-chemicals on sugar accumulation and quality in early and late maturing sugarcane cultivars. PhD Thesis, M. J. P. Rohilkhand University, Bareilly.

CHAPTER 7

MECHANISMS OF METALLOID UPTAKE, TRANSPORT, TOXICITY, AND TOLERANCE IN PLANTS

PALLAVI SHARMA,[1] VINEET SRIVASTAVA,[1] AMIT KUMAR GAUTAM,[1] ANISHA RUPASHREE,[1] RAJANI SINGH,[1] SANCHITA ABHIJITA,[1] SATISH KUMAR,[1] AMBUJ BHUSHAN JHA,[2] and AMARENDRA N. MISRA[1]

[1]*Centre for Life Sciences, Central University of Jharkhand, Brambe, Ranchi – 835205, Jharkhand, India, Tel.: +919470416688, +919437295915, E-mail: pallavi.sharma@cuj.ac.in, anm@cuj.ac.in*

[2]*Crop Development Centre, Department of Plant Sciences, University of Saskatchewan, 51 Campus Drive, Saskatoon, SK S7N 5A8, Canada*

CONTENTS

ABSTRACT

Metalloids are the group of elements that have physical and chemical properties in between metals and nonmetals. This group consists of arsenic (As), boron (B), silicon (Si), germanium (Ge), antimony (Sb), and tellurium (Te). Certain other elements, such as selenium (Se), are sometimes added to the list of metalloids. Arsenic, boron, selenium and antimony are known to cause toxicity in plants at higher concentrations. Natural occurrence and anthropogenic activities have led to widespread contamination of these metalloids in many parts of the world. Plants growing in these areas are exposed to various forms of metalloid species depending on the redox potential and pH of the soil. Toxic metalloid species cause yield reduction in crops. Moreover, consumption of these contaminated crops exposes a large section of the human population to potential metalloid poisoning. Thus, developing metalloid tolerance and reducing its concentration in edible parts of crops are of critical importance. A thorough molecular understanding of the mechanisms of metalloid uptake, transport, toxicity, and tolerance in plants will help in development of metalloid tolerant and safe plants that can sequester metalloids in non-edible plant tissues. Significant advancement has been made in the identification and characterization of genes and proteins responsible for uptake, movement and tolerance of metalloids within plants. In this chapter, we provide an overview of these findings and their application in development of transgenic plants with enhanced metalloid tolerance and ability to restrict metalloids in non-edible plant parts.

7.1   INTRODUCTION

Metalloids are the elements group that have physical and chemical characteristics intermediary between metals and non-metals. Elements such as arsenic (As), boron (B), silicon (Si), germanium (Ge), antimony (Sb), and tellurium (Te) are generally considered as metalloids. Although metalloids are considered as non-metal, selenium (Se) is commonly designated as metalloid due to similarity of its aqueous chemistry with those of arsenic and antimony [1]. Both natural sources such as erosion of metalloid bearing rocks, weathering and anthropogenic activities such as mining, discharge of industrial waste, smelting of metalloid containing ore, fertilizers, sewage sludge, fly ash, fossil fuel combustion especially coal, irrigation with metalloid contaminated water contribute to metalloid contamination of soil [2]. However, the contribution of natural sources to metalloid in soils and atmosphere is small in comparison to metalloid from human activity. Arsenic, antimony, boron, and selenium are usually found in the soil solution as either charged ions or un-dissociated (uncharged) molecules depending on the pH and the redox potential of the soil. Plant roots can take up bioavailable form of metalloids species along with beneficial elements from the soil. Metalloids are generally of intermediate size and therefore their uptake and transport were at first thought to occur by means of passive diffusion but over past decades numerous evidences prove that there are certain transporters present in the cell facilitating influx and efflux in cells [3]. Various forms of metalloid are taken up selectively through different pathway and transport mechanism.

Arsenic and antimony are non-essential element and has emerged as serious health issues for animals and threats to plant growth and development. Arsenic has been categorized as class I carcinogen by the International Agency of Research on Cancer [4]. In plants, high concentrations of arsenic may start a sequence of reactions that lead to growth inhibition, disruption of photosynthetic and respiratory systems, and stimulation of secondary metabolism [5]. Plants seem to be highly tolerant to antimony, in comparison to animals. However, high concentrations of antimony can lead to inhibition of growth and photosynthesis, reduction in the uptake of essential elements and synthesis of cellular metabolites. Low concentrations of boron and selenium are beneficial for human, animal and plants. In spite of a significant role of boron in plants, excess of boron causes considerable reduction in the yield of plants. Boron at high concentration reduces vigor, particularly of shoots, delays development causes leaf burn (chlorotic and necrotic patches in older leaves), chlorosis starting at the leaf tip and margins of mature leaves and decreases number, size and weight of fruits [6–9]. Although not essential, selenium is regarded as a beneficial element for higher plants at low concentration (1–10 μM), as it stimulates growth [10]. However, excess selenium is toxic to plants and shows the symptoms of chlorosis, stunted root growth, reduced biomass and withering of leaves and reduced photosynthesis efficiency due to damaged photosynthesis apparatus [11].

Plants, during evolution have become adapted for adverse situations by developing several distinct mechanisms for survival and defense system. Plants ability to survive in adverse stress condition is known as tolerance. Distinct tolerance mechanism exists for metalloid toxicity among different plant genotypes. The common defense strategies used by plants to alleviate the toxic effects of metalloids and increase survival ability are: efflux of the toxic metalloid, vacuolar compartmentalization, production of anti-oxidants, methylation, etc. Accumulation of metalloids in plants causes high exposure of these metalloids to humans and ruminants through intake with food, feed and fodder. Mechanistic insights into the pathways of toxic metalloid uptake, toxicity and tolerance can help in the development of metalloid excluding and/or tolerant crop plants with reduced concentration of metalloids in edible parts, thus limiting human and animal health risks.

7.2 TOXIC METALLOIDS IN SOILS

Metalloids occur naturally in soil. Both pedogenic and anthropogenic processes contribute to the high concentration of metalloids in the soil. Arsenic and antimony has no known function in living organisms whereas boron and selenium is beneficial for living organisms at low concentration. At high concentrations, all these metalloids become toxic to living organisms. Metalloids exist under the form of various chemical species depending on several factors including pH and the redox potential of soils. These species differ in bioavailability, mobility, toxicity, biotransformation and physicochemical behavior. Toxic metalloids (arsenic, boron, selenium and antimony), their concentration in Earth's crust, anthropogenic sources, current use, and speciation in soil are listed in Table 7.1.

7.2.1 ARSENIC (AS)

Arsenic, a naturally occurring ubiquitous element of earth's crust is placed in the group VA of periodic table. It possesses the properties of metalloids and is extremely toxic. Arsenic contamination of soil and water is among the major global problems. It has emerged as serious health issues for

TABLE 7.1 Toxic Metalloids: Their Position in Periodic Table, Concentration in Earth's Crust, Anthropogenic Sources, Industrial Uses and Speciation in Soil

Metalloid	Group in periodic table	Concentration in Earth's crust (mg/kg)	Anthropogenic Sources	Industrial uses	Specification in soil
Arsenic (As)	VA	0.20–40	Mining, discharge of industrial waste, smelting of arsenic containing ores, fossil fuel combustion especially coal and irrigation with arsenic contaminated water	Alloying agent in the manufacture of semi-conductors, transistors, lasers and processing of glass, pigments, textiles, paper, metal adhesives, wood preservatives, ammunition, hide tanning process, pesticides, feed additives and pharmaceuticals	Four oxidation states +5, +3, 0, and –3. Arsenic (III) is most abundant and toxic than other forms.
Boron (B)	IIIA	1–500	Use of fertilizers and irrigation waters high in boron	Synthesis of organic compounds, borosilicate glass, pyrotechnic flares, flame retardant and medicines	Oxidation state 0. Exists mainly as boric acid and borates.
Selenium (Se)	VIA	0.05–0.09	Coal and oil combustion facilities, selenium refining factories, base metal smelting and refining factories, mining and milling operations and end-product manufacturers including semiconductor manufacturers	Paint, pigment production, glass manufacturing, oil refinery, electricity generation, metallurgy and lately medicine	Four oxidation states –2, +2, +4, and +6. Selenite (IV) most abundant and selenate (VI) is potentially the most toxic.

TABLE 7.1 (Continued)

Metalloid	Group in periodic table	Concentration in Earth's crust (mg/kg)	Anthropogenic Sources	Industrial uses	Specification in soil
Antimony (Sb)	VA	0.20–0.50	Emissions from vehicles (where it is used as a fire-retardant in brake linings), waste disposal, incineration, fuel combustion, metal smelters and shooting activities	Production of semiconductors, infrared detectors and diodes, manufacture of lead storage batteries, solder, sheet and pipe metal, etc., fire-retardant, formulations for plastics, rubbers, textiles, paper, paints, explosives, pigments, antimony slats, ruby glass and medicine.	Four oxidation states, +5, +3, 0, and –3, with the antimonate (V) and the antimonite (III) being the most abundant forms. The antimonite (III) forms are more toxic than antimonate (V) forms.

animals' as well as one of the potent threats to plant growth and development. It has been categorized as class I carcinogen by the International Agency of Research on Cancer [5]. According to World Health Organization [12], arsenic concentration typically range from 0.2–40 mg/kg in non-contaminated soil while in contaminated soils, it is as high as 100–2500 mg/kg. The arsenic contamination is more prominent in South and South East Asia river deltas including the mining areas of India, China, Vietnam, Chile, Bangladesh, Mexico and Hungary. In the present scenario, the average concentration of total arsenic present in soils of West Bengal ranges from 105.4 to 1500 mg/kg [13].

Arsenic contamination in soil could be due to natural processes such as erosion of arsenic bearing rocks and weathering and anthropogenic activities such as mining, discharge of industrial waste, smelting of arsenic containing ore, fossil fuel combustion especially coal, irrigation with arsenic contaminated water and the use of arsenic based pesticides, herbicides and fertilizers [14]. Currently, arsenic is used commercially and industrially as alloying agents in the manufacture of lasers, semi-conductors and transistors, processing of glass, paper, wood preservatives, metal adhesives, pigments, textiles, and ammunition. They are also used in the hide tanning process and, to a limited degree, as pesticides, feed additives and medicines. All these processes contribute to arsenic pollution. The phyto-availability and phyto-toxicity of arsenic depends upon several factors. One of the important factors is the speciation of arsenic. The pH, redox potential, clay content, presence of other ions and soil microorganisms result in different forms of arsenic. Arsenic exists in four different oxidation states of –3, 0, 3, and 5. It is commonly distributed throughout Earth's crust, most frequently as arsenic sulfide or as metal arsenates and arsenides. In aerobic environments, As (V) is dominant, usually in the form of arsenate (AsO_4^{3-}) in various protonation states: H_3AsO_4, $H_2AsO_4^{3-}$, and $HAsO_4^{2-}$. Under reducing environments, As(III) dominates and exist as arsenite (AsO_3^{3-}), and its protonated forms H_3AsO_3, $H_2AsO_3^{-}$, and $HAsO_3^{2-}$. Organic forms of arsenic such as methylarsinic acid [MMA; $(CH_3)AsO_2H_2$)] and dimethylarsinic acid [DMA; $(CH_3)_2AsO_2H$] also exist. Each form has different interaction with environment and biological system. Inorganic forms are most toxic and among serious threats to animal health and plant growth. Toxic concentrations of arsenic can lead to poor seed germination, inhibition in root and shoot growth, wilting and necrosis of leaf margins, chlorosis, nutrient deficiencies and biomass inhibition in plants [15]. Since, few decades, arsenic toxicity and its interaction with different plant genotypes has remained major thrust area of research. Several investigations are conducted to reveal the mechanisms of arsenic toxicity, transport and tolerance in plants.

7.2.2   BORON (B)

Boron (B) is a widely occurring essential micronutrient metalloid in the earth's crust and belongs to group IIIA in the periodic table. It is referred to as an orphan of the periodic table because its chemical properties are very different from other elements such as Al or Ga in this group [16]. It is the 51st most common element and is found at an average concentration of 8 mg/kg (approximately 0.0008%) in the earth's crust. Boron toxicity can occur in soils where high concentration of boron has accumulated naturally, due to use of fertilizers and irrigation waters high in boron [7, 17–20]. Many boron-containing chemicals including boronic acids and borohydrides are commonly used in the synthesis of organic compounds. Boron is also used as rocket fire igniter and in manufacture of borosilicate glasses, pyrotechnic flares, flame retardant, medicines, etc. Among all plant mineral nutrients, boron is unusual as it is generally taken up as an uncharged molecule, boric acid (BA) at physiological pH by the root from the soil. More than 90% of the boron in plants is found in cell walls, forming borate ester cross-linked rhamnogalacturonan II (RG-II) dimers, which is essential for the cell wall porosity and tensile strength. In spite of its great importance in plants, high concentration of boron toxicity limits crop productivity in different regions of the world. Plants exposed to toxic levels of boron show root and shoot growth inhibition, reduced root cell division, lower leaf chlorophyll content and photosynthesis, decreased lignin and suberin levels and altered cellular metabolism [7, 21, 22]. Plant growth is inhibited by boron in the concentration range of 1–5 mM in the soil solution [8]. Boron toxicity is a considerable problem in semi-arid areas in the world including Mediterranean countries, Turkey, California, South Australia, and Chile. Boron toxicity in arid soils is often confounded with the associated problems of salinity.

7.2.3   SELENIUM (SE)

Selenium, a metalloid micronutrient belongs to group VIA of periodic table along with sulphur and tellurium. The Low concentration of selenium is essential nutrient for humans and animals, and beneficial for plants, archaea, bacteria green algae, and some other microorganisms. However, excess selenium is toxic [11, 23, 24]. Chemical properties, valence shells, electronic structures and atomic radii of selenium is similar to sulphur or tellurium. The Earth's crust contains an average selenium concentration of 0.05 and 0.09 ppm [25]. High concentrations of selenium are found in volcanic rock (upto 120 ppm), sand stones, uranium deposits and carbonaceous rocks. It is rarely seen in elemental form (Se0). Soil selenium concentrations are typically in the range 0.01–2 mg/kg with a world average of 0.4 mg/kg [26]. Extremely high concentration of selenium has been found in organic rich soils derived from black shales in Ireland (1200 mg/kg), England (3.1 mg/kg) [27] and volcanic soils of Hawaii (6–15 mg/kg) [28]. Main sources of selenium in Earth crust includes elenide associated with sulfide minerals [29], coal (coal combustion generate selenium dioxide), and volcanic emissions. Anthropogenic emission sources of atmospheric selenium include coal and oil combustion, selenium refining factories, base metal smelting and refining, mining and milling operations, and end-product manufacturers (e.g., some semiconductor manufacturers). Now a day's selenium is frequently used in paint, glass manufacturing, oil refinery, metallurgy, pigment production, electricity generation, and lately medicine, which contribute to selenium pollution [30]. Important properties of elements, e.g. their bioavailability and toxicity, depend on their chemical form or speciation. Selenium occurs in four redox states, Se(VI) SeO_4^{2-} selenite, Se(IV) SeO_3^{2-} selenite, Se(0) elemental Se, Se(-II) selenide and HSe hydrogen selenide. Based on thermodynamic considerations, alterations in pH and redox conditions may cause shifts in equilibrium distributions between molecular forms. Selenide and elementary selenium are favored in a reducing environment, selenite in a slightly oxidizing environment and selenate (analogous to sulfate) in an oxidizing environment. In humid regions and acid soils, the prevailing form is selenite, which is firmly adsorbed on sesquioxides and clay minerals and thus not readily available to plants [31]. Selenate is the dominant species of selenium in alkaline soils under well-aerated conditions. This form is neither adsorbed nor form insoluble salts and thus is readily available to plants [32]. Malformed selenoproteins and oxidative stress are two distinct types of stress that drive selenium toxicity in plants and could affect cellular processes in plants that have yet to be thoroughly explored. Selenium toxicity in plants shows the symptoms of chlorosis, stunted root growth, reduced biomass and withering of leaves and reduced photosynthetic efficiency due to damaged photosynthesis apparatus [11]. Plants growing in high selenium containing soils, accumulate toxic levels of selenium in food, feed and fodder [33].

7.2.4 ANTIMONY (SB)

Antimony is non-essential, potentially carcinogenic metalloid [34]. Natural occurrence of antimony in in the Earth's crust is as low as 0.2–0.3 mg/kg [35]. Background concentrations of antimony in soils range between 0.3 and 8.6 mg/kg [36, 37]. In general, they are below 1 mg/kg [38]. Higher concentrations are usually related to anthropogenic sources. Important anthropogenic sources of antimony in the environment are emissions from vehicles (where it is used as a fire-retardant in brake linings), waste disposal and incineration, fuel combustion, metal smelters, and shooting activities [39, 40]. Reuse of material from mining dumps also causes severe antimony contamination of agricultural land and residential areas [41]. Antimony is widely used and is the ninth most mined element. It is used in many industries such as production of semiconductors, infrared detectors and diodes, manufacture of lead storage batteries, solder, sheet and pipe metal etc., fire-retardant formulations for plastics, rubbers, textiles, paper and paints, explosives, pigments, antimony salts and ruby glass [39]. Antimony compounds have been used as medicines mainly in the treatment of two parasitic diseases, leishmaniasis and schistosomiasis. It exists in four oxidation states, +5, +3, 0, and −3, with the pentavalent and the trivalent being the most abundant forms. The trivalent state is more toxic than the pentavalent state. In the environment, antimony usually occurs as Sb(III) (antimonite) and Sb(V) (antimonate) in reducing and oxidizing conditions, respectively. In soils, Sb(III) is oxidized within hours to Sb(V)

[42]. Sb(III) is more toxic than Sb(V). In its trivalent form, antimony may have a level of genotoxicity similar to trivalent arsenic [43]. Soil mineral composition, its pH and redox potential, as well as phosphate fertilizer or EDTA supplies are among the factors influencing antimony mobility and availability to plants [44–47]. Moreover, clear differences in the ability to absorb, transport and tolerate antimony exist among species [48, 49]. Antimony can persist in the environment and bioaccumulate in organisms. Little fact is known about the biochemical effects of antimony; however, there is assumption that its mode of action is similar to that of arsenic. Antimony can damage plants, including growth retardation, inhibition of photosynthesis and decrease in the uptake of certain essential elements and decreases in the synthesis of certain metabolites. The excessive accumulation of antimony can be toxic to plants and can inhibit their growth. The levels of 5–10 mg/kg antimony in plant tissues have been suggested to be excessive or toxic [50]. In the past, soil contamination by antimony has been neglected as it mostly exists as a co-contaminant of more toxic elements such as lead or arsenic but it has become environmental problem of much concern in recent years, because of increasing mining and industrial use. A potentially important antimony exposure pathway of humans and animals to antimony in areas with contaminated soils is through food and feed plants [51]. Therefore, to minimize the health risks of antimony, it is necessary to control the level of antimony in the plant.

7.3 METALLOIDS UPTAKE AND TRANSPORT MECHANISMS IN PLANTS

Metalloids such as arsenic, antimony, boron, and selenium uptake and transport was initially believed to occur via passive diffusion but now it is well known that there are certain transporters, which facilitate influx and efflux in cells [3]. Figure 7.1 shows the transporters involved in uptake and transport of arsenic, boron, selenium and antimony. A better understanding of mechanism of metalloid uptake by plants and transporters involved in metalloids uptake from soils will help in development transgenic plants with less metalloid accumulation in their plant parts. Roots are primarily responsible for absorption in plants and xylems translocate arsenic from

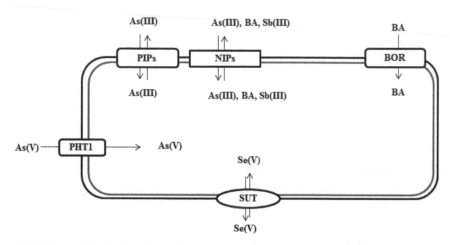

FIGURE 7.1   Transporters involved in uptake of metalloids in higher plants. In plant cells phosphate transporters (PHT1) play role in As(V) uptake whereas plasma membrane intrinsic protein subfamily (PIPs), Nodulin26-like intrinsic protein subfamily (NIPs) are involved in As(III) uptake. NIP transporters are also involved in uptake of arsenite, antimonite and boric acid/borate (BA). BOR is a boric acid/borate transporter for xylem loading and boron distribution within shoots. Selenate is taken up via sulphate transporters (SUT).

the root to the shoot and redistribute it in different tissues. Redox potential and pH determine metalloid solubility in soil. Various forms of metalloid species are available to plants from the soil. These forms are taken up selectively through different pathway and transport mechanism.

Plant species differ in their ability to accumulate and tolerate arsenic. Arsenic hyperaccumulator Chinese brake fern (*Pteris vitatta*) accumulated 20 times greater arsenic concentration than slender brake fern (*Pteris ensiformis*), a non-arsenic hyperaccumulator [52]. In another comparative study, castor plant (*Ricinus communis*) showed more arsenic than buckwheat (*Fagopyrum esculentum*) [53]. In case of arsenic, the most prevalent form in aerobic soil is arsenate (AsV). In plants, uptake kinetics of As(V), follow the Michaelis-Menten kinetics suggesting the involvement of a transporter [54, 55]. The Km value for As(V) has been reported to be 0.0157 mM in rice plants. As(V) is a chemical analog of phosphate with regards to membrane transport. The uptake mechanism involves co-transport of phosphate or As(V) and protons, with stoichiometry of at least $2H^+$ for each H_2PO_4 or H_2AsO_4 [56]. The oxyanion chemical structure of As(V) shows structure and chemical similarity with inorganic phosphorous and is thus readily taken up by phosphate transporters. As(V) has been shown to compete as a substrate for phosphate uptake system in a wide variety of plant species including non-hyperaccumulator plants such as Arabidopsis, rice and hyperaccumulator plants such as Chinese brake fern [57]. Asher and Reay [58] presented the first evidence of competitive inhibition of As(V) uptake by phosphate in plants. Among the nine high-affinity Pi transporters (*Pht1;1-Pht1;9*) encoded in *Arabidopsis thaliana*, *Pht1–5* and *Pht7–9* are known to be induced in roots under Pi-deficient conditions [59]. Shin et al. [60] demonstrated two high-affinity *Pht1* isoforms *AtPht1;1* (Pi transporter 1;1) and *AtPht1;4*, in *A. thaliana*, which promote and enhance the uptake of As(V) from soil with both low or high concentrations of Pi. Mutation and overexpression studies have indicated the role of Pi transporters in arsenic uptake. The null mutants of *pht1* moderately tolerate arsenate; while the double mutants of *pht1*, *pht4* are among significant arsenate tolerant. On the contrary, plants that overexpress the *Pht1* have shown to be arsenate-sensitive [61]. *Pht1;5* transporter in *A. thaliana* promotes the translocation of both Pi and As(V). Mutation of *Pht1;5* resulting in loss of *Pht1;5* mitigate As(V) toxicity in plants [62]. Overexpression of *AtPht1;7*, which is specifically expressed in reproductive tissues of Arabidopsis, enhances As(V) accumulation [63]. *AtPht1;8* and *AtPht1;9*, acquire Pi in the root [64]. *AtPHF1* (Pi Transporter Traffic Facilitator 1), an endoplasmic reticulum (ER)-localized protein has been shown to have an effect on the localization of *Pht1;1* transporters and As(V) tolerance in plants [65]. Like, *AtPHF1*, *OsPHF* affect the localization of *Pht1;1* transporters in *Oryza sativa* . Mutations in these ER localized protein affected the transport of *OsPht1;1* from the ER to the plasma membrane and thus decreased Pi and As(V) uptake [66, 67]. Role of several WRKY transcription factors in As(V) influx has also been demonstrated. WRKY6 and WRKY45, regulate the expression of AtPht1;1 and hence modulate As(V) uptake [68, 69]. In *O. sativa*, high affinity transporter *OsPht1;8*, transcription factor *OsPHR2* (Pi starvation response 2) which regulates the expression of *OsPht1;8* and a constitutively expressed high-affinity Pht1 transporter located in the plasma membrane *OsPht1;1*, have been shown to be involved in arsenic uptake

[67, 70–72]. In the flooded and anaerobic soil, As(III) is predominant form and hence Pi transport pathway is a minor route for arsenic uptake. At neutral to acidic pHs, the As(III) exist as a small uncharged molecule with a diameter of approximately 411 pm [73]. As(III) uptake in plant follows Michaelis-Menten kinetics suggesting that As(III) transport is an active process [54, 74, 75]. Rice, pea (*Pisumsativum*) and wheat plants shows Km values of 0.18, 0.34, and 0.51 mM, respectively, for As(III) [76]. Plant aquaporins, membrane channels that transport water and small neutral molecules, have been shown to participate in the transport of As(III) [77, 78]. Several NIPs (nodulin 26-like intrinsic proteins (NIP; subfamily of plant major intrinsic proteins; MIPs) protein are involved in As(III) uptake in roots of *A. thaliana* (*NIP5;1* and *NIP6;1*) and *O. sativa* (*NIP2;1*) [79]. These protein channels are also involved in the uptake of other metalloids such as boron and antimony. *NIP2:1* transporter, an important physiological silicon (Si) influx transporter also named as *OsLsi1* in rice, is involved in a major transport pathway responsible for As(III) uptake by root [73]. Zhao et al. [80] confirmed that *Lsi1* plays a role in arsenite efflux also in rice roots exposed to arsenate. The molecular mechanisms that regulate the pore selectivity of NIPs are still poorly understood [81]. Another plasma membrane intrinsic proteins (PIPs), a family of aquaporins, *OsPIP2;4*, *OsPIP2;6* and *OsPIP2;7*, have been shown to be involved in influx and efflux of As(III) through plasma membrane [82]. Also once absorbed into the roots of plants, As(V) is reduced to As(III) through arsenate reductase and then transported to shoot were its accumulation occur or is sequestered into the vacuoles. Presence of trace amount of organic arsenic species i.e. MMA and DMA in soils has been attributed to the use of arsenic containing pesticides and herbicides and methylation of inorganic arsenic by microorganisms [83, 84]. MMA and DMA enter plant roots using the same entry route as glycerol, which is transported into plant cells by aquaporins [85]. NIP proteins uptake MMA and DMA in undissociated form. However, their uptake mechanism is not clearly understood and very little is known about this. In rice, *OsNIP2;1* is the major transporter for diverse uncharged arsenic species. Rice mutant lacking *OsLsi1* showed 80% and 50% decrease in concentration of MMA and DMA, respectively, compared with the wild type [86].

Initially, general opinion was that the uptake of boron is a passive diffusion mechanism across lipid bilayers depending on the formation of boron complexes within the cell wall and plant water fluxes [16]. However, recent studies show that boron transport is an active process. Boron is absorbed by root and then it is translocated to the above ground plant parts. *Typha latifolia, Phragmites austalis,* and *Lemna* spp. accumulate high concentration of boron in their tissues. Usually, roots accumulate more boron than leaves and stems. Boron absorption by plants is influenced by various environmental factors including pH, the type of exchangeable ions present in the solution, amount of organic matter in soil and non-soil environments. One of the most important parameter affecting boron uptake is pH value. Plant uptake increases with decreasing pH because boric acid [$B(OH)_3$] pre-dominates in the solution rather than borax [$B(OH)_4$]. Boric acid exists as uncharged molecule at physiological pH and thus can pass directly across phospholipid bilayers. Aquaglyceroporins, the channel-like proteins further accelerate boron bi-directional movement of boron. Boron is transported from roots to stems and leaves into the xylem through the transpirational stream of water. Due to this an increased transpiration results in higher boron uptake. Phloem has also been shown to plays a role in mobility of boron into the plants. Inhibition of boron uptake by metabolic inhibitors and cold treatment in roots suggested active uptake of boron [87, 88]. BOR1 from *A. thaliana* is a boric acid/borate transporter involved in xylem loading and boron distribution within shoots. The bor1-1 mutant was found to be defective in boron translocation from roots to shoots and showed severe shoot growth inhibition and fertility reduction under boron-deficient conditions. In contrast to wild type *A. thaliana*, no preferential distribution of boron to young leaves was observed in bor1-1 mutant [89]. Over expression of *BOR1* led to enhanced shoot growth and fertility under low boron conditions. It has been proposed that *BOR1* may be involved in directing boron from the xylem to phloem for preferential delivery to young leaves. In rice, growing under boron-limited conditions, *OsBOR1* has been shown to be involved in the efficient uptake of boron into root cells [16, 90].

NIP5;1, (NIP subfamily) has been identified as boron transporter involved in efficient boron translocation under boron-deficient conditions [91]. Its mRNA is up-regulated under boron-limited conditions and a boron-deficiency tolerant plant was produced by increasing the expression of NIP5;1. NIP5;1 mutants showed severe reduction in growth of both roots and shoots under boron-limited conditions. NIP6;1, a gene like NIP5;1 is also present in the plasma membrane and expedites boric acid uptake into oocytes. In contrast to, NIP5;1 which is permeable to both boric acid and water, NIP6;1 does not show water channel activity [16]. NIP6;1 mutant lines showed defects in the expansion and low boron concentrations in young leaves under boron limitation. Strong NIP6;1 promoter activity was observed in the phloem region at nodes of the stem. It is proposed that NIP6;1 participates in xylem-phloem transfer for enhanced allocation of boron into young growing tissues [16]. Sutton et al. [92] showed the importance of HvBot1 and HvNIP2;1 in uptake of boron in barley. Over-expression of tonoplast aquaporin *AtTIP5;1* (*OxAtTIP5;1*) and *AtBOR4* gene considerably enhanced the tolerance of Arabidopsis to boron. Differential expression of these transporters can result in plants tolerant to high boron concentrations in the soil. The over-expression of tonoplast aquaporin *AtTIP5;1* (*OxAtTIP5;1*) gene considerably enhanced the tolerance to boron in Arabidopsis. Over-expression of *AtBOR4*, one of a six paralogues of *AtBOR1*, also provides tolerance to high boron concentrations in plants (3 mM). Expression of genes encoding for various boron transporters are finely regulated by boron availability in the environment for boron homeostasis. Altered expression of these transporters can lead to production of plants tolerant to stress produced by high boron in the environment.

Selinium uptake by plants is affected, by various factors such as pH, EC, temperature, soil organic matter, cation exchange capacity, nutrient balance, and concentration of other trace elements [93]. Selenium occurs naturally as Se^{2-}(selenide), Se^0(elemental selenium), $Se_2O_3^{2-}$(thioselenate), SeO_3^{2-} (selenite), and SeO_4^{2-} (selenate) [94]. Selenate is the abundant form of selenium in oxic soils and selenite is dominant in anoxic environments. Selenium is chemically similar to sulfur. The dominant forms of selenium and sulphur available to plants are sulphate, selenite, and selenite. Plants root mainly take selenate from soil but selenite and organic selenium compounds are also taken up readily [95–97]. Selenate absorption in plants from the soil solution is an active process. It competes with sulphate for absorption by roots due to their chemical similarity and transport of both anions occurs via sulphate transporter(s) in yeast [98, 99]. Transport of both selenate and sulphate occurs across the plasma membrane of root epidermal cells against their electrochemical gradients, which is driven by the cotransport of 3 protons for each ion [99, 100]. Selenate can be reduced to selenite, which can undergo further reduction to selenide (Se^{-2}). SeO_{3-}^2 taken up by roots is also rapidly transformed to other forms, including selenomethionine (SeMet) and selenomethionine Se-oxide hydrate (SeOMet), but commonly into unknown and water-insoluble forms. In chloroplast the chemical reduction of SeO_{4-}^2 to SeO_{3-}^2 and further conversion of SeCys occurs, while in cytosol the production of SeMet and methylation of SeMet takes place. Plants take up selenium from soil mainly in the form of selenite via sulphate transporters and metabolized via the sulphur assimilation pathway. Plants possess high affinity and low affinity transporters that are localized in root epidermal and cortical cells [101]. From plant, first sulfate transporter genes were isolated from *Sylosanthes hamate*, a tropical forage legume through functionally complementing selenate resistant mutant yeast that were defective in sulphate transport [102]. The high affinity transporter, with a Km for sulfate of approximately 9 μM, is expressed absolutely in roots and is thought to be the primary transporter of sulphate from the soil [99, 103]. On the other hand, the low affinity transporters, which are expressed in leaves and roots, exhibit a Km for sulphate of approximately 100 μM and are postulated to be involved in both uptake of sulphate from the soil solution into roots and intracellularly from the apoplast into the symplast. The selenate resistant *A. thaliana* sel mutants were found to contain lesions in the sulphate transporter gene Sultr1:2 characterized as being expressed in the root cortex, root tip and lateral roots, suggesting that this transporter mainly participates in importing sulphate and selenate from the soil into the root [99]. These results provide strong evidence to support selenate uptake via high-affinity sulphate transporters in plants. When plants are treated with mixtures of sulphate and selenate, the concentration ratio of sulphur:selenium in plant tissues rarely matches the concentration ratio in the rhizosphere solution [96, 104–106]. Parker et al. [106] provided evidence that selenium non-accumulators discriminate against selenate uptake relative to sulphate, whereas selenium hyperaccumulator species preferentially absorb selenium over sulphur. This implies that either the sulphate transporters in selenium hyperaccumulating and non-accumulating plants have different characteristics and these proteins are selective for selenate relative to sulphate in the hyperaccumulators or that hyperaccumulators have additional selenate transporters. Most of the selenium transport in selenite-fed plants was found to be in the form of selenate or as an unknown selenium compound, the relative proportions of these two forms varying both with time and with external selenite concentration. Asher [95] suggested metabolic involvement in the uptake and long distance transport of selenium supplied as selenite.

As chemical properties of arsenic and antimony are similar, it is expected that they share uptake, transport and tolerance mechanisms. In Arabidopsis, *NIP1* transporters participate in uptake of both arsenite and antimonite [107]. Two hypothesis have been proposed regarding Sb(V) transport in plants (i) antimonite Sb(V) enters root symplasm likely via anion transports such as those that transport Cl^- or NO_3^- with low selectivity; and ii) antimonite Sb (III) has been proposed to enter the xylem through apoplastic pathway. Unlike As(V), uptake of Sb (V) was not reduced with addition of phosphate (V) in maize or sunflowers [108] therefore it was concluded that antimony do not use phosphate transporter to enter plant roots. Tschan et al. [108] suggested that this could be because of the different structures of their pentavalent oxyanions. Arsenate (AsO_4^{3-}) is tetrahedral while $Sb(OH)_6$ is octahedral [109]. Linear uptake of antimony by plants further suggested that another selective pathway or nonselective apoplastic pathway exist for antimony uptake [110]. At low external concentrations, antimony enters root symplasm likely via anion transporters of low selectivity, such as Cl^- or NO_3^-. When external concentrations are 2–3 orders of magnitude higher than internal concentration, Sb(V) can overcome an electrical potential difference across the membrane and can be taken up passively by plants. Antimony uptakes by roots as well as translocation of the element to leaves were more significant when antimony was added to soil together with arsenic. This result suggested that the two metalloids utilizes different uptake pathway and that the alteration in membrane integrity by arsenic may be responsible for increased permeability and thus concentration of antimony in plants [111, 112]. The uptake of antimony in plants depends on various factors such as phyto-availability of antimony in soils, antimony speciation, pH, plant species, soil oxidation/reduction potential and coexisting ions in the soil [44, 46, 47, 113, 114]. Moreover, clear differences in the ability to absorb, transport and tolerate antimony exist among species [48]. At physiological pH, Sb(III) exist mainly as Sb(OH)$_3$, which is structurally similar to As(OH)$_3$ and glycerol. Sb(III) has been shown to compete with As(III) in plants suggesting that Sb (III) uptake in plants occurs passively through aquaporins and therefore is affected by water stream [76]. Sb(III) influx into rice roots is substantially faster than Sb(V) [115]. Rice and *Lolium perenne* L. show a high affinity for Sb(III), whereas *Holcus lanatus* L. has a high affinity for Sb(V) [115]. In rice roots, although Sb(III) was more efficiently taken up compared to Sb(V), however, most of the antimony was accumulated as Sb(V) in roots. It is possible that considerable amount of Sb(III) has been converted to Sb(V) in rice plants [115]. On the contrary, Okkenhaug [116] reported higher amount of Sb(III) in terrestrial plants from an active antimony mining area and suggested that reduction of pentavalent to trivalent antimony form occurred after plant uptake. Rice accumulated the highest antimony in the roots, followed by the stems and leaves. Rice is ineffective in translocating either Sb(III) or Sb(V) as the highest TF is 0.51. However, translocation of Sb(V) is ~3–4times higher than Sb(III) suggesting that Sb(V) is more mobile than Sb(III). Plant cell walls are considered key storage compartment for antimony in rice [115].

7.4 MECHANISMS OF METALLOID TOXICITY IN PLANTS

Toxicity due to excess metalloids in soil is a significant agricultural problem, which reduces crop yield in various parts of the world. Moreover, dietary intake of edible parts of the plants growing in metalloid contaminated soil is potential risks for human and animal health. Therefore, understanding the mechanisms of metalloid toxicity in plants is necessary for development of metalloid tolerant transgenic plants. Figure 7.2 shows the mechanisms of metalloid toxicity in plants.

Higher concentrations of arsenic has been shown to cause adverse effects on various key metabolic processes in plants such as photosynthesis, respiration, carbohydrate metabolism [117], nitrogen metabolism [118], phosphate metabolism [119], thiol metabolism [120, 121], RNA and protein metabolism [122] which leads to poor growth, reduced yield, and often death of plant [123, 124]. When arsenic non-tolerant plants are exposed to toxic concentrations of arsenic, they show toxicity symptoms

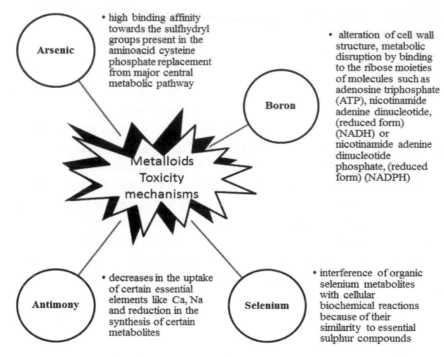

FIGURE 7.2   Mechanisms of metalloid toxicity in plants.

such as poor seed germination, marked reduction in root and shoot growth, decrease in plant height, decrease in tillering, inhibition of leaf formation, wilting and necrosis of leaf margins, chlorosis, nutrient deficiencies, biomass inhibition, lower fruit and grain yield, and sometimes death [15]. At high level of arsenic concentration, the process of photosynthesis is halted due to disorganization and damage of chloroplast membrane.

This affects CO_2 fixation rate and photosystem II (PSII) function [125]. The toxic effect of arsenic is greatly influenced by its nature. Studies on different plant species provides inconsistent phytotoxicity order of arsenic speciation indicating that arsenic interacts in a different way with different genotype and the existing nutrients [126]. Different forms of arsenic have distinct toxicity mechanism. Among the inorganic form, the arsenite is the most toxic. At cellular level, it shows high binding affinity towards the sulfhydryl groups present in the aminoacid cysteine. Thus, binding of arsenite with thiol groups of protein and enzyme and its cofactor results in disruption of protein structure and protein-protein interactions hence, many key metabolic processes in the cell get adversely affected [127]. Another important mechanism of arsenic toxicity is phosphate replacement from major central metabolic pathway. The oxidized arsenic form arsenate, being structure and chemical analog of phosphorous, substitutes inorganic phosphorous Pi from critical cellular and biochemical process. Initially it hijacks phosphate transporter while entering into plant roots, further in cell signaling it disrupts phosphorylation/dephosphorylation by replacing phosphate which cannot be revert back. Moreover, As(V) moves across inner mitochondrial membrane of plants through the Pi translocator [128] and the dicarboxylate carrier [129]. It is also known to pass through Pi transporters ATPHT family found on golgi and plastid [130]. Few enzymes use As(V) directly as substrate over Pi as most Pi liberating reactions are of irreversible nature [131]. Thus, it can compete with phosphate during phosphorylation reactions, leading to the formation of AsV adducts that are often unstable and short-lived. The production and quick auto hydrolysis of As(V)-ADP results in a futile cycle that disengage photophosphorylation and oxidative phosphorylation and results in decrease in the ability of cells to produce ATP and carry out normal metabolism [126]. Arsenic leads to excessive formation of ROS such as superoxide anion, hydroxyl radicals, hydrogen peroxide and lipid peroxide due to glutathione depletion and lipid peroxidation, consequently leading to cellular damage [132–134]. Lipid peroxidation reflects decreased membrane stability in arsenic sensitive plants [135]. Different plant species shows considerably different response when exposed to different arsenic concentration. For example the most common studied plants for arsenic, *P. vittata* (arsenic hyperaccumulator) and *P. ensiformis* (non-arsenic hyperaccumulator), especially at arsenic exposure of 267 μM when evaluated *P. ensiformis* shows higher arsenic-induced oxidative stress in plants than *P. vittata* [52].

Boron toxicity affects various processes in plants. Plants exposed to toxic levels of boron show reduced growth of roots and shoots, reduced root cell division, lower leaf chlorophyll contents and photosynthetic rates, decreased lignin and suberin levels and altered metabolism [7, 21, 22]. The toxicity of boron appears mainly due to its ability to form strong complexes with metabolites that have multiple hydroxyl groups in the cis-conformation such as ribose. Boron can bind compounds with two hydroxyl groups in the cis-configuration. Although the physiological basis for boron toxicity is not clear, three main cause have been proposed (i) alteration of cell wall structure; (ii) metabolic disruption by binding to the ribose moieties of molecules such as adenosine triphosphate (ATP), nicotinamide adenine dinucleotide (NADH) or nicotinamide adenine dinucleotide phosphate, (reduced form) (NADPH); and (iii) reduced cell division and growth by binding to ribose (free sugar or within RNA) [136]. Although boron can bind cis hydroxyls present on the ribose moiety of the energy-carrying molecules ATP, NADH, and NADPH, up to approximately 50 mM boron, photosynthesis and respiration has been shown to be largely insensitive [8]. Cell division and expansion are also inhibited only in the high concentration range of 1–5 mM boron [8]. In the case of boron toxicity, the cellular alterations in root meristems are related to a reduction of mitotic activity and modifications of the expression patterns of key core cell cycle genes [137]. Quantitative RT-PCR analysis revealed that expression of the negative cell cycle regulators WEE1 and SMR4 were enhanced considerably after 24 h of boric acid treatment. WEE1 codes for a kinase protein and is transcriptionally activated upon the termination of DNA replication or DNA damage, inhibiting growth of plant by arresting dividing cells in the G2-phase of the cell cycle [138]. Recently, boron toxicity mechanism has been suggested to involve breakage in DNA double-strand and probably blockage of replication triggered by a genotoxic stress caused by boric acid [139]. Boron has also been shown to inhibit one step of in vitro pre-mRNA splicing reaction [140], suggesting that boron toxicity is mainly because of disruption of RNA splicing [22]. A global expression analysis demonstrated that boron toxicity induces the expression of genes related with abscisic acid (ABA) signaling, ABA response and cell wall modifications, and represses genes that code for water transporters. These results indicate that boron toxicity cause a decrease in water and boric acid uptake, triggering a hydric stress response that causes root growth inhibition. High supply of boron can lead to over production of ROS, which in turn causes oxidative damage by peroxidation of lipid and accumulation of hydrogen peroxide in leaves [141, 142].

The toxic effects of selenium in plants are well documented and include stunted root growth, reduced biomass, chlorosis and reduced photosynthetic efficiency. It also disturbs the uptake and distribution of essential elements such as Ca, K, Na, Cu, Mg, Fe, Mn, and Zn from soil. In general selenite is more toxic to plants than selenate [143]. The toxicity of both selenate and selenite to most plants can be associated to three factors: (i) rapid absoption of selenate and selenite from the soil by roots and translocation to other parts of the plant; (ii) conversion of these anions into organic forms of selenium by metabolic reactions; and (iii) interference of organic selenium metabolites with cellular biochemical reactions because of their similarity to essential sulphur compounds. Incorporation into proteins of the amino acid analogues selenocysteine and selenomethionine, instead of equivalent sulphur amino acids cysteine and methionine results in small, but significant, changes in the biological properties of a selenium-substituted protein and is considered to be the main cause of selenium toxicity. The deformed selenoproteins hypothesis asserts that selenium toxicity occurs when at RNAcys inadvertently binds to selenocysteine instead of cysteine during translation to form non-specific and toxic selenoproteins. Brown and Shrift [144] first demonstrated that selenocysteine can be misincorporated into nonspecific selenoproteins in *Vigna radi*ata. Selenium enhances the accumulation of ROS and thereby induces oxidative stress in plants [145]. Both, selenate and selenite have been reported to induce oxidative in plants. Compared to Se-hyperaccumulator *Stanleya pinnata*, (higher glutathione content), *Stanleya albescens* plants showed enhanced superoxide and hydrogen peroxide accumulation when treated with selenate [146]. Glutathione is a tripeptide that plays important roles in plant cells including signaling and maintenance of cellular redox status in plants [147]. In fact, glutathione reduction is linked to disturbance of auxin homeostasis and reduced growth of root [148]. In Arabidopsis plants, levels of glutathione declined in both a dose- and time-dependent manner due to selenite treatment [149]. Arabidopsis mutant with a knockout in APR2, the dominant APR isozyme that completes sulphate reduction and probably mediates selenate reduction as well showed elevated levels of selenate, which decreased the content of glutathione and increased the accumulation of superoxide. ROS lead to damage of cellular components such as membranes and proteins. For example, *Triticum* *aestivum* seedlings treated with 100 mM selenate had a two- to three-fold increase in lipid peroxidation compared with control [150]. Compromised membrane integrity was observed in *Hordeum vulgare* (barley) along with increased activity of ROS-scavenging enzymes, including SOD, CAT and APX [151]. Selenate treatment is also likely to oxidize proteins in plants [152]. Similarly, increased activity of ROS-scavenging enzymes has also been reported in selenium-treated Ulva species [153]. Selenite treatment also led to accumulation of ROS such as hydrogen peroxide, superoxide, increased lipid peroxidation and cell mortality in plants [154, 155]. Further, oxidative stress can also be induced by inorganic selenium metabolites that are generated by the process of reduction of selenite to selenocysteine. and accumulation of selenoproteins which results in generation of superoxide due to damage and leakage in electron transport of mitochondria and the chloroplast [156]. The non-enzymatic reduction of selenite mediated by glutathione generates selenodiglutathione, which has been shown to be more toxic than selenite and capable of inducing mitochondrial superoxide [157].

Antimony in high concentrations can be toxic to plants and can inhibit their growth. It can cause growth reduction, inhibition of photosynthesis, reduction in the uptake of certain essential elements like Ca, Na and decrease in the synthesis of certain metabolites in plants [158–161, 162]. Relatively low antimony concentration in the medium (5 mg kg^{-1} antimony) significantly influenced root and shoot biomass of sunflower [163]. Antimony toxic mechanisms, in particular for its toxicity to plant root growth, are unclear. In its trivalent form, antimony may have a level of genotoxicity similar to trivalent arsenic [43]. However, Yang found no significant difference in the toxicity of Sb(III) and Sb(V) on root and shoot growth of rice grown in pots [159]. Antimony accumulation in the shoots is not required for phytotoxicity growth of barley was depressed in sand cultures at concentrations of 50–100 mg/L antimony in solution, although antimony was below the detection limit in the shoots (<2 mg/kg) [164]. When the antimony concentrations in the fronds of *Cyrtomium fortunei, Cyclosorus dentatus,* and *Microlepia hancei* reached 10.2 mg/kg, 27.2 mg/ kg and 53.3 mg/kg, respectively, the total biomasses of these plants were observed to decrease by 12.5%, 38.3%, and 35.0%, respectively [165]. Decrease in the concentrations of essential plant nutrients and significant increase in concentration of antimony correlate well with a decline in root and leaf biomass. The increase in antimony concentration led to significant decrease in Ca concentration in all plants parts. Calcium plays very important role in cell; it has structural role in plant cell walls and membrane, it can act as counter cation for anions in the vacuoles, and an intracellular messenger in the cytosol [166]. It has been suggested to function directly in various aspects of photosynthesis [167]. The concentration of K in the leaves of seedlings grown in presence of 100 and 150 mg/Lof antimony was lower ($P < 0.05$) compared to the K content in the leaves of the control plants. A decrease in plant K concentration has been reported to suppress of photosynthesis. K deficiency may be characterized by reduced plant growth with yellowing of the leaf edges [168, 169]. In seeds, a significant decrease in Ca and Na contents was registered after treatment of seedlings with 150 mg/L of antimony. Antimony concentration of 50 mg/L showed a considerable decrease in concentrations of K, Cu, and Pb in the roots. Higher antimony concentrations led to reduced Ca and Na content in the roots. Antimony led to decline in photosynthesis, transpiration, and content of photosynthetic pigments. This suggests that in spite of relatively low mobility of antimony in root-shoot system, antimony significantly alters physiological status in shoot and decreases plant growth. In addition, size of intercellular spaces in leaf decreased making leaf tissue more compact [163]. Antimony negatively influenced the content of chlorophyll a (chla) and chlorophyll b (chlb), which resulted in the decrease of total chlorophylls content in sunflower [163]. It was found that antimony could cause oxidative stress, increase the peroxidation of membrane lipids and stimulate antioxidant defense system in plants [161, 165, 170].

7.5   METALLOID TOLERANCE MECHANISMS IN PLANTS

During evolution, plants have become adapted for harsh environment resulting in development of several distinct mechanisms for survival and defense system. Plants capability to survive in adverse stress condition is known as tolerance. Metalloid tolerance varies greatly in tolerant and sensitive varieties of a plant. These differences have been used to reveal the tolerance mechanism for metalloid toxicity in plants. The common defense mechanisms used by plants to alleviate the toxic effect and increase survivility percentage are efflux of the toxic metalloid, complexation with phytochelatins, glutathione, and polyhydroxy metabolites, compartmentalization in vacuole and cell wall, production of antioxidants, methylation, etc. (Figure 7.3). A greater understanding of mechanisms to withstand toxic amount of metalloids in soils will help in development of metalloid tolerant transgenic plants. A brief description of tolerance mechanisms operating in response to metalloid toxicity in plants is given in this section.

7.5.1   EFFLUX MECHANISM

Efflux is a general detoxification mechanism in plants. In case of arsenic, majority of detoxification starts from reduction of As(V) into As(III) by arsenate reductase enzyme. It is then removed from cell cytosol through efflux mechanism or sequestered in vacuole. In arsenic hyperaccumulator plant *P. vittata*, ACR3 and ACR3;1, arsenate reductase gene encode

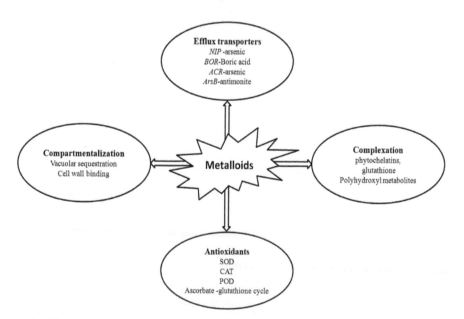

FIGURE 7.3   Mechanisms of metalloid tolerance in plants.

similar protein to arsenite effluxer of yeast ACR3. Lack of ACR3 expression but not ACR3;1, resulted in an arsenite-sensitive phenotype which indicated that ACR3 play an important role in arsenic tolerance in *P. vittata* [171]. NIP proteins which is responsible for As(III) uptake have been proposed to have bidirectional functional role in transport. NIP proteins from *A. thaliana*(NIP5;1), *L. japonicus* (NIP6:1) and *O. sativa* (NIP2;1 and NIP3;2) have role in both influx and efflux of arsenite in cell [172]. Arsenic efflux by NIP was demonstrated to be coupled with electrochemical potential gradient of protons generated by the plasma membrane H^+-ATPase [173]. However, uptake was more favored compared to efflux. The *Lsi1* transporter of silicon also mediates efflux of As(III) in rice through intercellular transport [73]. Logoteta et al. [174] observed that arsenite efflux is not adaptively increased in the tolerant phenotype of *H. lanatus* and suggested that efflux of arsenite could be a basal tolerance mechanism to reduce cellular arsenic in both tolerant and non-tolerant phenotypes. Tolerance to boron toxicity in plants has been associated with decreased accumulation of boron [8, 136]. BOR1, a boron transporter has been shown to be actively involved in the efflux of boron. Homologues of *A. thaliana* BOR1 play important role in tolerance of high boron in other plant species. Reid [175] identified boron-tolerance genes that encoded boron-efflux transporters in wheat and barley. He cloned *HvBor2* genes from barley and *Tabor2* from wheat which were strongly upregulated in roots of tolerant cultivars, and virtually undetectable in sensitive cultivars. Sutton et al. [92] performed a quantitative trait locus (QTL) analysis of two barley cultivars differing in their tolerance of high boron, and detected BOR1 homologue in the mapped region. Bot1 mRNA was detected in both roots and shoots, but was more in roots. The tolerant cultivar possesses multiple copies of Bot1 and has greater amounts of Bot1 mRNA than the susceptible cultivar. BOR1 homologues not only decrease boron uptake by roots as a primary boron tolerance mechanism, but also redistribute excess boron in leaves to confer further tolerance of high boron. In boron tolerant cultivars, Reid and Fitzpatrick [9] observed lower boron concentrations in leaf protoplasts, suggesting different partitioning of boron within the shoot. It is suggested that boron transporters export toxic boron out of cells into the apoplast and alleviate boron toxicity symptoms in the cytoplasm of leaf cells. Better antimony exclusion has also been suggested to be an important mechanism to alleviate antimony toxicity. In fact an *A. capillaries* ecotype with low capacity to efflux antimony did not perform well on antimony-rich mine soil, while the ecotype from the mining area efficiently restricted antimony transport to the shoots [176]. Antimonite has been suggested to be the state of antimony being extruded from the cell therefore, reduction of antimonate to antimonite in plants is very important. Integral membrane protein ArsB catalyzes the efflux of antimonite and, therefore, provides resistance against the toxic effects. ArsB forms a complex with the catalytic subunit ArsA and shows ATP-dependent activity. The efflux of antimonite is then coupled to an electrochemical gradient via proton exchange [176]. In spite of being arsenic analog, antimony is not hyperaccumulated and visual symptoms of injury was not observed in in *P. vittata*. It was observed that it possess efficient Sb(III) uptake and efflux ability, however, translocation and transformation capability of this plant was limited [177].

7.5.2   COMPLEXATION

Complexation of metalloids by phytochelatins, glutathione and polyhydroxyl metabolites are important detoxification strategy applied by plants in response to metalloid toxicity. Phytochelatin (PC) involvement in the detoxification of metalloids, such as, As, Se and Sb is well-documented [178–181]. Phytochelators are metal complexing short thiol rich peptides synthesized non translationally using glutathione as substrate by glutamate, cysteine, and glycine residues condensation in three sequential enzymatic reactions [182]. Higher concentration of PCs was synthesized in rice roots on exposure to arsenic toxicity [183]. PCs-AsIII as well as GSH-As(III) complex was transported and sequestered into vacuole. The PCs-AsIII complex is stable in acidic environment thus its sequesteration into vacuole is important for complete detoxification of As [184]. In Arabidopsis, the PC-deficient mutant was significantly more sensitive to As(V) than in the wild type [185]. Phytochelatins are believed to play an important role in antimony tolerance also [186]. Antimonite has strong affinity to sulphydryl groups and therefore forms complexes with phytochelatins and gets neutralized in plants. When PCS genes from the fission yeast

Schizosaccharomyces pombe (*SpPCS*) and *Triticum aestivum* (*TaPCS1*) were expressed in the strains of *S. cerevisiae*, antimonite hypersensitivity of the ycf1 mutant was totally suppressed. This suggested the involvement of PCs in antimonite resistance in this mutant [187]. However, buthionine-[S,R]-sulphoximine (BSO), (an inhibitor of PC synthesis) treatment did not enhance the susceptibility of clover to antimony toxicity suggesting no role of phytochelatins in the high antimony tolerance of clover [170]. In *A. thaliana* also, PCs do not significantly affect selenite detoxification. The wild-type of *A. thaliana* and its mutants (glutathione deficient Cad 2–1 and phytochelatins deficient Cad 1–3) when separately treated with varying concentrations of selenite and arsenate indicated that GSH and PCs affect arsenate detoxification but only GSH affects selenium detoxification. Selenium glutathione-complex prevents the production of detrimental selenopeptide [179]. Metallothionein gene plays role in selenium homeostasis, which in turn increases selenium tolerance in sugarcane [188]. qRT-PCR analysis indicated increase in expression of MT gene in leaf tissue with an increase in selenium supply. Polyhydroxyl metabolites such as malic acid, fructose, glucose, sucrose and citric acid can bind boron and this has been proposed as an important boron tolerance mechanism in plants [189]. Summed shoot concentration of the potentially boron-binding polyhydroxyl metabolites in Arabidopsis was found to be low in comparison to *Thellungiella halophila*, an *A. thaliana*-related 'extremophile' plant, generally found in B rich environment. In Thellungiella, boron-binding polyhydroxyl metabolites was over twofold higher than [B]int, and thus likely to allow appreciable 1:2 boron-metabolite complexation in the shoot.

7.5.3   COMPARTMENTALIZATION

Compartmentalization is an important means for detoxifying metalloids in plants. As(III)-PCs complex has been shown to be transported and sequestered into vacuole. The As(III)-PCs complex is stable in acidic environment thus its sequesteration into vacuole is important for complete detoxification of arsenic [184]. In-depth studies resulted in identification of vacuolar transporter, which is a remarkable achievement in understanding the tolerance mechanisms by Song et al. in 2010. He identified that ABCC (ATP Binding Cassette C) transporter of Mg-ATP dependent ABC family localized in tonoplast of vacuole plays important role in transportation across vacuolar membrane. PC synthesis and their complexation to As(III) are coordinated to the transport of the As(III)-PC complex to the vacoule subcellular compartment through the activity of two members of a subclass of ATP binding cassette (ABC) transporters: ABCC1 and ABCC-2. A novel vacuolar transporter named OsABCC1, is recently identified, is responsible for transportation of As(III)-PCs across the tonoplast [190]. The *osabcc1* knockout study in Arabidopsis plants exhibited that OsABCC1 has similar functions to AtABCC1 and AtABCC2. Plants which lack OsABCC1 were unable to transport PCs into vacuoles. Additional, arsenic transporters were also identified to participate in tonoplast transport. As described earlier PvACR3, a well characterized ACR3 As(III) efflux protein localized in *P. vittata* tonoplast shows the ability to sequester arsenic into vacuoles. *P. vittata*, an arsenic hyperaccumulating plants showed enhanced uptake of As(V) and efficient As(III) transportation into vacuoles by PvACR3 compared to non-hyperaccumulating plants [191]. *P. vittata* with non-functional PvACR3 exhibit an As(III)-sensitive phenotype [171].

Binding of antimony to cell wall has been suggested an important tolerance mechanism in plants in response to toxic concentrations of antimony. The cell walls are mainly composed of polyose(including cellulose, hemicellulose and pectin) and protein, providing carboxyl, hydroxyl, amino groups and aldehyde groups that can bind antimony and restrict its transport across the cytomembrane. *Ficus tikoua* shows a great tolerance for accumulating antimony. The concentrations of antimony were found to be higher in the roots of this plant than those in the stems and leaves, suggesting the excellent ability of this plant to accumulate antimony in roots [192]. Almost all the antimony was found to be bound to the root cell wall of *F. Tikoua*. With exposure to increasing concentrations of antimony in the solution, the proportion of antimony in the cell wall fraction increases, whereas the proportions of antimony in the cytoplasmic organelle and cytoplasmic supernatant fraction decreases suggesting that the cell wall functions as a barrier, protecting the protoplast from antimony toxicity [192].

7.5.4 ANTIOXIDANTS

It is now well-accepted fact that metalloids trigger the production of ROS which lead to the establishment of oxidative stress in plants [193–195]. Moreover, during oxidative stress, overproduction of ROS such as superoxide radical and hydroxyl radical which are strong oxidizers of lipids, proteins and nucleic acids causing membrane damage and in some cases cell death [196]. Plants possess antioxidants defense mechanisms comprising of enzymatic and non-enzymatic antioxidants to combat the oxidative stress caused by overproduction of ROS. Among these, superoxide dismutase (SOD) dismutates O_2^- to O_2 and which are further oxidized to molecular oxygen and H_2O by peroxidases (POX), catalase (CAT) and ascorbate-glutathione pathway enzymes like ascorbate peroxidase (APX) and glutathione reductase (GR) [193]. A tripeptide glutathione (γ-glutamyl-cysteinyl-glycine) is an abundant compound in plant tissues. It exists in both reduced (GSH) and oxidized (GSSG) states. GSH has the ability to regenerate ascorbic acid via the ascorbate-glutathione cycle and therefore plays a key role in antioxidative defense system [197]. Higher scavenging capacities of enzymic and nonenzymic antioxidants are considered to be responsible for metalloid tolerance in plants. Various antioxidant enzymes guiacol peroxidase (GPX), ascorbate peroxidase (APX), and superoxide dismutase (SOD) showed enhanced activity in presence of high concentration of As(V) in high arsenic accumulating cultivars compared to low arsenic accumulating cultivars and this has been suggested to protect a plant from the metalloid toxicity [198–200]. Ansari et al. [194] showed that high concentration of arsenic inhibits growth and causes oxidative stress and that the AsA-GSH cycle play key role in cellular defense against arsenic in *Brassica juncea* (L.). Under high boron supply also, activity of antioxidative enzymes increases and play a crucial role in conferring tolerance to boron stress in plants [195, 201–205]. In the roots of chickpea cultivar Gökce, better protection from boron-stress-induced oxidative stress compared to sensitive Küsmen cultivar was due to enhanced SOD, CAT and POX activities under high boron concentrations [204]. High boron concentration enhanced ascorbate pool size and activities of l-galactose dehydrogenase, an enzyme participating in ascorbate biosynthesis [202]. Wang et al. [205] also reported an increase in total GSH content in pear leaves under high boron supply. This pathway consists of enzymes GR, MDHAR, DHAR which are distributed in most cellular compartments [206]. In general, their activity is enhanced by boron toxicity [202, 205, 207]. In photosynthetic organisms, the accumulation of selenite-induced ROS can also be partially mitigated by an increase in the activity of ROS-scavenging enzymes, as reported in *Coffeaarabica* cells [208] and the cyanobacterium *Spirula platensis* [209]. The enzymes SOD, APX and CAT have been suggested to play important roles in rebalancing the excess ROS resulting from exposure to selenium in paddy wheat [210]. Compared with wild type plants, the *vtc1* mutant with a defect in ascorbic acid biosynthesis accumulates more superoxide and hydrogen peroxide during selenite treatment; this indicates that ascorbic acid help in reducing the accumulation of in plants challenged with selenite [211]. Recent studies have linked antimony-induced ROS accumulation to cell death in roots. Tolerance to antimony has been associated with increased activity of enzymes POD, CAT and APX in antimony tolerant fern *Pteris cretica*, whereas in paddy wheat, SOD, APX and CAT play important roles in rebalancing the excess ROS [165]. In Sb(III)-treated maize, POD activity was enhanced at low antimony levels but inhibited at high antimony levels; simultaneously, the SOD activity decreased, but the CAT activity increased with increasing concentrations of antimony in soils [160]. Under higher antimony stress, oxidative damage is prevented but, at extreme antimony contamination, the activities of these enzymes are impaired. Glutathione has been strongly correlated to antimony tolerance in Arabidopsis [212]. Plants treated with buthionine sulfoximine, an inhibitor of glutathione biosynthesis showed two-fold increased sensitivity to selenate.

7.5.5 METHYLATION

One of the major detoxification processes for selenium is methylation. Plants metabolize inorganic selenium to comparatively nontoxic, volatile forms (dimethyl selenide [DMSe] and dimethyl diselenide [DMDSe]), which escape to the atmosphere [213]. Methyltransferases catalyse the formation of these methylated compounds in biological systems. Selected members of the genus Astragulus (Fabaceae) are known for their ability to accumulate high levels of selenium, mainly in the form of Se-methyl-selenocysteine. Unlike non-selenium accumulating species, they do not unspecifically incorporate selenium into proteins. The enzyme selenocys-teinemethyltransferase which does not accept cysteine as a substrate is proposed to play a crucial role in conferring selenium tolerance [214]. Overexpression of this gene from hyperaccumulator *A. bisulcatus* in Arabidopsis and Indian mustard led to increased Se accumulation and volatilization and hence tolerance to selenium [215, 216].

Methylated pentavalent arsenic species arealsoknown to be less toxic than inorganic arsenic and are regularly found in plants. However, it is not known whether plants can methylate arsenic. It has been suggested that plants take up methylated arsenic produced by microorganisms [217].

7.6 PROGRESS IN DEVELOPING TRANSGENIC METALLOID TOLERANT PLANTS

Agriculturally important plants can be genetically engineered for metalloid extrusion, tolerance/sequestration and eventually for partitioning of metalloids in non-edible plant tissues. Genes introduced/disrupted in plants to develop metalloid tolerance have been listed in Table 7.2. Metalloid efflux transporter can be used to genetically engineer important food crops that can grow in metalloid contaminated sites without increasing the accumulation of toxic metalloids in the biomass or edible tissues [82]. Rice plasma membrane intrinsic protein (PIP) subfamily participates in arsenic tolerance and transport. Overexpression of bidirectional transporter genes *OsPIP2;4*, *OsPIP2;6*, and *OsPIP2;7* in Arabidopsis resulted in increased arsenite tolerance and more biomass accumulation. Further, no significant accumulation of arsenic in shoot and root tissues of transgenic plants was noticed in long-term uptake assays. In barley, Bot1, which functions as an efflux transporter, and the multifunctional aquaporin from the nodulin-26-like intrinsic protein (NIP) subfamily HvNIP2;1 are important determinant of boron toxicity tolerance. Overexpression of *AtBOR1* conferred tolerance to Arabidopsis under boron deficient conditions and plays a key role in xylem loading [218]. Overexpression of an AtBOR1 paralog, AtBOR4, in transgenic Arabidopsis plants also increased their tolerance to high boron levels [219]. Up-regulation of the *HvBot1* led to improved

TABLE 7.2   A Selected List of Genes Introduced/Disrupted in Plants to Develop Metalloid Tolerance

Gene name	Gene origin	Transgenic plant	Metalloid tolerance	Reference
PvACR3	*Pteris vittata*	*Arabidopsis thaliana*	Arsenic	[224]
γ-ECS and Ars C	*Escherichia coli*	*Arabidopsis thaliana*	Arsenic	[226]
γ-ECS	*Escherichia coli*	*Populus deltoids*	Arsenic	[63]
NtPCS1	*Nicotiana tobaccum*	*Nicotiana tobaccum*	Arsenic	[227]
PvGRX5	*Pteris vittata*	*Arabidopsis thaliana*	Arsenic	[237]
BET1	*Oryza sativa*	*Oryza sativa*	Boron	[222]
AtBOR1	*Arabidopsis thaliana*	*Arabidopsis thaliana*	Boron	[218]
AtBOR4	*Arabidopsis thaliana*	*Arabidopsis thaliana*	Boron	[219]
APS	*Arabidopsis thaliana*	*Brassica juncea*	Selenium	[230]
GR	*Escherichia coli*	*Brassica juncea*	Selenium	[231]
Se-cyslyase gene	*Mus musculus*	*Arabidopsis thaliana*	Selenium	[232]
PaAPR	*Pseudomonas aeruginosa*	*Arabidopsis thaliana*	Selenium	[94]
CcS–cystathionine-c-synthase	*Arabidopsis thaliana*	*Brassica juncea*	Selenium	[238]
SMT	*Astragalus bisulcatus*	*Arabidopsis thaliana*	Selenium	[215]
SMT	*Astragalus bisulcatus*	*Brassica juncea*	Selenium	[216]

tolerance to boron-toxicity [92]. The decreased expression of *HvNIP2;1* restricted boron uptake and thus made the rice plant tolerant to high soil boron [220]. Over-expression of tonoplast aquaporin AtTIP5;1 (OxAt-TIP5;1) gene significantly increases the toxicity tolerance to boron in Arabidopsis [221]. A gene responsible for tolerance to boron toxicity in rice, was identified and named BORON EXCESS TOLERANT1. The gene encodes a NAC (NAM, ATAF, and CUC)-like transcription factor and the function of the transcript is abolished in boron-toxicity-tolerant cultivars. Transgenic plants in which the expression of Os04g0477300 is abolished by RNA interference gain tolerance to boron toxicity [222]. BOR2 and BOR1 mutants had reduced root elongation under low B availability [223].

Chelation of metals and vacuolar sequestration are the main strategies for metalloids detoxification and tolerance in plants. PvACR3, a well characterized ACR3 As(III) efflux protein localized in *P. vittata* tonoplast shows the ability to sequester arsenic into vacuoles. PvACR3 non-functional exhibit an As(III)-sensitive phenotype [171] whereas overexpression of PvACR3 enhanced As(III) efflux into the external environment and increased tolerance [224]. Similarly, heterologous expression of yeast arsenite efflux transporter ACR3 into Arabidopsis endows plants with greater arsenic resistance, but does not lower arsenic tissue levels significantly [225]. Dhankher et al. [226] arsenic tolerant transgenic Arabidopsis plants expressing *E. coli* γ-glutamyl cysteine synthetase (γ-ECS) and arsenate reductase (Ars C) which could transport oxyanion arsenate to aboveground, reduce to arsenite and sequester it in thiol peptide complexes. *E. coli Ars C* gene encoding arsenate reductase catalyzes the GSH coupled electrochemical reduction of arsenate to the more toxic arsenite whereas γ-ECS catalyzes the first and rate-limiting step in the production of the cellular antioxidant GSH. Arabidopsis plants transformed with *Ars C* gene were hypersensitive to arsenate whereas plants expressing *E. coli γ-ECS* gene was moderately tolerant to arsenic compared to control plants. Plants expressing SRSIp/ArsC and ACT 2p/γ-ECS together showed higher tolerance to arsenic. These transgenic plants accumulated 4- to 17-fold greater fresh shoot weight and accumulated 2- to 3-fold more arsenic per gram of tissue than wild plants or transgenic plants expressing γ-ECS or ArsC alone. Eastern cottonwood plants expressing ECS had elevated thiol group levels, consistent with increased ECS activity. In addition, these ECS-expressing plants had enhanced growth on levels of arsenate toxic to control plants in vitro. Furthermore, roots of ECS-expressing plants accumulated significantly more arsenic than control roots (approximately twice as much), while shoots accumulated significantly less arsenic than control shoots (approximately two-thirds as much) [63]. Phytochelatins (PCs) and gluta-thione (GSH) are the main binding peptides involved in chelating heavy metal ions in plants and other living organisms. In general, dual-gene trans-formants exhibited significantly higher tolerance compared to single-gene transgenic lines and accumulated more arsenic. Simultaneous overexpres-sion of *AsPCS1* and *GSH1* led to elevated total PC production in transgenic Arabidopsis. These results indicate that such a stacking of modified genes is capable of increasing arseinc tolerance and accumulation in transgenic lines [130]. Phytochelatin synthase (PCS) catalyzes the synthesis of phy-tochelatins, which are involved in heavy metal detoxification in plants and other living organisms. Transgenic tobacco lines over-expressing *NtPCS1* in the sense or antisense direction showed increased tolerance to cadmium and arsenite, and growth retardation in the early stage, respectively, suggest-ing that NtPCS1 plays important roles in metal(loid) tolerance as well as in growth and development in tobacco [227]. Simultaneous overexpression of *AsPCS1* and *YCF1* (derived from garlic and baker's yeast) into *A. thaliana*, which is sensitive to heavy metals, leads to transgenic plants tolerant to arse-nic [228].

The assimilation of sulfate and selenate is activated by ATP sulfury-lase. Selenate is converted into adenosine phosphoselenate (ADP-Se), which is then reduced to selenite [229]. Overexpression of ATP sulfurylase gene (*APS*) from *A. thaliana* led to 4-times higher APS enzymatic activity and three fold higher selenium in transgenic *B. juncea* plants compared to wild plants [230]. Transgenics plants showed more tolerance towards sele-nium and grew faster than wild type [230]. Bacterial glutathione reductase was also expressed in the cytoplasm and the chloroplast of Indian mustard plants and both types grew at higher rate on agar medium containing toxic concentrations of selenate or selenite in comparison to wild type seedlings [231]. A mouse Se-cyslyase gene when transferred to Arabidopsis plant showed enhanced shoot selenium concentration up to 1.5-fold compared to wild type [232]. Another approach using genetic engineering to enhance phytoremediation potentials to transform fast-growing host plants with unique genes from natural hyperaccumulators.

One potential mitigation strategy is to genetically engineer plants to enable them to transform inorganic arsenic to methylated and volatile arsenic species. Transgenic *A. thaliana* plants showed increase arsenic tolerance by volatilization. When two ecotypes of *A. thaliana* were genetically engineered with the arsenite (As(III)) S-adenosylmethyltransferase (arsM) gene from the eukaryotic alga *Chlamydomonas reinhardtii*, most of the inorganic arsenic were converted into dimethylarsenate [DMA(V)] in the shoots [233]. However, the transgenic plants became more sensitive to As(III) in the medium, suggesting that DMA(V) is more phytotoxic than inorganic arsenic. Only small amounts of volatile arsenic were detected from the transgenic plants. Therefore it was suggested that *arsM* genes with a strong ability to methylate arsenic to volatile species is needed. Recently, Verma et al. [234] reported increased tolerance in *A. thaliana* expressing arsenic methyltransferase (*WaarsM*) gene from a fungus *Westerdykella aurantiaca* (*W. aurantiaca*). Similar to arsenic, selenium tolerance can also be developed in plants by expressing genes involved in selenium volatilization. Tagmount et al. [235] showed that S-adenosyl-L-methionine:L-methionine S-methyltransferase (MMT) is the enzyme responsible for the methylation of selenomethionine to Se-methylselenomethionine and that this enzyme was rate limiting with respect to the production of volatile selenium. Overexpression of MMT from Arabidopsis in *E. coli* (do not possess MMT activity) led to 10 times more volatile selenium compared to untransformed strain when Se-Met was supplied in the medium. Another, quite different approach is to use hyperaccumulator plant species as a source of plant genes to enhance tolerance. Milk vetch (*Astragalus bisulcatus*), a selenium hyperaccumulator accumulates selenium in excess of 4000 ppm in its leaves partly through the presence of the gene encoding selenocysteine methyltransferase (SMT). SMT converts SeCys to methylselenocysteine (MetSeCys), thereby diminishing selenium toxicity. SeCys can cause toxicity when incorporated into protein and non-protein amino acid. When *SMT* from *A. bisulcatus* was overexpressed in *A. thaliana*, 8-fold increase in foliar selenium accumulation and increased tolerance to selenite but not selenite was observed in transgenic plants [215]. Selenium-tolerant transgenic Arabidopsis plants overexpressing a *Brassica oleracea* methyltransferase had significantly lower amounts of hydrogen peroxide and superoxide when treated with selenite, despite accumulating the same concentration of total selenium as the wild type [236]. It is possible that the enhanced selenium tolerance could be due to significant reduction in ROS accumulation. Overexpression of SMT from *A. bisulcatus* in *B. juncea* led to 2- to 4-fold increase in total selenium accumulation and increased tolerance to selenite and selenate [216]. It was suggested that increased volatilization of dimethyl diselenide, might be responsible for the increased tolerance in transgenic plants as this will redirected selenocysteine away from protein misincorporation. Transgenic research to improve metalloid tolerance in plants has yielded promising results, which clarify the approaches to be taken to achieve considerable accomplishment, in terms of application of metalloid tolerant plants in the field.

7.7   CONCLUSIONS AND FUTURE PROSPECTS

Toxic metalloids are present in soil in form of various chemical species. It can enter soil through both natural processes and anthropological activities. Absorption of metalloids by plants is influenced by many factors including plants species, the concentration of arsenic in the soil, soil properties such as pH and clay content, and the presence of other ions. Metalloids can be taken up by means of passive diffusion as well as certain transporters present in the cell which facilitate their influx. Metalloids interfere with various metabolic processes of the cell. They interfere with cellular metabolism due to their high affinity for certain functional groups present on many enzymes, incorporation in biomolecules due to their analogy with functional groups and overproduction of ROS leading to oxidative stress. At elevated metalloid concentrations, biomass production and yields of various plant species are reduced significantly often leading to death. A variety of tolerance and resistance mechanisms including avoidance or exclusion, which minimizes the cellular accumulation of metalloids, and tolerance, which allows plants to survive while accumulating high concentrations of metals have been identified in plants. These include efflux of metalloids from plant roots back to the medium, complexation with thiols, particularly PCs and sequestration in vacuole, activation of antioxidant defense system and methylation. Tremendous progress has been made in understanding the mechanism of uptake, transport, toxicity and tolerance of metalloids in plants, however, much remain to be learned about the diversity of metalloid uptake, complex transportation behaviors and occurrence and regulation of arsenic chemical form and transformations in various plant species. A better understanding of the mechanisms responsible for metalloids uptake, transport, toxicity and tolerance in plants is needed for production of metalloid–tolerant plants for metalloid tolerance and safe cropping. Genetic modifications have been employed to enhance metalloid tolerance in plants. However, work to date has mostly focused on incorporation of a single or dual genes involved in sequestration and efflux, complexation, compartmentalization and detoxification of metalloid in plant, however, to make real progress, a multifaceted approach using multiple targets may be required.

KEYWORDS

- antimony
- arsenic
- boron
- metalloid
- selenium

REFERENCES

1. Uden, P. C. (2015). Speciation of Selenium, in Handbook of Elemental Speciation II - Species in the Environment, Food, Medicine and Occupational Health. Ed. by Cornelis R., Caruso, J. Crews H., & Heumann K., John Wiley and Sons Ltd., Chichester, UK, doi: 10. 1002/0470856009. Chapter 2, 346–365.
2. Alloway, B. J. (2013). Sources of heavy metals and metalloids in soils. Heavy metals in soils. *Environmental Pollution.* Springer, Dordrecht, *22*, pp. 11–50.
3. Zangi, R., & Filella, M. (2012). Transport routes of metalloids into and out of the cell, a review of the current knowledge. *Chemico-Biological Interactions.* *197*(1), 47–57.
4. IARC (1979). Monographs on the evaluation of the carcinogenic risk of chemicals to humans, vol. 20, IARC.

5. Sharma, P., Jha, A. B., & Dubey, R. S. (2014) Arsenic toxicity and tolerance mechanisms in crop plants. *Handbook of Plant and Crop Physiology*. 3rd edition, Edited by Pessarakli, M., CRC Press, Taylor and Francis Publishing Company, Florida, pp. 733–782.
6. Paull, J. G., Nable, R. O., Lake, A. W. H., Materne, M. A., & Rathjen, A. J. (1992). Response of annual medics (Medicago spp.) and field peas (*Pisum sativum*) to high concentration of boron, genetic variation and the mechanism of tolerance. *Crop and Pasture Science. 43*(1), 203–213.
7. Nable, R. O., Banuelos, G. S., & Paull, J. G. (1997). Boron toxicity. *Plant and Soil. 193*(1), 181–198.
8. Reid, R. J., Hayes, J. E., Post, A., Stangoulis, J. C. R., & Graham, R. D. (2004). A critical analysis of the causes of boron toxicity in plants. *Plant, Cell and Environment. 27*(11), 1405–1414.
9. Reid, R., & Fitzpatrick, K. L. (2009). Redistribution of boron in leaves reduces boron toxicity. *Plant Signaling and Behavior. 4*(11), 1091–1093.
10. Pilon-Smits, E. A. H., Quinn, C. F., Tapken, W., Malagoli, M., & Schiavon, M. (2009). Physiological functions of beneficial elements. *Current Opinion in Plant Biology. 12*(3), 267–274.
11. Wang, Y.-D., Wang, X., & Wong, Y.-s. (2012). Proteomics analysis reveals multiple regulatory mechanisms in response to selenium in rice. *Journal of Proteomics. 75*(6), 1849–1866.
12. World Health Organization (2001). Environmental health criteria 224, arsenic and arsenic compounds. 2nd edition, *World Health Organization*, Geneva, pp. 1–108.
13. Zandsalimi, S., Karimi, N., & Kohandel, A. (2011). Arsenic in soil, vegetation and water of a contaminated region. *International Journal of Environmental Science and Technology. 8*(2), 331–338.
14. Karimi, N., Ghaderian, S. M., Raab, A., Feldmann, J., & Meharg, A. A. (2009). An arsenic-accumulating, hypertolerant brassica, Isatis capadocica. *New Phytologist. 184*(1), 41–47.
15. Stoeva, N., & Bineva, T. (2003). Oxidative changes and photosynthesis in oat plants grown in As-contaminated soil. *Bulgarian Journal of Plant Physiology, 29*(1–2), 87–95.
16. Tanaka, M., & Fujiwara, T. (2008). Physiological roles and transport mechanisms of boron, perspectives from plants. *European Journal of Physiology, 456*(4), 671–677.
17. Chauhan, R. P. S., & Powar, S. L. (1978). Tolerance of wheat and pea to boron in irrigation water. *Plant and Soil. 50*(1), 145–149.
18. Severson, R. C., & Gough, L. P. (1983). Boron in mine soils and rehabilitation plant species at selected surface coalmines in western United States. *Journal of Environmental Quality, 12*(1), 142–146.
19. Branson, R. L. (1976). Soluble salts, exchangeable sodium, and boron in soils. In: Reisenauer H.M. (ed.) Soil and plant tissue testing in California, Division of Agricultural Sciences, Bulletin, University of California, Berkeley, 43–46.
20. Gupta, U. C. (1980). Boron nutrition of crops. *Advances in Agronomy, 31*, 273–307.
21. Nable, R. O., Lance, R. C., & Cartwright, B. (1990). Uptake of boron and silicon by barley genotypes with differing susceptibilities to boron toxicity. *Annals of Botany. 66*(1), 83–90.

22. Reid, R. (2007). Update on boron toxicity and tolerance in plants. *Advances in Plant and Animal Boron Nutrition.* Springer, Dordrecht, The Netherlands, pp. 83–90.

23. Saffaryazdi, A., Lahouti, M., Ganjeali, A., & Bayat, H. (2012). Impact of selenium supplementation on growth and selenium accumulation on spinach (*Spinacia oleracea* L.) plants. *Notulae Scientia Biologicae. 4*(4), 95.

24. Schild, F., Kieffer-Jaquinod, S., Palencia, A. s., Cobessi, D., Sarret, G. r., Zubieta, C., Jourdain, A. s., Dumas, R., Forge, V., & Testemale, D. (2014). Biochemical and biophysical characterization of the selenium-binding and reducing site in Arabidopsis thaliana homologue to mammals selenium-binding protein 1. *Journal of Biological Chemistry. 289*(46), 31765–31776.

25. Lakin, H. W. (1972). Selenium accumulation in soils and its absorption by plants and animals. *Geological Society of America Special Papers. 140*, 45–54.

26. Swaine, D. J. (1955). The trace element content of soils. Commonwealth Bur. Soil Sci., Tech. Communication No. 48, Commonwealth Agricultural Bureau, Harpenden, England, 91–99.

27. Thornton, I., Kinniburgh, D. G., Pullen, G., & Smith, C. A. (1983). Geochemical aspects of selenium in British soils and implications to animal health. In: Hemphill, D.D. (ed.), Trace Substances in Environmental Health, Proceedings of University of Missouri, 17th Annual Conference on trace substances in Environmental health, Missouri, 13–16th June, pp. 391–398.

28. Jacobs, L. W. (1990). Selenium in agriculture and the environment. *Soil Science. 149*(2), 121.

29. Sieprawska, A., Kornas, A., & Filek, M. (2015). Involvement of selenium in protective mechanisms of plants under environmental stress conditions, a review. *Acta Biologica Cracoviensia Series Botanica, 57(1),* 9–20. .

30. De Filippis, L. F. (2010). Biochemical and molecular aspects in phytoremediation of selenium. In: Ashraf, M., Ozturk, M,A, and Ahmad, M.S.A (eds.), *Plant Adaptation and Phytoremediation.* 1st edition, Springer, Netherlands, 193–226.

31. Elrashidi, M. A., Adriano, D. C., Workman, S. M., & Lindsay, W. L. (1987). Chemical equilibria of selenium in soils, A theoretical development. *Soil Science. 144*(2), 141–152.

32. Cary, E. E., Wieczorek, G. A., & Allaway, W. H. (1967). Reactions of selenite-selenium added to soils that produce low-selenium forages. *Soil Science Society of America Journal. 31*(1), 21–26.

33. Finley, J. W., Davis, C. D., & Feng, Y. (2000). Selenium from high selenium broccoli protects rats from colon cancer. *The Journal of Nutrition. 130*(9), 2384–2389.

34. Feng, R., Wei, C., Tu, S., Ding, Y, Wang, R., & Guo, J. (2013). The uptake and detoxification of antimony by plants, a review. *Environmental and Experimental Botany. 96*, 28–34.

35. Fowler, B. A., & Goering, P. L. (1991). Antimony. In E. Merian (ed.), Metals and their compounds in the environment: Occurrence, analysis and biological relevance. 2nd Volume, VCH publishers, Weinheim, pp. 939–944.

36. Kabata-Pendias, A., Pendias, H. (1984). Trace elements in soil and plants. 1st edition, CRC Press, Boca Raton, FL (USA), pp. 1–293 .

37. Johnson, C. A., Moench, H., Wersin, P., Kugler, P., & Wenger, C. (2005). Solubility of antimony and other elements in samples taken from shooting ranges. *Journal of Environmental Quality. 34*(1), 248–254.

38. Lintschinger, J., Michalke, B., Schulte-Hostede, S., & Schramel, P. (1998). Studies on speciation of antimony in soil contaminated by industrial activity. *International Journal of Environmental Analytical Chemistry. 72*(1), 11–25.

39. Cal-Prieto, M. J., Carlosena, A., Andrade, J. M., Martinez, M. L., Muniategui, S., Lopez-Mahia, P., & Prada, D. (2001). Antimony as a tracer of the anthropogenic influence on soils and estuarine sediments. *Water, Air, and Soil Pollution. 129*(1–4), 333–348.

40. Amereih, S., Meisel, T., Kahr, E., & Wegscheider, W. (2005). Speciation analysis of inorganic antimony in soil using HPLC-ID-ICP-MS. *Analytical and Bioanalytical Chemistry. 383*(7–8), 1052–1059.

41. Flynn, H. C., Meharg, A. A., Bowyer, P. K., & Paton, G. I. (2003). Antimony bioavailability in mine soils. *Environmental Pollution. 124*(1), 93–100.

42. Krachler, M., Emons, H., & Zheng, J. (2001). Speciation of antimony for the 21st century, promises and pitfalls. *Trends in Analytical Chemistry (TrAC). 20*(2), 79–90.

43. Gebel, T., (1997). Arsenic and antimony, comparative approach on mechanistic toxicology. *Chemico-biological interactions. 107*(3), 131–144.

44. Wan, X. M., Ty, S., Hockmann, K., & Schulin, R. (2013). Changes in Sb speciation with waterlogging of shooting range soils and impacts on plant uptake. *Environmental Pollution. 172*, 53–60.

45. Okkenhaug, G., Zhu, Y. G., Luo, L., Lei, M., Li, X., & Mulder, J. (2011). Distribution, speciation and availability of antimony (Sb) in soils and terrestrial plants from an active Sb mining area. *Environmental Pollution. 159*(10), 2427–2434.

46. Tighe, M., Lockwood, P. V., Ashley, P. M., Murison, R. D., & Wilson, S. C. (2013). The availability and mobility of arsenic and antimony in an acid sulfate soil pasture system. *Science of the Total Environment. 463*, 151–160.

47. Wilson, S. C., Lockwood, P. V., Ashley, P. M., & Tighe, M. (2010). The chemistry and behavior of antimony in the soil environment with comparisons to arsenic, a critical review. *Environmental Pollution. 158*(5), 1169–1181.

48. Shtangeeva, I., Steinnes, E., & Lierhagen, S. (2012). Uptake of different forms of antimony by wheat and rye seedlings. *Environmental Science and Pollution Research. 19*(2), 502–509.

49. Tschan, M., Robinson, B., Johnson, C. A., Burgi, A., & Schulin, R. (2010). Antimony uptake and toxicity in sunflower and maize growing in SbIII and SbV contaminated soil. *Plant and Soil. 334*(1–2), 235–245.

50. Kabata, A., & Pendias, H. (2001). Trace elements in soils and plants. 3rd edition   CRC Press, Boca Raton, FL, pp: 1–252.

51. Ainsworth, N., Cooke, J. A., & Johnson, M. S. (1990). Distribution of antimony in contaminated grassland: 1 – vegetation and soils. *Environmental Pollution. 65*(1), 65–77.

52. Singh, N., & Ma, L. Q. (2006). Arsenic speciation, and arsenic and phosphate distribution in arsenic hyperaccumulator *Pteris vittata* L. and non-hyperaccumulator *Pteris ensiformis* L. *Environmental Pollution. 141*(2), 238–246.

53. Mahmud, R., Inoue, N., Kasajima, S. Y., & Shaheen, R. (2008). Assessment of potential indigenous plant species for the phytoremediation of arsenic-contaminated areas of Bangladesh. *International Journal of Phytoremediation. 10*(2), 119–132.

54. Abedin, M. J., Feldmann, J., & Meharg, A. A. (2002). Uptake kinetics of arsenic species in rice plants. *Plant Physiology. 128*(3), 1120–1128.

55. Abbas, M. H. H., & Meharg, A. A. (2008). Arsenate, arsenite and dimethyl arsenic acid (DMA) uptake and tolerance in maize (*Zea mays* L.). *Plant and Soil. 304*(1), 277–289.

56. Ullrich-Eberius, C. I., Sanz, A., & Novacky, A. J. (1989). Evaluation of arsenate- and vanadate-associated changes of electrical membrane potential and phosphate transport in Lemna gibba G1. *Journal of Experimental Botany. 40*, 119–128.

57. Wang, J., Zhao, F.-J., Meharg, A. A., Raab, A., Feldmann, J., & McGrath, S. P. (2002). Mechanisms of arsenic hyperaccumulation in *Pteris vittata*. Uptake kinetics, interactions with phosphate, and arsenic speciation. *Plant Physiology. 130*(3), 1552–1561.

58. Asher, C. J., & Reay, P. F. (1979). Arsenic uptake by barley seedlings. *Functional Plant Biology. 6*(4), 459–466.

59. Mudge, S. R., Rae, A. L., Diatloff, E., & Smith, F. W. (2002). Expression analysis suggests novel roles for members of the Pht1 family of phosphate transporters in Arabidopsis. *The Plant Journal. 31*(3), 341–353.

60. Shin, H., Shin, H. S., Dewbre, G. R., & Harrison, M. J. (2004). Phosphate transport in Arabidopsis, Pht1; 1 and Pht1; 4 play a major role in phosphate acquisition from both low and high phosphate environments. *The Plant Journal. 39*(4), 629–642.

61. Catarecha, P., Segura, M. D., Franco-Zorrilla, J. M., Garcia-Ponce, B., Lanza, M. n., Solano, R., Paz-Ares, J., & Leyva, A. (2007). A mutant of the Arabidopsis phosphate transporter PHT1; 1 displays enhanced arsenic accumulation. *The Plant Cell. 19*(3), 1123–1133.

62. Nagarajan, V. K., Jain, A., Poling, M. D., Lewis, A. J., Raghothama, K. G., & Smith, A. P. (2011). Arabidopsis Pht1; 5 mobilizes phosphate between source and sink organs and influences the interaction between phosphate homeostasis and ethylene signaling. *Plant Physiology. 156*(3), 1149–1163.

63. LeBlanc, M. S., McKinney, E. C., Meagher, R. B., & Smith, A. P. (2013). Hijacking membrane transporters for arsenic phytoextraction. *Journal of Biotechnology. 163*(1), 1–9.

64. Remy, E., Cabrito, T. R., Batista, R. A., Teixeira, M. C., Sa-Correia, I., & Duque, P. (2012). The Pht1; 9 and Pht1; 8 transporters mediate inorganic phosphate acquisition by the Arabidopsis thaliana root during phosphorus starvation. *New Phytologist. 195*(2), 356–371.

65. González, E., Solano, R., Rubio, V., Leyva, A., & Paz-Ares, J. (2005). Phosphate Transporter Traffic Facilitator1 is a plant-specific SEC12-related protein that enables the endoplasmic reticulum exit of a high-affinity phosphate transporter in Arabidopsis. *The Plant Cell. 17*(12), 3500–3512.

66. Chen, J., Liu, Y., Ni, J., Wang, Y., Bai, Y., Shi, J., Gan, J., Wu, Z., & Wu, P. (2011). OsPHF1 regulates the plasma membrane localization of low-and high-affinity inorganic phosphate transporters and determines inorganic phosphate uptake and translocation in rice. *Plant Physiology. 57*(1), 269–278.

67. Wu, Z., Ren, H., McGrath, S. P., Wu, P., & Zhao, F.-J. (2011). Investigating the contribution of the phosphate transport pathway to arsenic accumulation in rice. *Plant Physiology.* *157*(1), 498–508.

68. Castrillo, G., Sanchez-Bermejo, E., de Lorenzo, L., Crevillen, P., Fraile-Escanciano, A., Mohan, T. C., Mouriz, A., Catarecha, P., Sobrino-Plata, J., & Olsson, S. (2013). WRKY6 transcription factor restricts arsenate uptake and transposon activation in Arabidopsis. *The Plant Cell.* *25*(8), 2944–2957.

69. Wang, H., Xu, Q., Kong, Y.-H., Chen, Y., Duan, J.-Y., Wu, W. H., & Chen, Y.-F. (2014). Arabidopsis WRKY45 transcription factor activates Phosphate Transporter; 1 expression in response to phosphate starvation. *Plant Physiology.* *164*(4), 2020–2029.

70. Jia, H., Ren, H., Gu, M., Zhao, J., Sun, S., Zhang, X., Chen, J., Wu, P., & Xu, G. (2011). The phosphate transporter gene OsPht1; 8 is involved in phosphate homeostasis in rice. *Plant Physiology.* *156*(3), 1164–1175.

71. Sun, S., Gu, M., Cao, Y., Huang, X., Zhang, X., Ai, P., Zhao, J., Fan, X., & Xu, G. (2012). A constitutive expressed phosphate transporter, OsPht1; 1, modulates phosphate uptake and translocation in phosphate-replete rice. *Plant Physiology.* *159*(4), 1571–1581.

72. Kamiya, T., Islam, R., Duan, G., Uraguchi, S., & Fujiwara, T. (2013). Phosphate deficiency signaling pathway is a target of arsenate and phosphate transporter OsPT1 is involved in As accumulation in shoots of rice. *Soil Science and Plant Nutrition.* *59*(4), 580–590.

73. Ma, J. F., Yamaji, N., Mitani, N., Xu, X.-Y., Su, Y.-H., McGrath, S. P., & Zhao, F.-J. (2008). Transporters of arsenite in rice and their role in arsenic accumulation in rice grain. Proceedings of the *National Academy of Sciences.* *105*(29), 9931–9935.

74. Chen, Z., Zhu, Y. G., Liu, W. J., & Meharg, A. A. (2005). Direct evidence showing the effect of root surface iron plaque on arsenite and arsenate uptake into rice (*Oryza sativa*) roots. *New Phytologist.* *165*(1), 91–97.

75. Zhang, X., Zhao, F. J., Huang, Q., Williams, P. N., Sun, G. X., & Zhu, Y. G. (2009). Arsenic uptake and speciation in the rootless duckweed Wolffia globosa. *New Phytologist.* *182*(2), 421–428.

76. Meharg, A. A., & Jardine, L. (2003). Arsenite transport into paddy rice (Oryza sativa) roots. *New Phytologist.* *157*(1), 39–44.

77. Li, G., Santoni, V. R., & Maurel, C. (2014). Plant aquaporins, roles in plant physiology. *Biochimica et Biophysica Acta.* *1840*(5), 1574–1582.

78. Mukhopadhyay, R., Bhattacharjee, H., & Rosen, B. P. (2014). Aquaglyceroporins, generalized metalloid channels. Biochimica et Biophysica Acta (BBA) *General Subjects.* *1840*(5), 1583–1591.

79. Bienert, G. P., SchÃ¼ssler, M. D., & Jahn, T. P. (2008). Metalloids, essential, beneficial or toxic? Major intrinsic proteins sort it out. *Trends in Biochemical Sciences* *33*(1), 20–26.

80. Zhao, F. J., Ago, Y., Mitani, N., Li, R. Y., Su, Y. H., Yamaji, N., McGrath, S. P., & Ma, J. F. (2010). The role of the rice aquaporin Lsi1 in arsenite efflux from roots. *New Phytologist.* *186*(2), 392–399.

81. Pommerrenig, B., Diehn, T. A., & Bienert, G. P. (2015). Metalloido-porins, Essentiality of Nodulin 26-like intrinsic proteins in metalloid transport. *Plant Science.* *238*, 212–227.

82. Mosa, K. A., Kumar, K., Chhikara, S., McDermott, J., Liu, Z., Musante, C., White, J. C., & Dhankher, O. P. (2012). Members of rice plasma membrane intrinsic proteins subfamily are involved in arsenite permeability and tolerance in plants. *Transgenic Research. 21*(6), 1265–1277.

83. Ye, J., Rensing, C., Rosen, B. P., & Zhu, Y.-G. (2012). Arsenic biomethylation by photosynthetic organisms. *Trends in Plant Science 17*(3), 155–162.

84. Huang, J.-H. (2014). Impact of microorganisms on arsenic biogeochemistry, a review. *Water, Air and Soil Pollution. 225*(2), 1–25.

85. Rahman, M. A., Kadohashi, K., Maki, T., & Hasegawa, H. (2011). Transport of DMAA and MMAA into rice (*Oryza sativa* L.) roots. *Environmental and Experimental Botany. 72*(1), 41–46.

86. Li, R.-Y., Ago, Y., Liu, W.-J., Mitani, N., Feldmann, J. r., McGrath, S. P., Ma, J. F., & Zhao, F. J. (2009). The rice aquaporin Lsi1 mediates uptake of methylated arsenic species. *Plant Physiology. 150*(4), 2071–2080.

87. Pfeffer, H., Dannel, F., & Romheld, V. (1999). Are there two mechanisms for boron uptake in sunflower? *Journal of Plant Physiology. 155*(1), 34–40.

88. Dannel, F., Pfeffer, H., & Romheld, V. (2000). Characterization of root boron pools, boron uptake and boron translocation in sunflower using the stable isotopes 10 B and 11 B. *Functional Plant Biology. 27*(5), 397–405.

89. Takano, J., Noguchi, K., Yasumori, M., Kobayashi, M., Gajdos, Z., Miwa, K., Hayashi, H., Yoneyama, T., & Fujiwara, T. (2002). Arabidopsis boron transporter for xylem loading. *Nature. 420*(6913), 337–340.

90. Nakagawa, Y., Hanaoka, H., Kobayashi, M., Miyoshi, K., Miwa, K., & Fujiwara, T. (2007). Cell-type specificity of the expression of *OsBOR1*, a rice efflux boron transporter gene, is regulated in response to boron availability for efficient boron uptake and xylem loading. *The Plant Cell. 19*(8), 2624–2635.

91. Takano, J., Wada, M., Ludewig, U., Schaaf, G., Von Wiren, N., & Fujiwara, T. (2006). The Arabidopsis major intrinsic protein NIP5; 1 is essential for efficient boron uptake and plant development under boron limitation. *The Plant Cell. 18*(6), 1498–1509.

92. Sutton, T., Baumann, U., Hayes, J., Collins, N. C., Shi, B.-J., Schnurbusch, T., Hay, A., Mayo, G., Pallotta, M., & Tester, M. (2007). Boron-toxicity tolerance in barley arising from efflux transporter amplification. *Science. 318*(5855), 1446–1449.

93. Kabata-Pendias, A. (2010). Trace elements in soils and plants. 4th edition, CRC Press/Taylor & Francis Group, Boca Raton, FL, USA, pp. 1–548.

94. Sors, T. G., Ellis, D. R., Na, G. N., Lahner, B., Lee, S., Leustek, T., Pickering, I. J., & Salt, D. E. (2005). Analysis of sulfur and selenium assimilation in Astragalus plants with varying capacities to accumulate selenium. *The Plant Journal. 42*(6), 785–797

95. Asher, C. J., Butler, G. W., & Peterson, P. J. (1977). Selenium transport in root systems of tomato. *Journal of Experimental Botany. 28*(2), 279–291.

96. White, P. J., Bowen, H. C., Parmaguru, P., Fritz, M., Spracklen, W. P., Spiby, R. E., Meacham, M. C., Mead, A., Harriman, M., Trueman, L. J. (2004). Interactions between selenium and sulphur nutrition in *Arabidopsis thaliana*. *Journal of Experimental Botany. 55(404),* 1927–1937.

97. El-Ramady, H., Abdalla, N., Alshaal, T., El-Henawy, A., Salah, E. D. A. F., Shams, M. S., Shalaby, T., Bayoumi, Y., Elhawat, N., & Shehata, S. (2015). Selenium and its role in higher plants. In: Lichtfouse, E., Schwarzbauer, J. & Robert, D. (eds.) *Pollutants in Buildings, Water and Living Organisms*. Environmental Chemistry for a Sustainable World, vol 7. Springer, Cham: Springer International Publishing Switzerland, pp. 235–296.

98. Ulrich, J. M., & Shrift, A. (1968). Selenium absorption by excised Astragalus roots. *Plant Physiology. 43*(1), 14–20.

99. Shibagaki, N., Rose, A., McDermott, J. P., Fujiwara, T., Hayashi, H., Yoneyama, T., & Davies, J. P. (2002). Selenate resistant mutants of Arabidopsis thaliana identify SULTR1; 2, a sulfate transporter required for efficient transport of sulfate into roots. *The Plant Journal. 29*(4), 475–486.

100. Hawkesford, M. J., & Zhao, F.-J. (2007). Strategies for increasing the selenium content of wheat. *Journal of Cereal Science. 46*(3), 282–292.

101. Leustek, T. (2002). Sulfate metabolism. In Somerville, C.R. & Meyerowitz E.M., (ed.)The Arabadopsis Book, e0017 Ed. by Society of Plant Biologists, Rockville, MD.

102. Smith, F. W., Ealing, P. M., Hawkesford, M. J., & Clarkson, D. T. (1995). Plant members of a family of sulfate transporters reveal functional subtypes. Proceedings of the *National Academy of Sciences. 92*(20), 9373–9377.

103. Smith, F. W., Hawkesford, M. J., Ealing, P. M., Clarkson, D. T., Berg, P. J., Belcher, A. R., & Warrilow, A. G. S. (1997). Regulation of expression of a cDNA from barley roots encoding a high affinity sulphate transporter. *The Plant Journal. 12*(4), 875–884.

104. Mikkelsen, R. L., & Wan, H. F. (1990). The effect of selenium on sulfur uptake by barley and rice. *Plant and Soil. 121*(1), 151–153.

105. Kopsell, D. A., & Randle, W. M. (1997). Selenate concentration affects selenium and sulfur uptake and accumulation byGranex 33'onions. *Journal of the American Society for Horticultural Science. 122*(5), 721–726.

106. Parker, D. R., Page, A. L., & Bell, P. F. (1992). Contrasting selenate-sulfate interactions in selenium-accumulating and nonaccumulating plant species. *Soil Science Society of America Journal. 56*(6), 1818–1824.

107. Kamiya, T., & Fujiwara, T. (2009). Arabidopsis NIP1; 1 transports antimonite and determines antimonite sensitivity. *Plant and Cell Physiology. 50*(11), 1977–1981.

108. Tschan, M., Robinson, B., & Schulin, R. (2008). Antimony uptake by Zea mays (L.) and Helianthus annuus (L.) from nutrient solution. *Environmental Geochemistry and Health. 30*(2), 187–191.

109. Baes, C. F., & Mesmer, R. E. (1986). Arsenic, antimony, bismuth. The hydrolysis of cations. 2nd edition, pp. 366–375.

110. Bell, P. F., McLaughlin, M. J., Cozens, G., Stevens, D. P., Owens, G., & South, H. (2003). Plant uptake of 14C-EDTA, 14C-Citrate, and 14C-Histidine from chelator-buffered and conventional hydroponic solutions. *Plant and Soil. 253*(2), 311–319.

111. Feng, R., Wei, C., Tu, S., Tang, S., & Wu, F. (2011). Detoxification of antimony by selenium and their interaction in paddy rice under hydroponic conditions. *Microchemical Journal. 97*(1), 57–61.

112. Muller, K., Daus, B., Mattusch, J., Vetterlein, D., Merbach, I., & Wennrich, R. (2013). Impact of arsenic on uptake and bio-accumulation of antimony by arsenic hyperaccumulator Pteris vittata. *Environmental Pollution. 174*, 128–133.

113. Okkenhaug, G., Amstatter, K., Lassen Bue, H., Cornelissen, G., Breedveld, G. D., Henriksen, T., & Mulder, J. (2012). Antimony (Sb) contaminated shooting range soil, Sb mobility and immobilization by soil amendments. *Environmental Science and Technology. 47*(12), 6431–6439.

114. Wan, X.-m., Tandy, S., Hockmann, K., & Schulin, R. (2013). Effects of water-logging on the solubility and redox state of Sb in a shooting range soil and its uptake by grasses, a tank experiment. *Plant and Soil. 371*(1–2), 155–166.

115. Ren, J. H., Ma, L. Q., Sun, H. J., Cai, F., & Luo, J. (2014). Antimony uptake, translocation and speciation in rice plants exposed to antimonite and antimonate. *Science of the Total Environment. 475*, 83–89.

116. Okkenhaug, G., Zhu, Y.-G., Luo, L., Lei, M., Li, X., & Mulder, J. (2011). Distribution, speciation and availability of antimony (Sb) in soils and terrestrial plants from an active Sb mining area. *Environmental Pollution. 159*(10), 2427–2434.

117. Miteva, E., & Merakchiyska, M. (2002). Response of chloroplasts and photosynthetic mechanism of bean plants to excess arsenic in soil. *Bulgarian Journal of Agricultural Science. 8*, 151–156.

118. Singh, N., Ma, L. Q., Vu, J. C., & Raj, A. (2009). Effects of arsenic on nitrate metabolism in arsenic hyperaccumulating and non-hyperaccumulating ferns. *Environmental Pollution. 157*(8), 2300–2305.

119. Dixon, H. B. F. (1996). The biochemical action of arsonic acids especially as phosphate analogues. *Advances in Inorganic Chemistry. 44*(c), 191–227.

120. Srivastava, S., Srivastava, A. K., Suprasanna, P., & D'Souza, S. F. (2009). Comparative biochemical and transcriptional profiling of two contrasting varieties of *Brassica juncea* L. in response to arsenic exposure reveals mechanisms of stress perception and tolerance. *Journal of Experimental Botany. 60*(12), 3419–3431.

121. Tripathi, P., Mishra, A., Dwivedi, S., Chakrabarty, D., Trivedi, P. K., Singh, R. P., & Tripathi, R. D. (2012). Differential response of oxidative stress and thiol metabolism in contrasting rice genotypes for arsenic tolerance. *Ecotoxicology and Environmental Safety. 79*, 189–198.

122. Mishra, S., & Dubey, R. S. (2006). Inhibition of ribonuclease and protease activities in arsenic exposed rice seedlings, role of proline as enzyme protectant. *Journal of Plant Physiology. 163*(9), 927–936.

123. Meharg, A. A., & Hartley-Whitaker, J. (2002). Arsenic uptake and metabolism in arsenic resistant and nonresistant plant species. *New Phytologist. 154*(1), 29–43.

124. Tu, C., & Ma, L. Q. (2002). Effects of arsenic concentrations and forms on arsenic uptake by the hyperaccumulator ladder brake. *Journal of Environmental Quality. 31*(2), 641–647.

125. Garg, N., & Singla, P. (2011). Arsenic toxicity in crop plants, physiological effects and tolerance mechanisms. *Environmental Chemistry Letters 9*(3), 303–321.

126. Finnegan, P., & Chen, W. (2012). Arsenic toxicity, the effects on plant metabolism. *Frontiers in Physiology. 3*, 182.

127. Gochfeld, M. (1997). Factors influencing susceptibility to metals. *Environmental Health Perspectives. 105*(Suppl 4), 817.

128. De Santis, A., Borraccino, G., Arrigoni, O., & Palmieri, F. (1975). The mechanism of phosphate permeation in purified bean mitochondria. *Plant and Cell Physiology. 16*(5), 911–923.

129. Palmieri, L., Picault, N., Arrigoni, R., Besin, E., Palmieri, F., & Hodges, M. (2008). Molecular identification of three Arabidopsis thaliana mitochondrial dicarboxylate carrier isoforms, organ distribution, bacterial expression, reconstitution into liposomes and functional characterization. *Biochemical Journal. 410*(3), 621–629.

130. Guo, J., Dai, X., Xu, W., & Ma, M. (2008). Overexpressing GSH1 and AsPCS1 simultaneously increases the tolerance and accumulation of cadmium and arsenic in Arabidopsis thaliana. *Chemosphere. 72*(7), 1020–1026.

131. Tawfik, D. S., & Viola, R. E. (2011). Arsenate replacing phosphate, alternative life chemistries and ion promiscuity. *Biochemistry. 50*(7), 1128–1134.

132. Gunes, A., Pilbeam, D. J., & Inal, A. (2009). Effect of arsenic-phosphorus interaction on arsenic-induced oxidative stress in chickpea plants. *Plant and Soil. 314*(1–2), 211–220.

133. Mishra, S., Jha, A. B., & Dubey, R. S. (2011).Arsenite treatment induces oxidative stress, upregulates antioxidant system, and causes phytochelatin synthesis in rice seedlings. *Protoplasma. 248*(3), 565–577.

134. Chou, W. C., Jie, C., Kenedy, A. A., Jones, R. J., Trush, M. A., & Dang, C. V. (2004). Role of NADPH oxidase in arsenic-induced reactive oxygen species formation and cytotoxicity in myeloid leukemia cells. Proceedings of the *National Academy of Sciences of the United States of America. 101*(13), 4578–4583.

135. Smith, S. E., Christophersen, H. M., Pope, S., & Smith, F. A. (2010). Arsenic uptake and toxicity in plants, integrating mycorrhizal influences. *Plant and Soil. 327*(1–2), 1–21.

136. Hayes, J. E., & Reid, R. J. (2004). Boron tolerance in barley is mediated by efflux of boron from the roots. *Plant Physiology. 136*(2), 3376–3382.

137. West, G., Inze, D., & Beemster, G. T. S. (2004). Cell cycle modulation in the response of the primary root of Arabidopsis to salt stress. *Plant Physiology. 135*(2), 1050–1058.

138. De Schutter, K., Joubes, J. r. m., Cools, T., Verkest, A., Corellou, F., Babiychuk, E., Van Der Schueren, E., Beeckman, T., Kushnir, S., & Inze, D. (2007). Arabidopsis WEE1 kinase controls cell cycle arrest in response to activation of the DNA integrity checkpoint. *The Plant Cell. 19*(1), 211–225.

139. Sakamoto, T., Inui, Y. T., Uraguchi, S., Yoshizumi, T., Matsunaga, S., Mastui, M., Umeda, M., Fukui, K., & Fujiwara, T. (2011). Condensin II alleviates DNA damage and is essential for tolerance of boron overload stress in Arabidopsis. *The Plant Cell. 23*(9), 3533–3546.

140. Shomron, N., & Ast, G. (2003). Boric acid reversibly inhibits the second step of pre-mRNA splicing. *FEBS letters. 552*(2–3), 219–224.

141. Karabal, E., Yucel, M., & Oktem, H. S. A. (2003). Antioxidant responses of tolerant and sensitive barley cultivars to boron toxicity. *Plant Science. 164*(6), 925–933.

142. Molassiotis, A., Sotiropoulos, T., Tanou, G., Diamantidis, G., & Therios, I. (2006).Boron-induced oxidative damage and antioxidant and nucleolytic responses in shoot tips culture of the apple rootstock EM 9 (Malus domestica Borkh). *Environmental and Experimental Botany. 56*(1), 54–62.

143. Geoffroy, L., Gilbin, R., Simon, O., Floriani, M., Adam, C., Pradines, C., Cournac, L., & Garnier-Laplace, J. (2007). Effect of selenate on growth and photosynthesis of Chlamydomonas reinhardtii. *Aquatic Toxicology. 83*(2), 149–158.

144. Brown, T. A., & Shrift, A. (1980). Identification of selenocysteine in the proteins of selenate-grown Vigna radiata. *Plant Physiology 66*(4), 758–761.

145. Van Hoewyk, D., (2013). A tale of two toxicities, malformed selenoproteins and oxidative stress both contribute to selenium stress in plants. *Annals of Botany. 112*(6), 965–972.

146. Freeman, J. L., Tamaoki, M., Stushnoff, C., Quinn, C. F., Cappa, J. J., Devonshire, J., Fakra, S. C., Marcus, M. A., McGrath, S. P., & Van Hoewyk, D. (2010). Molecular mechanisms of selenium tolerance and hyperaccumulation in *Stanleya pinnata. Plant Physiology. 153*(4), 1630–1652.

147. Noctor, G., Mhamdi, A., Chaouch, S., Han, Y. I., Neukermans, J., Marquez-Garcia, B., Queval, G., & Foyer, C. H. (2011). Glutathione in plants, an integrated overview. *Plant, Cell and Environment 35*(2), 454–484.

148. Koprivova, A., Mugford, S. T., & Kopriva, S. (2010). Arabidopsis root growth dependence on glutathione is linked to auxin transport. *Plant Cell Reports. 29*(10), 1157–1167.

149. Hugouvieux, V. r., Dutilleul, C., Jourdain, A. s., Reynaud, F., Lopez, V. R., & Bourguignon, J. (2009). Arabidopsis putative selenium-binding protein expression is tightly linked to cellular sulfur demand and can reduce sensitivity to stresses requiring glutathione for tolerance. *Plant Physiology. 151*(2), 768–781.

150. Labanowska, M., Filek, M., Koscielniak, J., Kurdziel, M., Kulis, E., & Hartikainen, H. (2012). The effects of short-term selenium stress on Polish and Finnish wheat seedlings-EPR, enzymatic and fluorescence studies. *Journal of Plant Physiology. 169*(3), 275–284.

151. Akbulut, M., & Cakir, S. (2010). The effects of Se phytotoxicity on the antioxidant systems of leaf tissues in barley (*Hordeum vulgare* L.) seedlings. *Plant Physiology and Biochemistry. 48*(2), 160–166.

152. Sabbagh, M., & Van Hoewyk, D. (2012). Malformed selenoproteins are removed by the ubiquitin-proteasome pathway in *Stanleya pinnata. Plant and Cell Physiology. 53*(3), 555–564.

153. Schiavon, M., Moro, I., Pilon-Smits, E. A. H., Matozzo, V., Malagoli, M., & Dalla Vecchia, F. (2012). Accumulation of selenium in Ulva sp. and effects on morphology, ultrastructure and antioxidant enzymes and metabolites. *Aquatic Toxicology. 122*, 222–231.

154. Lehotai Lehotai, N. r., Kolbert, Z., Peto, A., Feigl, G. b., Ordog, A., Kumar, D., Tari, I., & Erdei, L. (2012). Selenite-induced hormonal and signalling mechanisms during root growth of *Arabidopsis thaliana* L. *Journal of Experimental Botany. 63*(15), 5677–5687.

155. Mroczek-Zdyrska, M., & Wojcik, M. G. (2012). The influence of selenium on root growth and oxidative stress induced by lead in Vicia faba L. minor plants. *Biological Trace Element Research. 147*(1–3), 320–328.

156. Staniek, K., Gille, L., Kozlov, A. V., & Nohl, H. (2002). Mitochondrial superoxide radical formation is controlled by electron bifurcation to the high and low potential pathways. *Free Radical Research. 36*(4), 381–387.

157. Wallenberg, M., Olm, E., Hebert, C., Bjornstedt, M., & Fernandes, A. P. (2010). Selenium compounds are substrates for glutaredoxins, a novel pathway for selenium metabolism and a potential mechanism for selenium-mediated cytotoxicity. *Biochemical Journal. 429*(1), 85–93.

158. Tschan, M., Robinson, B. H., & Schulin, R. (2009). Antimony in the soil-plant system, a review. *Environmental Chemistry. 6*(2), 106–115.

159. He, M., & Yang, J. (1999). Effects of different forms of antimony on rice during the period of germination and growth and antimony concentration in rice tissue. *Science of the Total Environment. 243*, 149–155.

160. Pan, X., Zhang, D., Chen, X., Bao, A., & Li, L. (2011). Antimony accumulation, growth performance, antioxidant defense system and photosynthesis of Zea mays in response to antimony pollution in soil. *Water, Air and Soil Pollution. 215*(1–4), 517–523.

161. Vaculikova, M., Vaculik, M., Simkova, L., Fialova, I., Kochanova, Z., Sedlakova, B., & Luxova, M. (2014). Influence of silicon on maize roots exposed to antimony-growth and antioxidative response. *Plant Physiology and Biochemistry. 83*, 279–284.

162. Shtangeeva, I., Singh, B., Bali, R., Ayrault, S., & Timofeev, S. (2014). Antimony accumulation in wheat seedlings grown in soil and water. *Communications in Soil Science and Plant Analysis. 45*(7), 968–983.

163. Vaculik, M., Pavlovic, A., & Lux, A. (2015). Silicon alleviates cadmium toxicity by enhanced photosynthetic rate and modified bundle sheath's cell chloroplasts ultrastructure in maize. *Ecotoxicology and Environmental Safety. 120*, 66–73.

164. Davis, R. D., Beckett, P. H. T., & Wollan, E. (1978). Critical levels of twenty potentially toxic elements in young spring barley. *Plant and Soil. 49*(2), 395–408.

165. Feng, R., Wei, C., Tu, S., Wu, F., & Yang, L. (2009). Antimony accumulation and antioxidative responses in four fern plants. *Plant and Soil 317*(1–2), 93–101.

166. White, P. J., & Broadley, M. R. (2003). Calcium in plants. *Annals of Botany. 92*(4), 487–511.

167. Brand, J. J., & Becker, D. W. (1984). Evidence for direct roles of calcium in photosynthesis. *Journal of Bioenergetics and Biomembranes. 16*(4), 239–249.

168. Terry, N., & Ulrich, A. (1973). Effects of phosphorus deficiency on the photosynthesis and respiration of leaves of sugar beet. *Plant Physiology. 51*(1), 43–47.

169. Behboudian, M. H., & Anderson, D. R. (1990). Effects of potassium deficiency on water relations and photosynthesis of the tomato plant. *Plant and Soil. 127*(1), 137–139.

170. Corrales, I., Barcelo, J., Bech, J., & Poschenrieder, C. (2014). Antimony accumulation and toxicity tolerance mechanisms in Trifolium species. *Journal of Geochemical Exploration. 147*, 167–172.

171. Indriolo, E., Na, G., Ellis, D., Salt, D. E., & Banks, J. A. (2010). A vacuolar arsenite transporter necessary for arsenic tolerance in the arsenic hyperaccumulating fern Pteris vittata is missing in flowering plants. *The Plant Cell. 22*(6), 2045–2057.

172. Bienert, G. P., Thorsen, M., Schüssler, M. D., Nilsson, H. R., Wagner, A., Tams, M. J., & Jahn, T. P. (2008). A subgroup of plant aquaporins facilitate the bi-directional diffusion of As $(OH)_3$ and Sb $(OH)_3$ across membranes. *BMC Biology. 6*(1), 1.

173. Maciaszczyk-Dziubinska, E., Wawrzycka, D., & Wysocki, R. (2012). Arsenic and antimony transporters in eukaryotes. *International Journal of Molecular Sciences. 13*(3), 3527–3548.

174. Logoteta, B., Xu, X. Y., Macnair, M. R., McGrath, S. P., & Zhao, F. J. (2009). Arsenite efflux is not enhanced in the arsenate tolerant phenotype of Holcus lanatus. *New Phytologist. 183*(2), 340–348.

175. Reid, R. (2007). Identification of boron transporter genes likely to be responsible for tolerance to boron toxicity in wheat and barley. *Plant and Cell Physiology. 48*(12), 1673–1678.

176. Bech, J., Corrales, I., Tume, P., Barcelo, J., Duran, P., Roca, N., & Poschenrieder, C. (2012). Accumulation of antimony and other potentially toxic elements in plants around a former antimony mine located in the Ribes Valley (Eastern Pyrenees). *Journal of Geochemical Exploration. 113*, 100–105.

177. Tisarum, R., Lessl, J. T., Dong, X., & de Oliveira, L. M. (2014). Antimony uptake, efflux speciation in arsenic hyperaccumulator *Pteris vittata*. *Environmental Pollution. 186*, 110–114.

178. Grill, E., Winnacker, E.-L., & Zenk, M. H. (1987). Phytochelatins, a class of heavy-metal-binding peptides from plants, are functionally analogous to metallothioneins. *Proceedings of the National Academy of Sciences. 84*(2), 439–443.

179. Aborode, F. A., Raab, A., Voigt, M., Costa, L. M., Krupp, E. M., & Feldmann, J. (2016). The importance of glutathione and phytochelatins on the selenite and arsenate detoxification in Arabidopsis thaliana. *Journal of Environmental Sciences. 49*, 150–161.

180. Vazquez Reina, S. l., Esteban, E., & Goldsbrough, P. (2005). Arsenate induced phytochelatins in white lupin, influence of phosphate status. *Physiologia Plantarum. 124*(1), 41–49.

181. Spain, S. M., & Rabenstein, D. L. (2004). Characterization of the selenotrisulfide formed by reaction of selenite with end-capped phytochelatin-2. *Analytical and Bioanalytical Chemistry. 378*(6), 1561–1567.

182. Pal, R., & Rai, J. P. N. (2010). Phytochelatins, peptides involved in heavy metal detoxification. *Applied Biochemistry and Biotechnology. 160*(3), 945–963.

183. Batista, B. L., Nigar, M., Mestrot, A., Rocha, B. A., Junior, F. B., Price, A. H., Raab, A., & Feldmann, J. (2014). Identification and quantification of phytochelatins in roots of rice to long-term exposure, evidence of individual role on arsenic accumulation and translocation. *Journal of Experimental Botany. 65*(6), 1467–1479.

184. Park, J., Song, W. Y., Ko, D., Eom, Y., Hansen, T. H., Schiller, M., Lee, T. G., Martinoia, E., & Lee, Y. (2012). The phytochelatin transporters AtABCC1 and AtABCC2 mediate tolerance to cadmium and mercury. *The Plant Journal. 69*(2), 278–288.

185. Kamiya, T., & Fujiwara, T. (2011). A novel allele of the Arabidopsis phytochelatin synthase 1 gene conferring high sensitivity to arsenic and antimony. *Soil Science and Plant Nutrition. 57*(2), 272–278.

186. Filella, M., Belzile, N., & Lett, M. C. (2007). Antimony in the environment, a review focused on natural waters. III. Microbiota relevant interactions. *Earth-Science Reviews. 80*(3), 195–217.

187. Wysocki, R., Clemens, S., Augustyniak, D., Golik, P., Maciaszczyk, E., Tamas, M. J., & Dziadkowiec, D. (2003). Metalloid tolerance based on phytochelatins is not functionally equivalent to the arsenite transporter Acr3p. *Biochemical and Biophysical Research Communications. 304*(2), 293–300.

188. Jain, R., Verma, R., Singh, A., Chandra, A., & Solomon, S. (2015). Influence of selenium on metallothionein gene expression and physiological characteristics of sugarcane plants. *Plant Growth Regulation. 77*(2), 109–115.

189. Lamdan, N. L., Attia, Z., Moran, N., & Moshelion, M. (2012). The Arabidopsis related halophyte Thellungiella halophila, boron tolerance via boron complexation with metabolites? *Plant, Cell and Environment. 35*(4), 735–746.

190. Song, W.-Y., Yamaki, T., Yamaji, N., Ko, D., Jung, K.-H., Fujii-Kashino, M., An, G., Martinoia, E., Lee, Y., & Ma, J. F. (2014). A rice ABC transporter, OsABCC1, reduces arsenic accumulation in the grain. *Proceedings of the National Academy of Sciences. 111*(44), 15699–15704.

191. Poynton, C. Y., Huang, J. W., Blaylock, M. J., Kochian, L. V., & Elless, M. P. (2004). Mechanisms of arsenic hyperaccumulation in Pteris species, root As influx and translocation. *Planta. 219*(6), 1080–1088.

192. Wang, Y., Chai, L., Yang, Z., Hussani, M., Xiao, R., & Tang, C. (2017). Subcellular distribution and chemical forms of antimony in Ficus tikoua. *International Journal of Phytoremediation. 19(2),* 97–103.

193. Sharma, P., Jha, A. B., Dubey, R. S., & Pessarakli, M. (2012). Reactive oxygen species, oxidative damage, and antioxidative defense mechanism in plants under stressful conditions. *Journal of Botany. Article ID 217037*, 26 pp.

194. Ansari, M. K. A., Zia, M. H., Ahmad, A., Aref, I. M., Fatma, T., Iqbal, M., & Owens, G. (2016). Status of antioxidant defense system for detoxification of arsenic in *Brassica juncea* (L.). *Ecoprint, An International Journal of Ecology. 22*, 7–19.

195. Ayvaz, M., Guven, A., Blokhina, O., & Fagerstedt, K. V. (2016). Boron stress, oxidative damage and antioxidant protection in potato cultivars (*Solanum tuberosum* L.). *Acta Agriculturae Scandinavica, Section B-Soil and Plant Science. 66*(4), 302–316.

196. Del Rio, D., Stewart, A. J., & Pellegrini, N. (2005). A review of recent studies on malondialdehyde as toxic molecule and biological marker of oxidative stress. *Nutrition, Metabolism and Cardiovascular Diseases. 15*(4), 316–328.

197. Foyer, C. H., & Noctor, G. (2011). Ascorbate and glutathione, the heart of the redox hub. *Plant physiology. 155*(1), 2–18.

198. Cao, X., Ma, L. Q., & Tu, C. (2004). Antioxidative responses to arsenic in the arsenic-hyperaccumulator Chinese brake fern (*Pteris vittata* L.). *Environmental Pollution, 128*(3), 317–325.

199. Srivastava, M., Ma, L. Q., Singh, N., & Singh, S. (2005). Antioxidant responses of hyper-accumulator and sensitive fern species to arsenic. *Journal of Experimental Botany. 56*(415), 1335–1342.

200. Dave, R., Singh, P. K., Tripathi, P., Shri, M., Dixit, G., Dwivedi, S., Chakrabarty, D., Trivedi, P. K., Sharma, Y. K., & Dhankher, O. P. (2013). Arsenite tolerance is related to proportional thiolic metabolite synthesis in rice (*Oryza sativa* L.). *Archives of Environmental Contamination and Toxicology. 64*(2), 235–242.

201. Gunes, A., Soylemezoglu, G., Inal, A., Bagci, E. G., Coban, S., & Sahin, O. (2006). Antioxidant and stomatal responses of grapevine (*Vitis vinifera* L.) to boron toxicity. *Scientia Horticulturae. 110*(3), 279–284.

202. Cervilla, L. M., Blasco, B. a., RÃos, J. J., Romero, L., Ruiz, & J. M. (2007). Oxidative stress and antioxidants in tomato (*Solanum lycopersicum*) plants subjected to boron toxicity. *Annals of Botany. 100*(4), 747–756.

203. Eraslan, F., Inal, A., Savasturk, O., & Gunes, A. (2007). Changes in antioxidative system and membrane damage of lettuce in response to salinity and boron toxicity. *Scientia Horticulturae. 114*(1), 5–10.

204. Ardic, M., Sekmen, A. H., Tokur, S., Ozdemir, F., & Turkan, I. (2009). Antioxidant responses of chickpea plants subjected to boron toxicity. *Plant Biology. 11*(3), 328–338.

205. Wang, J. Z., Tao, S. T., Qi, K. J., Wu, J., Wu, H. Q., & Zhang, S. L. (2011). Changes in photosynthetic properties and antioxidative system of pear leaves to boron toxicity. *African Journal of Biotechnology. 10*(85), 19693–19700.

206. Ishikawa, T., Dowdle, J., & Smirnoff, N. (2006). Progress in manipulating ascorbic acid biosynthesis and accumulation in plants. *Physiologia Plantarum. 126*(3), 343–355.

207. Lopez-Gomez, E., San Juan, M. A., Diaz-Vivancos, P., Beneyto, J. M., García-Legaz, M. F., & Hernandez, J. A. (2007). Effect of rootstocks grafting and boron on the antioxidant systems and salinity tolerance of loquat plants (*Eriobotrya japonica* Lindl.). *Environmental and Experimental Botany. 60*(2), 151–158.

208. Gomes-Junior, R. A., Gratao, P. L., Gaziola, S. A., Mazzafera, P., Lea, P. J., & Azevedo, R. A. (2007). Selenium-induced oxidative stress in coffee cell suspension cultures. *Functional Plant Biology. 34*(5), 449–456.

209. Chen, T. F., Zheng, W. J., Wong, Y. S., & Yang, F. (2008). Selenium-induced changes in activities of antioxidant enzymes and content of photosynthetic pigments in Spirulina platensis. *Journal of Integrative Plant Biology. 50*(1), 40–48.

210. Feng, R., Liao, G., Guo, J., Wang, R., Xu, Y., Ding, Y., Mo, L., Fan, Z., & Li, N. (2016). Responses of root growth and antioxidative systems of paddy rice exposed to antimony and selenium. *Environmental and Experimental Botany. 122*, 29–38.

211. Tamaoki, M., Freeman, J. L., & Pilon-Smits, E. A. H. (2008). Cooperative ethylene and jasmonic acid signaling regulates selenite resistance in Arabidopsis. *Plant Physiology. 146*(3), 1219–1230.

212. Grant, K., Carey, N. M., Mendoza, M., Schulze, J., Pilon, M., Pilon-Smits, E. A. H., & Van Hoewyk, D. (2011). Adenosine 5'-phosphosulfate reductase (APR2) mutation in Arabidopsis implicates glutathione deficiency in selenate toxicity. *Biochemical Journal. 438*(2), 325–335.

213. Terry, N., Zayed, A. M., De Souza, M. P., & Tarun, A. S. (2000). Selenium in higher plants. *Annual Review of Plant Biology. 51*(1), 401–432.

214. Neuhierl, B., & Bock, A. (1996). On the mechanism of selenium tolerance in selenium accumulating plants. *European Journal of Biochemistry. 239*(1), 235–238.

215. Ellis, D. R., Sors, T. G., Brunk, D. G., Albrecht, C., Orser, C., Lahner, B., Wood, K. V., Harris, H. H., Pickering, I. J., & Salt, D. E. (2004). Production of Se-methylselenocysteine in transgenic plants expressing selenocysteine methyltransferase. BMC *Plant Biology. 4*(1), 1.

216. LeDuc, D. L., Tarun, A. S., Montes-Bayon, M., Meija, J., Malit, M. F., Wu, C. P., AbdelSamie, M., Chiang, C.-Y., Tagmount, A., & Neuhierl, B. (2004). Overexpression of selenocysteine methyltransferase in Arabidopsis and Indian mustard increases selenium tolerance and accumulation. *Plant Physiology. 135*(1), 377–383.

217. Lomax, C., Liu, W. J., Wu, L., Xue, K., Xiong, J., Zhou, J., McGrath, S. P., Meharg, A. A., Miller, A. J., & Zhao, F. J. (2012). Methylated arsenic species in plants originate from soil microorganisms. *New Phytologist. 193*(3), 665–672.

218. Miwa, K., Takano, J., & Fujiwara, T. (2006). Improvement of seed yields under boronâ-limiting conditions through overexpression of BOR1, a boron transporter for xylem loading, in Arabidopsis thaliana. *The Plant Journal. 46*(6), 1084–1091.

219. Miwa, K., Takano, J., Omori, H., Seki, M., Shinozaki, K., & Fujiwara, T. (2007). Plants tolerant of high boron levels. *Science. 318*(5855), 1417–1417.

220. Schnurbusch, T., Hayes, J., Hrmova, M., Baumann, U., Ramesh, S. A., Tyerman, S. D., Langridge, P., & Sutton, T. (2010). Boron toxicity tolerance in barley through reduced expression of the multifunctional aquaporin HvNIP2; 1. *Plant Physiology. 153*(4), 1706–1715.

221. Pang, Y., Li, L., Ren, F., Lu, P., Wei, P., Cai, J., Xin, L., Zhang, J., Chen, J., & Wang, X. (2010). Overexpression of the tonoplast aquaporin AtTIP5; 1 conferred tolerance to boron toxicity in Arabidopsis. *Journal of Genetics and Genomics. 37*(6), 389–397.

222. Ochiai, K., Shimizu, A., Okumoto, Y., Fujiwara, T., & Matoh, T. (2011). Suppression of a NAC-like transcription factor gene improves boron-toxicity tolerance in rice. *Plant Physiology. 156*(3), 1457–1463.

223. Miwa, K., Wakuta, S., Takada, S., Ide, K., Takano, J., Naito, S., Omori, H., Matsunaga, T., & Fujiwara, T. (2013). Roles of BOR2, a boron exporter, in cross-linking of rhamnogalacturonan II and root elongation under boron limitation in Arabidopsis. *Plant Physiology. 163*(4), 1699–1709.

224. Chen, Y., Xu, W., Shen, H., Yan, H., Xu, W., He, Z., & Ma, M. (2013). Engineering arsenic tolerance and hyperaccumulation in plants for phytoremediation by a PvACR3 transgenic approach. *Environmental Science and Technology. 47*(16), 9355–9362.

225. Ali, W., Isner, J. C., Isayenkov, S. V., Liu, W., Zhao, F. J., & Maathuis, F. J. M. (2012). Heterologous expression of the yeast arsenite efflux system ACR3 improves *Arabidopsis thaliana* tolerance to arsenic stress. *New Phytologist. 194*(3), 716–723.

226. Dhankher, O. P., Li, Y., Rosen, B. P., Shi, J., Salt, D., Senecoff, J. F., Sashti, N. A., & Meagher, R. B. (2002). Engineering tolerance and hyperaccumulation of arsenic in plants by combining arsenate reductase and γ-glutamylcysteine synthetase expression. *Nature Biotechnology. 20*(11), 1140–1145.

227. Lee, B. D., & Hwang, S. (2015). Tobacco phytochelatin synthase (NtPCS1) plays important roles in cadmium and arsenic tolerance and in early plant development in tobacco. *Plant Biotechnology Reports. 9*(3), 107–114.

228. Guo, J., Xu, W., & Ma, M. (2012). The assembly of metals chelation by thiols and vacuolar compartmentalization conferred increased tolerance to and accumulation of cadmium and arsenic in transgenic *Arabidopsis thaliana*. *Journal of Hazardous Materials. 199*, 309–313.

229. de Souza, M. P., Lytle, C. M., Mulholland, M. M., Otte, M. L., & Terry, N. (2000). Selenium assimilation and volatilization from dimethylselenoniopropionate by Indian mustard. *Plant Physiology. 122*(4), 1281–1288.

230. Pilon-Smits, E. A. H., Hwang, S., Lytle, C. M., Zhu, Y., Tai, J. C., Bravo, R. C., Chen, Y., Leustek, T., & Terry, N. (1999). Overexpression of ATP sulfurylase in Indian mustard leads to increased selenate uptake, reduction, and tolerance. *Plant Physiology. 119*(1), 123–132.

231. Pilon-Smits, E. A. H., Zhu, Y. L., Sears, T., & Terry, N. (2000). Overexpression of glutathione reductase in *Brassica juncea*, effects on cadmium accumulation and tolerance. *Physiologia Plantarum. 110*(4), 455–460.

232. Pilon Pilon, M., Owen, J. D., Garifullina, G. F., Kurihara, T., Mihara, H., Esaki, N., & Pilon-Smits, E. A. H. (2003). Enhanced selenium tolerance and accumulation in transgenic Arabidopsis expressing a mouse selenocysteine lyase. *Plant Physiology. 131*(3), 1250–1257.

233. Tang, Z., Lv, Y., Chen, F., Zhang, W., Rosen, B. P., & Zhao, F. J. (2016). Arsenic methylation in Arabidopsis thaliana expressing an algal arsenite methyltransferase gene increases arsenic phytotoxicity. *Journal of Agricultural and Food Chemistry. 64*(13), 2674–2681.

234. Verma, S., Verma, P. K., Pande, V., Tripathi, R. D., & Chakrabarty, D. (2016). Transgenic Arabidopsis thaliana expressing fungal arsenic methyltransferase gene (WaarsM) showed enhanced arsenic tolerance via volatilization. *Environmental and Experimental Botany. 132*, 113–120.

235. Tagmount, A., Berken, A., & Terry, N. (2002). An essential role of S-adenosyl-l-methionine, l-methionineS-methyltransferase in selenium volatilization by plants. Methylation of selenomethionine to selenium-methyl-l-selenium-methionine, the precursor of volatile selenium. *Plant Physiology. 130*(2), 847–856.

236. Zhou, Y. J., Zhang, S. P., Liu, C. W., & Cai, Y. Q. (2009). The protection of selenium on ROS mediated-apoptosis by mitochondria dysfunction in cadmium-induced LLC-PK 1 cells. *Toxicology in Vitro. 23*(2), 288–294.

237. Sundaram, S., Wu, S., Ma, L. Q., & Rathinasabapathi, B. (2009). Expression of a Pteris vittata glutaredoxin PvGRX5 in transgenic Arabidopsis thaliana increases plant arsenic tolerance and decreases arsenic accumulation in the leaves. *Plant, Cell and Environment. 32*(7), 851–858.

238. Van Huysen, T., Abdel-Ghany, S., Hale, K. L., LeDuc, D., Terry, N., & Pilon-Smits, E. A. H. (2003). Overexpression of cystathionine-Î-γ--synthase enhances selenium volatilization in *Brassica juncea*. *Planta. 218*(1), 71–78.

CHAPTER 8

MORPHO-PHYSIOLOGICAL RESPONSES OF WHEAT GENOTYPES UNDER HIGH TEMPERATURE STRESS

SHARAD K. DWIVEDI, SANTOSH KUMAR, and K. K. RAO

ICAR Research Complex for Eastern Region, Patna – 80 (0014). India, E-mail: sharad.dwivedi9736@gmail.com

CONTENTS

ABSTRACT

Delayed harvesting of rice forced late sowing of wheat particularly in Indo-Gangetic plain that causes severe yield penalty due to terminal heat stress. Terminal heat causes series of changes in plant system in term of physiology, growth, and yield. Various studies have indicated that the average maximum temperature more than 32°C during reproductive phase negatively influenced wheat grain yield and average yield loss upto 30% was reported. Physiological and biochemical traits like relative water content (RWC), membrane stability index (MSI), photosynthetic rate, chlorophyll content, various osmolytes, anti-oxidants, and some molecular chaperones (HSPs) showed variations under high temperature condition. Moreover, genotypic variation is obvious in all traits regarding thermal susceptibility. Stay green and canopy temperature depression has been found to be significant co-relation with grain yield. Terminal heat stress led to reduced grain filling duration, thousand-grain weight, grain number per year, and ultimately final yield. Furthermore, elevated temperature driven interruption in the transport of photosynthate from green foliage (source) to anther tissues (sink) leads to high pollen mortality and thereby decreases grain yield. Some wheat genotypes like NW (1014). Halna, Raj (3765). WH760, GW273 and HD2987 has been found to be suitable for late sown terminal heat stress condition and produce optimum yield.

8.1   INTRODUCTION

Production of food grains in modern decade is not keeping pace with growing population demand, which in turn leads to inflation and a risk to food and nutritional security in India and other developing countries. Furthermore, the spreading of urbanization has forced agriculture into more harsh situations and marginal lands, while the global food requirements has been projected to increase by 70% by the end of (2050). necessitate improvement in agricultural productivity with a lesser amount of resources like land and water [1]. Wheat is one of the staple food crops and the cheapest source of carbohydrate and proteins in most part of the world. It is grown on approximately 30% of cereal area of the world and around 220 million hectares of global wheat cultivating area suffers from high temperature stress [2]. South East Asia comprising of India, Bangladesh, Nepal, and Pakistan are the most populous regions of the world with around 1.5 billion population [3]. Together with rice, wheat is also a primary food crop of this region and therefore, it is of vital significance for food security of these developing nations [4]. Around 36 million ha (approx. 16% of the global wheat area) South Asia is under wheat cultivation, which contribute around 15% of the world's wheat production [5]. The present wheat production in South Asia is around 98 million tons but the projected demand till 2020 has been estimated to be around 137 million tons. The demand of wheat is rising but there is no additional increase in cultivating land chiefly by reason of growing urbanization and diversification [6].

In most parts of the world, global climate change due to rising ambient temperature is considered as one of the most negative actors for agricultural productivity. Global temperature is expected to be increased by 3 to 5°C by the end of this century [7]. Temperature accelerates the developmental process in plants leading to the induction of earlier senescence and shortens the growth cycle [8]. Terminal heat stress is a key abiotic stress severely affecting wheat growth and yield [4, 9]. A major part of wheat cultivation in South East Asia including India has been found to be under threat of high temperature stress [4]. Heat stress is more prevalent in Eastern Indo-Gangetic Plains (EIGP), central and peninsular India, and Bangladesh; and moderate in north western parts of IGP. Delayed sowing of wheat due to late harvesting of rice is one of the main reasons for terminal heat stress in eastern part of India. In India, wheat is grown under subtropical environment during mild winters, which warms up towards the grain filling stage of the crop. North-West Plain Zone (NWPZ) contributes about 80% of total wheat production. Forced crop maturity and yield reduction of wheat in NWPZ is due to high temperature in the months of February and March, which causes stressful conditions for growth. The problem of heat stress is likely turn out to be even worse in near future under global climate change, which has become one of the utmost challenges that humanity will face to feed the growing population. Lower yields are obtained in dry and semi-dry environments as a result of continual rise in temperature that coincide during anthesis and grain filling periods of crops [9, 10].

Hence, development and identification of terminal heat stress tolerance wheat germplasm may be a noble strategy to resolve the imminent crucial problem caused by global warming. Besides, it is crucial to develop genotypes that are early in maturity so as to escape the terminal heat stress [4]. In recent decade, Indian wheat program has released a few varieties having moderate level of heat tolerance, viz., RAJ (3765). UP (2425). and DBW 16 for the North Western Plain Zone and, NW (1014). Halna, WH760, HD (2987). HD (2643). HP (1744). DBW 14, and NW 2036 for the North Eastern Plain Zone of India.

8.2   MORPHO-PHYSIOLOGICAL TRAITS INFLUENCED BY HIGH TEMPERATURE

Late sowing of wheat in India particularly in EIGP is quite common practice due to late harvesting of long duration rice varieties. High temperature induces considerable variations in gas exchange processes and membrane thermo stability. Disturbance of protection system under high temperature stress caused a alteration in normal carbon metabolic process, which in turn negatively regulate starch granule deposition in developing endosperm. Studies also show that rise in temperature causes metabolic changes in wheat senescence programme [11]. Furthermore, heat stress also lead to the inhibition of chlorophyll biosynthesis which trigger the senescence programme [12]. Heat stress during terminal phase of the crop inhibits the starch biosynthesis which in turn reduces the normal grain size. Several studies [13, 14] suggested that the estimation of membrane stability in terms of ion leakage as act as an index for screening wheat genotypes against heat. Apart from these, production of active oxygen species (AOS) is another criteria to measure the impact of heat stress [15]. However, plants have developed a chain of scavenging mechanisms that convert highly AOS to H_2O under heat stress conditions [16]. The protection mechanism at cellular and sub-cellular level is manifested by various anti-oxidents like superoxide dismutase, ascorbate peroxidase, glutathione reductase, catalase, and metabolites like glutathione, ascorbic acid, a-tocopherol, and carotenoids produce regularly inside plant system [17, 18].

8.2.1   MORPHO-ANATOMICAL AND PHENOLOGICAL RESPONSES

High temperature alters plant morphology and phenology in terms of leaf area, plant height, and pattern of plant development. Previous studies showed that long-term impact of high temperature stress on growing seeds causes' poor germination and vigor; in turn affect emergence and seedling establishment. The harmful effect of heat stress has also been observed in terms of scorching of leaves and twigs, leaf senescence, lesser canopy growth and poor yield [19]. One of the severe effects of high temperature is the premature death of plants [20]. Moreover, few informations are available regarding the anatomical changes of plant system under high temperatures. Reduction in cell size, partial closure of stomata for restriction of water loss and increased number of xylem vessels in root and shoot has been observed at whole plant level in response to heat stress [21]. At the sub-cellular level, significant alterations occur in chloroplast structure that causes variation in normal photosynthetic process. Heat stress negatively regulates the stacking of grana and structural organization of thylakoids [22]. The combined effects of all these alterations due to heat stress may result in poor canopy growth, physiological functions and plant productivity. Further depiction of variations in terms of plant phenology arise due to high temperature stress provides a clue for better understandings of interaction between the plant and its surrounding atmosphere. Different stages of plant growth varies in their sensitivity to rise in temperature; however, it is species and genotypic specific [23]. Moreover, reproductive stage is highly vulnerable to heat stress. During reproduction, a spell of heat stress lead to the pollen sterility abortion of open flower and this may vary between species [24–26].

8.2.2   PHYSIOLOGICAL RESPONSES UNDER HEAT STRESS

Physiological responses of wheat to terminal heat stress have been found to be well determined by genotype resistance or susceptibility [27]. Increase in temperature due to late sowing significantly decreases leaf ascorbic acid content, relative water content (RWC) and lipid peroxidation in wheat after anthesis[17].

8.2.2.1   Waters Relations and Heat Stress

Plant water status is one of the important parameter under changing temperatures. In general, with availability of sufficient moisture plants try to maintain tissue water status regardless of temperature; yet, water status is severely hampered due to scarcity of water under high temperatures [28]. Ref. [29] has shown that high temperature stress leads to reduced water availability under field conditions. In general, during day time high rate of transpiration in plants creates water insufficiency, causing a lower water potential and leading to disturbances of many physiological processes [30]. RWC of wheat cultivars were studied in leaves under normal and late sowing conditions at ICAR-RCER, Patna during the dry seasons of (2014). (2015). and 2016. There was significant reduction in RWC (10.3 and 13.7%) across the cultivars under late and very late sown situations as compared to the normal sown condition.

8.2.2.2   Accumulation of Compatible Osmolytes Under Heat Stress

A major adaptive mechanism to extreme temperatures is accumulation of certain low molecular weight organic compounds called osmolytes like sugars, polyols, proline, glycine betaine (GB), and tertiary sulphonium [31, 32]. Accumulation of such osmolytes provide tolerance to the plant under high temperature stress. One important osmolyte GB, plays an significant role in plants under salinity or high temperature [31]. Another compatible solute proline also act as an osmoprotectant and widely reported in various plant in response to abiotic stresses [33]. GB and proline act as a buffering agent thus buffer cellular redox potential under high temperature stress [34].

8.2.2.3   Photosynthetic Characteristics Under Heat Stress

Alterations in various gas exchange traits under high temperature stress are important indicators for thermo-tolerance of the plant as they show correlation with plant growth. Any obstacle in photosynthetic process can limit plant growth under heat stress. Ref. [35] reported that photochemical reactions in thylakoid lamellae and carbon metabolism in the stroma of chloroplast are the main sites of heat injury. Chlorophyll fluorescence has been shown to be correlated with thermo-tolerance [36]. Under high temperatures photo system II activity is greatly reduced as it is highly thermo-sensitive, [37]. Heat stress modifies the energy distribution and also the carbon metabolism enzymes activities, mainly the rubisco, thereby altering the rate of RuBP regeneration through disruption of electron transport and inactivation of the oxygen evolving enzymes of PSII [38]. Moreover, concentration of photosynthetic pigments [39], different soluble proteins and rubisco binding proteins (RBP) decreases due to heat shock. Heat stress also affects large and small sub units of rubisco in darkness however induces them in light, indicating their functions as chaperones and heat shock proteins (HSP) [40]. Further, the starch or sucrose synthesizes greatly influenced by heat stress as noticed from decreased activities of sucrose phosphate synthase [41], ADP glucose pyrophosphorylase and invertase [42]. A familiar impact of increasing temperature in plants is the damage caused by heat-induced imbalance in photosynthesis is and respiration generally the photosynthesis rate decreases while dark and photo-respiration rates increases significantly under rising temperatures. In addition, it has been determined that the assimilation rate of photosynthetic CO_2 is not much influenced by heat stress in developing leaves than in completely developed leaves.

Heat-exposed plants exhibit a collective decrease in chlorophyll content. The chlorophyll degradation in plants may be attributed to the over accumulation of ROS in plants due to high temperature, which in turn, results in down-regulation of photosynthetic rate. Alternatively, reduction in photosynthetic rate can also been explained through restrictions in stomatal and non-stomatal conductance under high temperature stress [43]. According to Ref. [44] limitation in non-stomatal conductance may be induced by the ROS-induced membrane injury as depicted from the membrane stability index. Transpiration rate can also be correlated with stomatal conductance and unavailability of enough water due to lower conductance may be a reason for diminished transpiration.

8.2.2.4 Assimilate Partitioning Inside Plant System Under Heat Stress

The activity of source and sink was highly reduced under high temperature stress leading to the economic loss in terms of grain yield. Assimilate partitioning in plant system take place via two major modes, i.e., symplastic and apoplastic pathways, and these plays significant role in assimilate transfer and partitioning under high temperature stress [45]. However, lot of variation exist among wheat genotypes for assimilate partitioning under high temperature stress [46]. Ref. [47] reported that three main reasons for lower grain filling rate in wheat under heat stress are source (flag leaf blade), sink (ear), and transport pathway (peduncle). The transport and portioning of assimilates (phloem loading) is optimum upto 30°C, however, the moment through the shoot is independent of temperature. In case of wheat, the transport and mobilization of assimilates is highly temperature dependent. It suggests that improved mobilization efficiency from the source to the sink act as a key strategy for better grain filling and yield in wheat under high temperature stress. However, limited knowledge is available on assimilate partitioning under heat stress thus in depth study is required to improve production efficiency of the crop.

8.2.2.5   Cell Membrane Thermostability and Heat Stress

Persistent role of cellular membranes under different stresses is critical for photosynthetic and respiration mechanisms [48]. Heat stress hastens the kinetic energy and movement of molecules across membranes which make the lipid bilayer of membranes further fluid by either proteins denaturation or more unsaturated fatty acids [49]. Such modifications increase the membrane permeability, as evident from increased loss of electrolytes. The decreased cell membrane thermo-stability (CMT) indicated by increased solute leakage, has long been used as an indirect measurement of high temperature tolerance in various plant species including wheat [50].

8.2.2.6   Hormonal Changes Inside Plant System Under Heat Stress

Plants have the capability to sense and acclimatize to unfavorable climatic conditions, though the degree of adaptability or tolerance differs across the species and genotypes. Hormonal regulation plays a significant role in this aspect. It was observed that the hormonal homeostasis, stability, content, biosynthesis, and compartmentalization are altered under heat stress [51]. Different environmental stresses including high temperature resulted in increased levels of ABA. ABA helps in adaptation of plants to desiccation by modulating the regulation of numerous abiotic genes [52]. Ref. [51] reported that the induction of ABA is an important component of thermotolerance. Other studies also suggest that ABA-mediated induction of several HSPs (e.g., HSP70) may be probable mechanism for conferring thermo-tolerance [53].

Among other plant hormones, salicylic acid (SA) is also involved in high temperature responses evoke by plants. SA act as stabilizer for the heat shock transcription factors (trimers) and aids them in binding heat shock elements to the promoter of heat shock related genes. Ca^{2+} homeostasis and antioxidant systems are thought to be involved in the long-term thermo-tolerance induced by SA [54]. In dwarf wheat variety, heat stress-induced decrease in cytokinin content has been found to be accountable for decreased grain filling and its dry weight [55]. The possible functions of other phytohormones in plant heat-tolerance are yet to be revealed.

8.2.3   OXIDATIVE STRESS AND ANTIOXIDANTS POTENTIAL OF PLANT SYSTEM UNDER HEAT STRESS

Oxidative stress also known as secondary stress arises as a consequence of different kind of abiotic stresses. Heat stress induces oxidative stress along with dehydration of tissue. Under heat stress various kind of AOS like singlet oxygen ($^1O_2$), superoxide radical (O^{2-}), hydroxyl radical (OH^-), and hydrogen peroxide (H_2O_2) are produced in plant system, which causes cellular injury [56]. Reactive oxygen species induces peroxidation of lipid which can easily be estimated in terms of malondialdehyde (MDA) or 2-thiobarbituric acid reactive substances (TBARS) content, thus diminishing the semi-permeability of cellular membrane and altering its functions [57]. For scavenging ROS, plant produces various anti-oxidants like catalase (CAT), super oxide dismutase (SOD) and ascorperoxidase (APX). The scavenging of O^{2-} by superoxide dismutase (SOD) results in the production of H_2O_2, which is removed by APX or CAT. Indeed further research is needed to detect the signaling molecules, which improve the production of antioxidants in cells subjected to high temperature.

8.2.3.1   Role of Stress Proteins Under Heat Stress

Plant synthesizes various stress proteins in response to abiotic stresses, which performs a vital role in survival mechanism of the plants [34].

Further, HSPs are entirely involved in high temperature response; certain other proteins also play significant role in this background. It has been found that with sudden or steady raise in temperature plants experience the increased production of HSPs [58]. HSP-triggered thermo-tolerance by (a) induction coincides with the organism under stress; (b) HSPs biosynthesis extremely rapid and intensive. Particular HSPs have been identified in response to raising temperatures, in different crop species. There is substantial evidence that attainment of heat-tolerance is directly related to the synthesis and accumulation of HSPs [59].

8.3   CRITERIA FOR SCREENING HEAT TOLERANT WHEAT GENOTYPES

Various criteria have been reported by many researchers to identify heat tolerant wheat genotypes. Traits like heat susceptibility index (HSI) [60], membrane thermo-stability [61] canopy temperature depression (CTD) [61] chlorophyll content; the normalized difference vegetation index, stay-green trait [62] and stomatal conductance [61] have been reported as the marker traits to differentiate heat susceptible and tolerant wheat genotypes. CTD is considered to be the most efficient to assess heat tolerance since one single reading integrates scores of leaves [61], CTD is highly heritable and easy to measure using a hand-held infrared thermometer on sunny days [61]. Even though an association between the stay green trait and yield and yield traits has been reported in various crops, published studies on a possible association between the stay green trait and CTD in different crops are scarce. Under hot, irrigated conditions infrared thermometry has been successfully used to screen wheat genotypes for their performance by measuring the difference between canopy and air temperature [61]. Ref. [63] reported yield loss upto 3 to 5% for 1°C rise in temperature which also influences physiology, growth and yield traits. High temperature at anthesis decreases the grain number per spike [64] and grain size [65], both of which have significant effects on grain yield. The grain yield affected by decreasing size of individual grains due to high temperature at the grain filling stage. Ref. [66] reported that in wheat, both number of grains and grain weight seems to be sensitive to heat stress, as at maturity there is a decline in the number of grains per year with rising temperature. Reproductive processes are clearly affected by high temperatures in most plants, which eventually affect fertilization and post-fertilization processes leading to reduced crop yield. Recommended wheat cultivars for sowing under delayed sowings of the Indo-Gangetic Plains are PBW 373, UP (2425). and RAJ 3765 for the NWPZ and NW (1014). HD (2643). HUW 510, HUW 234, HW (2045). DBW 14, NW (2036). and HP 1744 for the NEPZ. Although the release of a massive amount of varieties over the last two decades, HUW 234 (released in 1986) is still the dominant variety in the EIGP [4]. A key reason behind this success is the wider adaptation of this variety to low resource environments prevalent in eastern India together with its ability to perform well under abiotic stresses such as heat, limited irrigation, and variable nutrient doses [67].

8.4 MECHANISM OF HEAT TOLERANCE

Heat tolerance mechanism can be understood by investigating the physiological responses of tolerant and susceptible genotypes at various stages of plant development, especially during grain filling stage of wheat. Plants adopt various mechanisms to survive under high temperature condition, *viz.*, phonological and morphological adaptations and avoidance or acclimation responses such as changing leaf angle, cooling through transpirational system, or altering in membrane lipid compositions. As per the studies, smaller yield losses in different plants correlated with early maturity under high temperature which indicates one of the escape mechanisms [68]. Plants experience many types of environmental stresses at different growth stages and their mechanisms of response may vary at tissue level [69]. The initial stress signals in terms of ionic effects or membrane composition triggers downstream signaling processes and transcriptional control, which led to the activation of stress-responsive mechanisms and create homeostasis through protecting and repairing damaged proteins and membranes [70]. Several key tolerance mechanisms, including osmolyte accumulation and compartmentalization, ROS scavengers, late embryogenesis abundant proteins and factors involved in signaling process and gene level regulation are major drivers to counter-act the heat stress effect [71]. The tolerance process begins with sensing of heat stress, their signaling and production of many metabolites that enable the plant to counteract the ill effect of high temperature stress. The ROS scavengers like CAT, SOD, POX, APX, and ascorbic acid are also the important players in tolerance mechanism of heat stress [51]. Further-more, at molecular level the induction and expression of HSPs is highly correlated with the thermo tolerance mechanism of the plant. HSPs act as a molecular chaperone and provide protection to the cellular machinery. Many studies pointed out the role of HSPs in various stress responsive mechanisms [71].

8.5   CONCLUSION

From the various studies it is clear that average maximum temperature more than 32°C during grain filling negatively influenced wheat grain yield. Heat stress caused series of physiological and biochemical changes in traits like RWC, membrane stability index (MSI), photosynthetic rate, chlorophyll content, various osmolytes, anti-oxidants and some molecular chaperones (HSPs) were highly influenced. However, wheat genotypes like NW (1014). Halna, Raj (3765). WH760, GW273, and HD2987 has been found to be promising for late sown terminal heat stress condition and produce optimum yield.

KEYWORDS

- **anti-oxidants**
- **grain yield**
- **heat stress**
- **HSPs**
- **physiological changes**
- **wheat**

REFERENCES

1. Fischer, R. A., Byerlee, D., & Edmcades, G. O. (2014). Crop yields and global food security: will yield increase continue to feed the world? *ACIAR Monograph No. 158.* Australian Centre for International Agricultural Research, Canberra, 634 pp.

2. Cossani, C. M., & Reynolds, M. P. (2012). Physiological traits for improving heat tolerance in wheat. *Plant Physiol. 160,* 1710–1718. http://dx.doi.org/10.1104/pp.112.207753.

3. Witcombe, J. R., & Virk, D. S. (2001). Number of crosses and population size for participatory and classical plant breeding. *Euphytica. 122,* 451–462.

4. Joshi, A. K., Chand, R., Arun, B., Singh, R. P., & Ortiz, R. (2007a). Breeding crops for reduced-tillage management in the intensive, rice-wheat systems of south Asia. *Euphytica. 153,* 135–151

5. FAO (2007). Statistical database. http://www.faostat.fao.org.

6. Anonymous (2007). Vision 2025. Directorate of Wheat Research, Indian Council of Agricultural Research, Karnal, India.

7. IPCC. (2014). Climate change 2014: impacts, adaptation, and vulnerability. Part A: global and sectoral aspects. Contribution of working group II to the fifth assessment report of the inter-governmental panel on climate change. Cambridge University Press, Cambridge.

8. Bita, C. E., & Gerats, T. (2013). Plant tolerance to high temperature in a changing environment: scientific fundamentals and production of heat stress-tolerant crops. *Front. Plant Sci. 4,* 1–18.

9. Dwivedi, S. K., Kumar, S., & Prakash, V. (2015). Effect of late sowing on yield and yield attributes of wheat genotypes in Eastern Indo Gangetic Plains. *Journal of Agrisearch. 2*(4), 304–306.

10. García del Moral, L. F., Rharrabti, Y., Villegas, D., & Royo C. (2003). Evaluation of grain yield and its components in durum wheat under Mediterranean conditions: an ontogenic approach. *Agron. J. 95,* 266–274

11. Farooq, M., Bramley, H., Palta, J. A., & Siddique, K. H. M. (2011). Heat stress in wheat during reproductive and grain-filling phases. *Critical Reviews in Plant Sciences. 30,* 1–17.

12. Gupta, N. K., Gupta, S., & Kumar, A. (2000). Cytokinin application increases cell membrane and chlorophyll stability in wheat (*Triticum aestuvum* L.). *Cereal Research Communication. 28,* 287–291.

13. Kushwaha, S. R., Deshmukh, P. S., Sairam, R. K., & Singh, M. K. (2011). Effect of high temperature stress on growth, biomass and yield of wheat genotypes. *Indian Journal of Plant Physiology. 16,* 93–97.

14. Deshmukh, P. S., Sairam, R. K., & Shukla, D. S. (1991). Measurement of ion leakage as a screening technique for drought resistance in wheat genotypes. *Indian Journal of Plant Physiology. 34,* 89–91.

15. Liu, X., & Huang, B. (2000). Heat stress injury in relation to membrane lipid peroxidation in creeping bentgrass. *Crop Science. 40,* 503–510

16. Larkindale, J., & Huang, B. (2004). Thermo-tolerance and antioxidant systems in Agrostisstoloifera: involvement of salicylic acid, abscisic acid, calcium, hydrogen peroxide, and ethylene. *Journal of Plant Physiology. 161,* 405–413.

17. Sairam, R. K., Srivastava, G. C., & Saxena, D. C. (2000). Increased antioxidant activity under elevated temperature: a mechanism of heat stress tolerance in wheat genotypes. *Biologia Plantarum. 43*, 245–251.

18. Bansal, R., & Srivastava, J. P. (2012). Antiodative defense system in pigeon-pea roots under water logging stress. *Acta Physiologiae Plantarum. 34*, 515–522.

19. Vollenweider, P., & Gunthardt-Goerg, M. S. (2005). Diagnosis of abiotic and biotic stress factors using the visible symptoms in foliage. *Environ. Pollut. 137*, 455–465.

20. Hall, A. E. (1992). Breeding for heat tolerance. *Plant Breed. Rev. 10*, 129–168.

21. Anon, S., Fernandez, J. A., Franco, J. A., Torrecillas, A., Alarcon, J. J., & Sanchez-Blanco, M. J., (1992). Effects of water stress and night temperature precondition-ing on water relations and morphological and anatomical changes of Lotus creticus plants. *Sci. Hortic. 101*, 333–342.

22. Karim, M. A., Fracheboud, Y., & Stamp, P. (1997). Heat tolerance of maize with reference of some physiological characteristics. *Ann. Bangladesh Agri. 7*, 27–33.

23. Howarth, C. J. (2005). Genetic improvements of tolerance to high temperature. In: Ashraf, M., Harris, P. J. C. (Eds.), Abiotic Stresses: Plant Resistance Through Breed-ing and Molecular Approaches. Howarth Press Inc., New York. pp. 277–300.

24. Guilioni, L., Wery, J., & Tardieu, F. (1997). Heat stress-induced abortion of buds and flowers in pea: is sensitivity linked to organ age or to relations between reproduc-tive organs? *Ann. Bot. 80*, 159–168.

25. Young, L. W.,Wilen, R. W., & Bonham-Smith, P. C. (2004). High temperature stress of *Brassica napus* during flowering reduces micro- and mega gametophyte fertility, induces fruit abortion, and disrupts seed production. *J. Exp. Bot. 55*, 485–495.

26. Sato, S., Kamiyama, M., Iwata, T., Makita, N., Furukawa, H., & Ikeda, H. (2006). Moderate increase of mean daily temperature adversely affects fruit set of *Lycopersi-con esculentum* by disrupting specific physiological processes in male reproductive development. *Ann. Bot. 97*, 731–738.

27. Almeselmani, M., Deshmukh, P. S., & Chinnusamy, V. (2012). Effect of prolong high temperature stress on respiration, photosynthesis and gene expression in wheat (*Triticum aestivum* L.) varieties differing in their thermo-tolerance. *Plant Stress. 6*, 25–32.

28. Machado, S., & Paulsen, G. M., (2001). Combined effects of drought and high tem-perature on water relations of wheat and sorghum. *Plant Soil. 233*, 179–187.

29. Simoes-Araujo, J. L., Rumjanek, N. G., & Margis-Pinheiro, M. (2003). Small heat shock proteins genes are differentially expressed in distinct varieties of common bean. *Braz. J. Plant Physiol. 15*, 33–41.

30. Tsukaguchi, T., Kawamitsu, Y., Takeda, H., Suzuki, K., & Egawa, Y. (2003). Water status of flower buds and leaves as affected by high temperature in heat tolerant and heat-sensitive cultivars of snap bean (*Phaseolus vulgaris* L.). *Plant Prod. Sci. 6*, 4–27.

31. Sakamoto, A., & Murata, N. (2002). The role of glycine betaine in the protection of plants from stress: clues from transgenic plants. *Plant Cell Environ. 25*, 163–171.

32. Sairam, R. K., & Tyagi, A. (2004). Physiology and molecular biology of salinity stress tolerance in plants. *Curr. Sci. 86*, 407–421.

33. Kavi Kishore, P. B., Sangam, S., Amrutha, R. N., Laxmi, P. S., Naidu, K. R., Rao, K. R. S. S., Rao, S., Reddy, K. J., Theriappan, P., & Sreenivasulu, N. (2005). Regulation

of proline biosynthesis, degradation, uptake and transport in higher plants: its implications in plant growth and abiotic stress tolerance. *Curr. Sci. 88*, 424–443.

34. Wahid, A., & Close, T. J. (2007). Expression of dehydrins under heat stress and their relationship with water relations of sugarcane leaves. *Biol. Plant. 51*, 104–109.

35. Wise, R. R., Olson, A. J., Schrader, S. M., & Sharkey, T. D. (2004). Electron transport is the functional limitation of photosynthesis in field-grown Pima cotton plants at high temperature. *Plant Cell Environ. 27*, 717–724.

36. Yamada, M., Hidaka, T., & Fukamachi, H. (1996). Heat tolerance in leaves of tropical fruit crops as measured by chlorophyll fluorescence. *Sci. Hortic. 67*, 39–48.

37. Camejo, D., Rodriguez, P., Morales, M. A., Dell'amico, J. M., Torrecillas, A., & Alarcon, J. J. (2005). High temperature effects on photosynthetic activity of two tomato cultivars with different heat susceptibility. *J. Plant Physiol. 162*, 281–289.

38. Salvucci, M. E., & Crafts-Brandner, S. J. (2004b). Inhibition of photosynthesis by heat stress: the activation state of Rubisco as a limiting factor in photosynthesis. *Physiol. Plant. 120*, 179–186.

39. Todorov, D. T., Karanov, E. N., Smith, A. R., & Hall, M. A. (2003). Chlorophyllase activity and chlorophyll content in wild type and *eti 5* mutant of *Arabidopsisthaliana* subjected to low and high temperatures. *Biol. Plant. 46*, 633–636.

40. Kepova, K. D., Holzer, R., Stoilova, L. S., & Feller, U. (2005). Heat stress effects on ribulose-1,5-bisphosphate carboxylase/oxygenase, Rubiscobindind protein and Rubisco activase in wheat leaves. *Biol. Plant. 49*, 521–525.

41. Chaitanya, K. V., Sundar, D., & Reddy, A. R. (2001). Mulberry leaf metabolism under high temperature stress. *Biol. Plant. 44*, 379–384

42. Vu, J. C. V., Gesch, R. W., Pennanen, A. H., Allen, L. H. J., Boote, K. J., & Bowes, G., (2001). Soybean photosynthesis, Rubisco and carbohydrate enzymes function at supra-optimal temperatures in elevated CO_2. *J. Plant Physiol. 158*, 295–307.

43. Misson, L., Limousin, J., Rodriguez, R., & Letts, M. G. (2010). Leaf physiological responses to extreme drought sin Mediterranean *Quercus ilex* forest. *Plant Cell Environ. 33*, 1898–1910.

44. Shao, H., Chu, L., Shao, M., Li, S., & Yao, J. (2008). Bioengineering plant resistance to abiotic stresses by the global calcium signal system. Biotech. *Advances. 26*, 503–510.

45. Taiz, L., & Zeiger, E., (2006). Book: *Plant Physiology 4th edition*. Sinauer Associates Inc. Publishers, Massachusetts, USA.

46. Yang, J., Sears, R. G., Gill, B. S., & Paulsen, G. M. (2002). Genotypic differences in utilization of assimilate sources during maturation of wheat under chronic heat and heat shock stresses. *Euphytica. 125*, 179–188

47. Wardlaw, I. F., (1974). Temperature control of translocation. In: Bielske, R. L., Ferguson, A. R., Cresswell, M. M. (Eds.), *Mechanism of Regulation of Plant Growth*. Bull. Royal Soc., New Zealand, Wellington. pp. 533–538.

48. Dwivedi, S. K., Basu, S., Kumar, S., Kumar, G., Prakash V., Kumar, S., Mishra, J. S., Bhatt, B. P., Malviya, N., Singh, G. P., & Arora, A. (2017). Heat stress induced impairment of starch mobilisation regulatespollen viability and grain yield in wheat: Study in Eastern Indo-Gangetic Plains. *Field Crops Research. 206*, 106–114.

49. Blum, A. (1988). *Plant Breeding for Stress Environments*. CRC Press Inc., Boca Raton, Florida. pp. 223.

50. Savchenko, G. E., Klyuchareva, E. A., Abrabchik, L. M., & Serdyuchenko, E. V. (2002). Effect of periodic heat shock on the membrane system of etioplasts. *Russ. J. Plant Physiol. 49*, 349–359.

51. Blum, A., Klueva, N., & Nguyen, H. T. (2001). Wheat cellular thermo tolerance is related to yield under heat stress. *Euphytica. 117*, 117–123.

52. Maestri, E., Klueva, N., Perrotta, C., Gulli, M., Nguyen, H. T., & Marmiroli, N., (2002). Molecular genetics of heat tolerance and heat shock proteins in cereals. *Plant Mol. Biol. 48*, 667–681.

53. Xiong, L., Lee, H., Ishitani, M., & Zhu, J.-K. (2002). Regulation of osmotic stress responsive gene expression by LOS6/ABA1 locus in *Arabidopsis*. *J. Biol. Chem. 277*, 8588–8596.

54. Pareek, A., Singla, S. L., & Grover, A. (1998). Proteins alterations associated with salinity, desiccation, high and low temperature stresses and abscisic acid application in seedlings of Pusa 169, a high-yielding rice (*Oryza sativa* L.) cultivar. *Curr. Sci. 75*, 1023–1035.

55. Wang, L.-J., & Li, S.-L. (2006b). Salicylic acid-induced heat or cold tolerance in relation to Ca^{2+} homeostasis and antioxidant systems in young grape plants. *Plant Sci., 170*, 685–694.

56. Banowetz, G. M., Ammar, K., & Chen, D. D. (1999). Temperature effects on cytokinin accumulation and kernel mass in dwarf wheat. *Ann. Bot. 83*, 303–307.

57. Liu, X., & Huang, B. (2000). Heat stress injury in relation to membrane lipid peroxidation in creeping bent grass. *Crop Sci. 40*, 503–510.

58. Xu, S., Li, J., Zhang, X., Wei, H., & Cui, L. (2006). Effects of heat acclimation pretreatment on changes of membrane lipid peroxidation, antioxidant metabolites, and ultrastructure of chloroplasts in two cool-season turfgrass species under heat stress. *Environ. Exp. Bot. 56*, 274–285.

59. Nakamoto, H., & Hiyama, T. (1999). Heat-shock proteins and temperature stress. In: Pessarakli, M. (Ed.), *Handbook of Plant and Crop Stress*. Marcel Dekker, New York, pp. 399–416.

60. Bowen, J., Michael, L.-Y., Plummer, K. I. M., & Ferguson, I. A. N. (2002). The heat shock response is involved in thermotolerance in suspension-cultured apple fruit cells. *J. Plant Physiol. 159*, 599–606.

61. Mason, R. E., Mondal, S., Beecher, F. W., Pacheco, A., Jampala, B., Ibrahim, A. M. H., & Hays, D. B. (2010). QTL associated with heat susceptibility index in wheat (*Triticumaestivum,* L.) under short-term reproductive stage heat stress. *Euphytica. 174*, 23–436.

62. Reynolds, M. P., Balota, M., Delgado, M. I. B., Amani, I., & Fischer, R. A. (1994). Physiological and morphological traits associated with spring wheat 1125 yield under hot, irrigated conditions. *Aust. J. Plant Physiol. 21*, 717–730.

63. Harris, K., Subudhi, P. K., Borrell, A., Jordan, D., Rosenow, D., Nguyen, H. T., Klein, P., Klein, R., & Mullet, J. (2007). Sorghum stay-green QTL individually reduce post-flowering drought-induced leaf senescence. *J. Exp. Bot. 58*, 327–338.

64. Gibson, L. R., & Paulsen, G. M. (1999). Yield components of wheat grown under high temperature stress during reproductive growth. *Crop Sci. 39*, 1841–1846.

65. Prasad, P. V. V., Pisipati, S. R., Ristic, Z., Bukovnik, U., & Fritz, A. K. (2008a). Impact of nighttime temperature on physiology and growth of spring wheat. *Crop Sci. 48,* 2372–2380.

66. Viswanathan, C., & Khanna-Chopra, R. (2001). Effect of heat stress on grain growth, starch synthesis and protein synthesis in grains of wheat (*Triticumaestivum* L.) varieties differing in grain weight stability. *J. Agron. Crop Sci. 186,* 1–7.

67. Ferris, R., Ellis, R. H., Wheeler, T. R., & Hadley, P. (1998). Effect of high temperature stress at anthesis on grain yield and biomass of field-grown crops of wheat. *Ann. Bot. 82,* 631–639.

68. Arun, B., Joshi, A. K., Chand, R., & Singh, D. (2003). Wheat somaclonal variants showing, earliness, improved spot blotch resistance and higher yield. *Euphytica. 132,* 235–241.

69. Adams, S. R., Cockshull, K. E., & Cave, C. R. J. (2001). Effect of temperature on the growth and development of tomato fruits. *Ann. Bot. 88,* 869–877.

70. Queitsch, C., Hong, S. W., Vierling, E., & Lindquest, S. (2000). Heat shock protein 101 plays a crucial role in thermotolerance in *Arabidopsis. Plant Cell. 12,* 479–492.

71. Bohnert, H. J., Gong, Q., Li, P., & Ma, S. (2006). Unraveling abiotic stress tolerance mechanisms—getting genomics going. *Curr. Opin. Plant Biol. 9,* 180–188.

72. Wang, W., Vinocur, B., Shoseyov, O., & Altman, A. (2004). Role of plant heat-shock proteins and molecular chaperones in the abiotic stress response. *Trends Plant Sci. 9,* 244–252.

CHAPTER 9

CONTROL OF MICRORNA BIOGENESIS AND ACTION IN PLANTS

MOHD. ZAHID RIZVI

Department of Botany, Shia Post Graduate College,
Sitapur Road, Lucknow – 226020, Uttar Pradesh, India,
E-mail: zahid682001@gmail.com

CONTENTS

ABSTRACT

MicroRNAs are endogenous, noncoding, regulatory small RNA molecules in the size range of 21–24 nucleotides that regulate gene expression. miRNAs are processed from single stranded precursors containing stem-loop structures, by the endonuclease activity of RNAse III; DICER-LIKE1, later on generating a double stranded miRNA. One of the strands of mature double stranded miRNA thus formed, designated as guide strand is preferentially incorporated into an ARGONAUTE protein containing effector complex; RNA induced Silencing Complex to repress the expression of target RNA containing their complementary sequence in plants. miRNAs, regulate gene expression either by target mRNA degradation or translational repression. But plants have been reported to employ mostly mRNA degradation mechanism for silencing expression of genes. The biogenesis, stability and activities of miRNAs are tightly regulated to ensure their normal functions in various plant developmental processes, metabolism as well as plant's responses to abiotic and biotic stresses. The focus of the present review will be on advances in molecular machinery and mechanisms involved in regulation of miRNA biogenesis and regulation of their role in controlling gene expression in plants. The underlying process of regulating miRNA levels through their turnover involving their stability and degradation are also briefly discussed.

9.1   INTRODUCTION

Small RNAs of 21- to 24-nucleotides in length have vital regulatory roles in eukaryotic genomes [21, 44]. Current researches have realized the critical role of small RNAs in the protection against viruses, genomic alterations and control of gene expression [21, 44]. The genesis of these miniature RNAs is from RNA precursors formed after transcription of *miRNA* genes. Later on mature small RNAs are formed by a RNA silencing machinery mediated processing of these precursor molecules. MicroRNAs (miRNAs) and small interfering RNAs (siRNAs) are two common types of small RNAs. Genetic screening of the lin-4 and let-7 mutants in the worm *Caenorhabditis elegans* led to the discovery of miRNAs [86, 131]. Screening of small RNAs in *Arabidopsis* resulted in discovery of plant miRNAs [96, 119, 132]. Later on it was revealed that miRNAs are almost common components of different gene regulatory systems in higher eukaryotes. miRNAs are endogenous, miniature RNAs having one strand of 21–24 nucleotides (nts) size range. The process of genesis of mature miRNAs initiates through primary transcripts (pri-miRNAs), formed by transcription of *miRNA* genes (*MIR*). Subsequently, precursor-miRNAs (pre-miRNAs) are formed from pri-miRNAs. Pre-miRNAs are single-stranded miRNA precursors that form self-complementary stem-loops. The mature miRNAs are subsequently generated from pre-miRNAs through two RNAse III-mediated steps. A RNA induced silencing complex (RISC) which contains an ARGONAUTE (AGO) protein further interacts with mature miRNAs and mature miRNAs are loaded in this ribonucleo-protein complex. The miRNA-loaded RISC guides mRNA cleavage or translational repression of target mRNAs having sites with complementarity to the loaded miRNAs thus negatively regulating the expression of protein-coding mRNAs [5, 69, 70]. A different form of small RNAs, siR-NAs, are associated with a phenomenon designated as RNA interference (RNAi); [91]. The RNAi is a mechanism for inhibiting gene expression which operates at post-transcriptional level and commonly called as post transcriptional gene silencing (PTGS). It occurs mostly in plants. The similarities between miRNAs and siRNAs is that both miRNAs and siRNAs are processed by the RNaseIII enzymes and later on loaded into the RISC machinery. This active RISC machinery subsequently controls expression of target genes. While the fact that separates miRNAs from siRNAs is that while miRNAs are derived from one-strand stem-loop molecules, siRNAs are generated by lenthy, double-stranded forms [5, 69]. At functional level, the difference between miRNAs and siRNAs is that siRNAs can act on their parent genes whereas miRNAs do not down-regulate the parent genes. *Trans*-acting siRNAs (ta-siRNAs), [122, 161], are RNAs formed from defined genetic loci named as *TRANS-ACTING SIRNAs* (*TAS*). Ta-siRNAs do not code for proteins. The breaking of *TAS* precursors such as *TAS 1, 2 and TAS 3*, guided by the miRNA (miR173 and miR390, respectively) leads to the production of ta-siRNAs from the cleaved precursors, which later on, negatively control their protein-coding mRNA targets [2, 21, 122, 161, 177]. Plant miRNAs have an enhanced complementary matching to the target mRNAs as compared to animal miRNAs [68, 137]. Most of the miRNAs in plants are evolutionarily conserved and their targets are the genes involved in developmental timing and patterning of various plant processes such as proliferation of stem cell, development of vegetative and reproductive organs and reactions to environmental changes, that is, abiotic and biotic stresses [26, 42, 104, 123, 169].

The focus of the present review will be studies related to molecular mechanisms involved in regulation of miRNA biogenesis and their activities in plants. Mechanism of controlling levels of miRNAs through their turnover are also briefly discussed.

9.2   OVERVIEW OF BIOGENESIS, MATURATION, AND MECHANISM OF ACTION OF MIRNAS

In *Arabidopsis*, miRNA biogenesis has been studied extensively. Biogenesis and processing of miRNA involves many stages and many proteins help in these processes (Table 9.1; Fig. 1). Control of target gene expression involving miRNAs is also assisted by many proteins (Table 9.1; Fig. 1). *miRNA* genes upon transcription generate pri-miRNAs. Later on shorter length pre-miRNAs are generated from pri-miRNAs. Two ribonuclease III (RNase III)-type endonucleases are involved in genesis of mature miRNAs from pri-miRNAs. Drosha, a RNAse III, assists in processing of pri-miRNAs to produce pre-miRNAs in animals in nucleus. After that protein exportin 5 helps in exporting pre-miRNAs to cytoplasm. Dicer, second RNAase III in the cytoplasm further acts on exported pre-miRNAs to form a duplex made up of one miRNA strand (miRNA) and another antisense miRNA strand (miRNA*). On the other hand in plants, the above-mentioned two miRNA processing steps are possibly performed in the nucleus by a RNAse III enzyme, DICER LIKE1 (DCL1), which is similar to Animal Dicer [82]. A double-stranded RNA (dsRNA) binding protein HYPONASTY LEAVES1 (HYL1) and another protein SERRATE (SE), a C2H2 zinc finger protein, help DCL1 in the processing of pri-miRNAs to pre-miRNAs [55, 81, 99, 182]. CBP20 and CBP80 subunits of a cap-binding complex (CBP) protein in nucleus also have important role in miRNA processing [47, 84]. DCL1, HYL1 and SE mutually co-operate in the miRNA processing and they are found together in the nuclear bodies, termed Dicing bodies or D-bodies [38, 40, 148]. DAWDLE (DDL), a forkhead-associated domain (FHA) possessing RNA binding protein of *Arabidopsis*, helps in recognition of pri-miRNAs by DCL1 [186].

TABLE 9.1   Proteins Involved in Biogenesis, Processing and Action of miRNAs

Proteins	Nature	Function	References
DICER LIKE1 (DCL1)	Ribonuclease III (RNase III)-like domain containing protein	Helps in formation of miRNA/miRNA* duplex by processing of pri-miRNAs	[82, 107]
HYPONASTY LEAVES1 (HYL1)	Double stranded RNA (dsRNA) binding protein	Helps DCL1 in processing of pri-miRNA to pre-miRNA, role in strand selection during assembly of miRNA-AGO1complex	[33, 55, 81, 105]
SERRATE (SE)	C2H2 zinc finger protein	Assists DCL1 in processing of pri-miRNA to pre-miRNA, involved in alternative splicing	[98, 128, 182]
DAWDLE (DDL)	A forkhead-associated domain (FHA) containing RNA binding protein	helps in recognization or access of pri-miRNA by DCL1	[136, 186]
Tough (TGH)	G-patch domain single stranded RNA (ssRNA) binding protein	Interacts with DCL1 and required for the efficient processing of pri-miRNA	[134, 136]
CDC5	Cell division cycle 5 DNA binding protein	Improves the DCL1 activity and possibly modulates pri-miRNA processing	[192, 193]
CBC	Nuclear cap binding complex RNA binding protein	Needed for proper pri-miRNA processing and pre-mRNA splicing	[74, 84, 127]
HUA EN-HANCER1 (HEN1)	Small RNA methyltransferase	Methylates miRNA/miRNA* duplex	[57, 185]
HASTY	Nucloecytoplasmic transporter protein	Proposed to be involved in export of miRNA/miRNA* duplex or miRNA-RISC complex from nucleus to cytoplasm	[10, 118]
ARGONAUT1 (AGO1) protein	PAZ domain protein	The catalytic action of AGO1, in the plant RISC assembly, cleaves the target mRNA, which is complementary to miRNA	[6, 162]

TABLE 9.1   (Continued)

Proteins	Nature	Function	References
Hsp90	Heat shock protein (molecular chaperon)	Involved in the interaction of a miRNA with AGO in plants	[61, 62]

The 3'-ends of double stranded miRNA/miRNA* complex generated through DCL1 mediated processing of pre-miRNAs, possess two projecting nucleotides [176]. Thereafter, a small RNA methyltransferase HUA ENHANCER1 (HEN1) comes into action. It adds a 2'-O-methyl group on the 3'-terminal nucleotide of double stranded plant miRNA/miRNA* complex, thus stopping uridylation by HEN1 SUPPRESSOR1 (HESO1) and decay of miRNAs by 3'-5' exoribonucleases; small-RNA-degrading nuclease (SDN1, SDN2 and SDN3) and protect them against degradation [52, 57]. Biogenesis of miRNAs in plants differes from that of animal miRNAs in this step also.

In plants, after the formation of the miRNA/miRNA* duplex, a protein HASTY which is similar to exportin5 protein found in metazoans, has been proposed to carry the duplex from the nucleus to the cytoplasm [10, 118]. Thereafter one strand of double stranded miRNA complex ("guide" strand) is incorporated into miRNA-protein assembly, designated as RNA induced silencing complex (RISC); [132]. Another strand of the duplex ("passenger" strand), named as miRNA*, is degenerated [21, 108]. The RISC assembly possesses ARGONAUTE1 (AGO1) protein. Besides plants [9], AGO proteins are core components of the RNA silencing machinery of all eukaryotes and some prokaryotes also. These proteins have been proposed to evolve from translation initiation factors [58]. Different AGO paralogs found in species have been linked with functions of many types of small RNAs. Within the 10 paralogs in *Arabidopsis*, AGO1 is associated with function of miRNAs in *Arabidopsis* [6, 162]. The catalytic activity of AGO1, associated with the plant RISC complex, cleaves the target mRNA, which is complementary to miRNA [6]. Similar to animals, ATP and chaperone Hsp90 are thought to be involved in the interaction of a miRNA with AGO in plants also [62]. PAZ, MID and PIWI are the domains of AGO proteins. The MID (middle) domain attaches to the phosphate group on the 5'-end of the miRNA, therefore, miRNA can be used in mRNA-target digestion. The PIWI domain of AGO protein bears structural similarity to the ribonuclease catalytic domain of RNase H [187]. The PAZ domain attaches to the 3'-terminal nucleotide of the miRNA having methylated 2'-OH group [44]. Later on the PIWI domain of AGO protein helps in breaking of the phosphodiester bond between the tenth and eleventh nucleotide of the mRNA-target region thus generating products with 5'-phosphate and 3'-hydroxyl groups (Figure 9.1).

Two post-transcriptional mechanisms; mRNA cleavage or translational repression are used by miRNAs to regulate gene expression through their incorporation into the RISC complex thus forming the functional miRNA-RISC assembly [5, 69]. mRNA cleavage is selectively used to regulate target gene expression whose mRNA exhibits perfect or almost-perfect complementarity with miRNA; while in cases of mismatches between miRNA and target mRNA, the translational repression of target mRNA is preferred [18, 68]. In animals, miRNA bind to a site in the 3'-untranslated region (3'-UTR) of mRNAs which is having mismatched regions in nucleotides 10 and 11, hence there is not full complementarily of target mRNA with the miRNAs in animals, due to which in them, the mechanism employed for posttranslational regulation is translational repression [54]. Although central mismatches in regions of nucleotides 10 and 11 are tolerated to some extent, but perfect complementarily of the "seed" region (nucleotides 2-7) is indispensable for target recognition [87]. In animals, proposed mechanism for repression of translation by miRNA is competition of AGO protein with translation initiation factor eIF-4E, which plays important role in translation. This theory is supported by the fact that AGO protein attaches to the cap of mRNA [58, 77]. Evolutionarily conserved miRNAs that did not show perfect complementarily to their target mRNAs during binding and cleavage of mRNAs, were reported in plants also [68]. Most plant miRNAs with high complementarily to target mRNAs predominantly employ mRNA cleavage mechanism to post-transcriptionally regulate gene expression. Active AGO possessing RISC assembly guided by miRNA breaks one phosphodiester bond of the target mRNA within the miRNA-binding site [96, 97, 154]. The broken mRNA fragments separate from RISC complex and they are subsequently destroyed involving many mechanisms. The exonuclease XRN4 and some XRN4-independent mechanisms have been reported to be involved in degradation of products

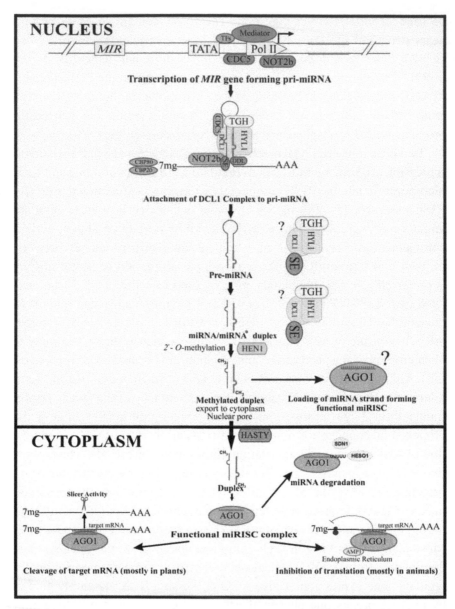

FIGURE 9.1   Biogenesis, processing, and mechanism of action of microRNA.

of 3' cleavage of some miRNA targets. Instances, where mRNA fragments are formed after 5' cleavage, the addition of uridines at the 3' ends of

5' cleavage products causes 5'-3' exonucleolytic decay of the 5' cleavage products [144]. Either translational repression or translational repression and mRNA cleavage mechanisms both are also employed by plant miRNAs to regulate gene expression [14]. Support for these observations comes from miR172 which controls flowering time and floral organ identity [4, 24], and miR 156/157 and miR854 [3, 41] that usually employ translational repression mechanism to regulate gene expression. On the other hand miR172 can also direct the cleavage of the target mRNAs [71, 142]. Hence miR172 exhibits coexistence of both translational repression and mRNA cleavage mechanism to regulate target gene expression.

9.3 REGULATION OF MIRNA BIOGENESIS, PROCESSING, EXPRESSION, AND ACTION

There are many different ways through which regulatory control can be imposed on expression and action of miRNAs. Transcriptional control and control of miRNA processing are important regulatory mechanisms.

9.3.1 REGULATION OF TRANSCRIPTION OF MIRNA GENES

Variable transcription of different *miRNA* genes is one of the different factors, that controls the levels of different types of miRNAs. Differential spatially and temporally controlled expression patterns of miRNA in plant is the reason for variable levels of the same miRNA among different types of tissue or at different developmental stages [69].

miRNAs arise by transcription of *miRNA* genes. Most of plant *miRNA* genes exist in the non-protein coding intergenic regions (IGR); [48]. But as an exception some miRNAs are found in intronic sequences of some protein coding genes [129]. Plant *miRNA* genes are transcribed by RNA polymerase II (Pol II) and this has been supported by several studies. First evidence indicating about this is, 5' cap and a 3' poly (A) tail, which are characteristic of a Pol II transcript, are found in primary transcripts of plant miRNAs. 5' CAP and 3' poly (A) tail stabilize pri-miRNAs [68, 176, 191]. The cyclin dependent kinase F1 (CDKF1) controls phosphorylation of the C-terminal domain of Pol II. The mutants lacking CDKF1 are without CAP structure of pri-miRNAs and levels of pri-miRNAs are reduced, which reinforces the observation that CAP structure stabilizes pri-miRNAs [53]. Protein factors have significant role in stabilization of pri-miRNAs. Dawdle (DDL), a forkhead-associated domain (FHA)-containing protein binds pri-miRNA and in its prescence, accumulation of pri-miRNAs is reported. Therefore, Dawdle (DDL) also plays important role in pri-miRNA stability although it possibly has no effect on transcription of *miRNA* genes [186]. Another observation pointing towards the transcription of *miRNA* genes by RNA Pol II, is the presence of TATA box and transcription initiator (INR) promoter elements in *miRNA* loci which are important aspects of Pol II genes [176, 196]. In addition to TATA box, *miRNA* promoters also contain 21 *cis*-regulatory motifs, therefore, *miRNA* expression may be regulated at transcriptional level [109, 176, 194], involving *miRNA* transcription factors. A general transcriptional coactivator mediator (multisubunit complex) regulates *MIR* transcription by helping attachment of Pol II to *miRNA* promoters. Absence of mediator, causes decreased *miRNA* promoter activities, which is subsequently responsible for decreased levels of pri-miRNAs and miRNAs [76]. Two homologous proteins, Not2a and Not2b, which are members of the conserved carbon catabolite repression 4 (CCR4)-NOT complex, also regulate *miRNA* transcription in *Arabidopsis*. Besides at transcriptional level, they also regulate levels of mRNAs at post-transcriptional level [28, 164]. NOT2b through interaction with the Pol II C-terminal domain also affects *miRNA* transcription [64]. The cell division cycle5 (CDC5) protein is a conserved DNA-binding protein in animals and plants [117]. CDC5 is a positive transcription factor of *miRNAs*. Absence of CDC5 adversely affects *miRNA* promoter activity and the occupancy of Pol II at *miRNA* promoters [193]. A transcription factor, SANT-domain-possessing protein Powerdress (PWR) specifically promotes the transcription of some *MIR172* family members, while it has no effect on other *miRNAs* [188].

Primary miRNA transcript, named as pri-miRNA have characteristic hairpin-like imperfect stem-loop. Generally from a hairpin-like structure of a pri-miRNA only one unique mature miRNA type is generated. But sometimes from one pri-miRNA, two or more hairpins, each of which generates a different mature miRNA species, are formed [110]. Transcripts formed from some transposable elements (TEs), which may code miRNA, and siRNA have been reported in both *Arabidopsis* and rice [124].

9.3.2   REGULATION OF PROCESSING OF MIRNA TRANSCRIPTS: PROTEIN FACTORS

Contrary to animals [112], in plants, the control of processing of miRNAs is not clearly worked out. The knowledge regarding regulation of biogenesis and miRNA processing in plants is obtained from some experimental evidences especially in *Arabidopsis* [69, 115, 162].

DICER-LIKE1 (DCL1), a multi-domain ribonuclease III (RNase III)-like protein is involved in miRNA biogenesis. DCL1 is a member of group of four *Arabidopsis* DCL proteins [69]. In hypomorphic alleles of the *DCL1* mutants, unusual scheme of flower development, defects in timing of flowering and leaf development were observed while in null alleles lethality was observed and the levels of most miRNAs were reduced in mutants of the *DCL1* gene; [45, 119, 141]. Importance of DCL1 enzyme in the miRNA biogenesis has been shown in these studies.

Transcription of *miRNA* gene(s), leads to formation of pri-miRNAs which are changed to pre-miRNAs, containing a stem-loop structure with 2-nt 3' overhangs at the end of stem, and subsequently to double stranded miRNA/miRNA* complex with 2-nt 3' overhang and a 5' phosphate at each strand through DCL1 [106, 107]. Plant pri-miRNA hairpin structures exhibit variability in length and structure with differential positioning of the miRNA/miRNA* duplex. The activity of DCL1 enzyme which plays important role in processing of pri-miRNAs, is subjected to control by structure of pri-miRNAs [107, 147, 168, 199]. An imperfectly paired lower stem of ~15 base pair (bp) below the miRNA/ miRNA* duplex is important for the initial loop-distal cleavage of pri-miRNAs while the loop plays a critical role in processing [107, 147, 168]. The secondary structures of pri-miRNAs are important in miRNA processing. In pri-miRNAs having lengthy upper stem structures such as pri-miR159a and pri-miR319a, after cleavage of the loop, the miRNAs is released from the stem with additional cuts [11, 30]. On the other hand in some pri-miRNAs with multibranched end loops, processing by DCL1 is in both directions. There are observations, which emphasize upon the importance of direction of processing. Processing in the direction of base-to-loop has positive effect on miRNA production, while cleavage in the opposite direction represses the generation of miRNAs from pri-miRNAs with multibranched terminal loops [199].

Many protein factors are also required for the efficient processing of pri-miRNAs. The double-stranded RNA (dsRNA)-binding protein HYL1, the zinc finger protein SE and the G-patch domain protein tough (TGH) are important protein factors. By their interaction with DCL1, they play an important role in processing of pri-miRNAs [38, 40, 81, 98, 134, 148, 160, 182]. HYL1 is a ds RNA binding protein with two RNA-binding domains [170, 184]. HYL1 helps in accurate pri-miRNA processing by possibly binding the miRNA/miRNA* duplex region as a dimer [184]. The N-terminal domain of SE binds to single-stranded RNAs (ssRNAs) [64, 100, 136]. While the zinc finger domain of SE interacts with DCL1 and optimizes DCL1 activity [64]. Both HYL1 and SE enhance the efficiency and accuracy of pri-miRNA processing [32]. These observations find support in the results in which *hyl1* mutants show impaired cleavage of products of pri-miRNAs and mutations in the helicase and RNase III domains of DCL1, which play important role in cleavage site selection and catalytic activity of DCL1, respectively, reverse the abnormalities caused by mutations in *hyl1* [93]. SE protein possibly plays important role in interaction between a miRNA precursor and DCL1 catalytic site [100]. A ssRNA-binding protein,TGH binds with pre-miRNAs and pri-miRNAs *in vivo* [134, 136]. Besides DCL1, TGH interacts with HYL1 and SE, therefore, it may be a component of the DCL1 complex [134]. TGH positively affects the efficiency of cleavage and/or the interaction of pri-miRNAs with DCL1, which is reinforced by reports of reduced DCL1 activity as well as the association of pri-miRNAs with the HYL1 complex in loss-of-function mutations in *TGH* [134]. A different protein factor CDC5 can modify pri-miRNA processing by improving the DCL1 activity [193]. SE besides its role in pri-miRNA processing, controls alternative splicing [128]. PRL1, which is a WD-40 protein, interacts with the DCL1, which plays vital part in processing of pri-miRNAs; and DCL3 and DCL4, which process the dsRNAs; and positively influences the processing of pri-miRNAs and dsRNAs. Therefore, PRL1 may promote the production of miRNAs and siRNAs and thus result in their accumulation [192]. CDC5 and PRL1 mutually co-ordinate to possibly enhance DCL1 activity and thus they have positive effect on pri-miRNA levels and miRNA accumulation [192]. High osmotic stress gene expression 5 (HOS5) and two serine/arginine-rich splicing factors RS40 and RS41, interact with HYL1 and SE, having important role in miRNA biogenesis. These proteins might have important roles in correct miRNA strand selection and the maintenance of miRNA levels and therefore may be involved in biogenesis of a group of miRNAs besides their role in pre-mRNA splicing [23]. The accessory factors of DCL1 exert variable effects on individual miRNAs [85]. Due to specefic spatial–temporal expression schemes, effects of protein factors on different miRNAs vary, for example, CDC5 exerts more influence on the miRNAs expressed in the proliferating cells because of its preferentially higher expression in these cells [92]. Proteins SICKLE (SIC) and Receptor for activated C kinase1 (RACK1) are possibly required by above mentioned protein factors for their optimal activity. SIC, a proline-rich protein important in plant development and adaptation to abiotic stresses, in conjunction with HYL1 is needed in processing of some pri-miRNAs and subsequently accumulation of miRNAs [190]. RACK1 along with SE regulates the DCL1 activity, which is indicated by the observation of reduced accumulation of miRNAs and the processing precision of some pri-miRNAs on lack of RACK1 [150].

9.3.2.1 Control of Localization, Levels and Action of DCL1 and HYL1

HYL1, TGH, SE and DCL1 are found in the subnuclear Dicer-body (D-body). Pri-miRNA processing or storage possibly occurs in D bodies, which is indicated by their association with pri-miRNAs [38, 40, 134, 148]. CDC5 and NOT2 are found in the DCL1 containing subnuclear loci but without association with HYL1, hence there is not much clarity about whether CDC5 and NOT2 are constituents of D-bodies [164, 193]. miRNA biogenesis is dependent very much on correct localization of D-body. Appropriate D-body pattern formation in *Arabidopsis* needs NOT2 protein and a RNA-binding protein MOS2 [164, 174]. MOS2 interacts with pri-miRNAs *in vivo* but it does not interact with DCL1, HYL1 or SE [174]. In mutants *mos2*, the localization of HYL1 in D-bodies is improper, the interaction of pri-miRNA with HYL1 is negatively affected and the levels of miRNAs are decreased indicating that MOS2 may help in the D-body formation and the recruitment of pri-miRNAs to the D-bodies [174]. NOT2s has direct interaction with DCL1. NOT2s possibly have a role in assembly of D-body, which is indicated by occurrence of increased numbers of DCL1-containing loci without affecting the localization of HYL1, in case of abnormalities in NOT2s [164].

Because DCL1, HYL1, and SE are involved in miRNA processing, therefore, miRNA processing can be controlled by regulating the transcription of *DCL1*, *HYL1* and *SE* genes themselves through many transcription factors. The miRNA processing can therefore be controlled by balance between enhancement of expression of *DCL1* by Stabilized1 (STA1), a pre-mRNA processing factor [7] and blocking transcription of *HYL1* and *SE* and thus repressing miRNA production by Histone acetyltransferase GCN5 [75]. Tissue- or developmental stage-specific expression of stem loops of short interspaced elements (SINE) RNA in *Arabidopsis*, which are similar to the hairpin structures of miRNA precursors and bind to HYL1 protein, might modify the production of plant miRNA by competing with HYL1 [125].

The phosphorylation of HYL1 and DCL1 also affects pri-miRNA processing. DDL interacts with the phosphothreonine-containing helicase and RNAse III domains of DCL1 [101]. The phosphorylation of DCL1 takes place *in vivo* [36]. Absence of DDL negatively affects miRNA maturation, which indicates that the interaction of DDL with phosphorylated DCL1 has possibly important roles in pri-miRNA processing [186]. C-terminal domain phosphatase-like1 (CPL1) is a protein phosphatase and can dephosphorylate a serine motif in the C-terminal heptad repeat domain (CTD) of RNA polymerase II [105]. CPL1 supports the maintenance of the hypophosphorylated state of phosphorylated HYL1, needed for optimal action of HYL1, and thereby causing accurate and efficient primiRNA processing [105]. SE interacts with CPL1. Lack of SE impairs the CPL1-HYL1 interaction and dephosphorylation of HYL1, indicating that SE mediates CPL1 interaction with HYL1 [105].

9.3.2.2 Role of Splicing in Processing of Pri-miRNAs: Nuclear Cap-Binding Complex (CBC)

Pri-miRNAs also contain introns [29, 116]. Splicing of introns can change the stem-loop structures of pri-miRNAs and thus regulate miRNA maturation [80, 139, 153]. For example, upon changes in stemloop structures of pri-miR162a and pri-miR842-miR846 dicistron by alternative splicing, their processing is reduced in *Arabidopsis* [56, 66]. Besides changing pri-miRNA structure, splicing may enhance pri-miRNA processing. For example, the processing efficiency of pri-mi163 and pri-miR161 can be enhanced by the splicing of the 3' introns following their stem-loops [8, 143]. Positive effect of introns and active 5' splice sites was observed on the accumulation of miRNAs generated from genes having intron. In examples of exonic *MIR*161, *MIR*163 and *MIR*172a pri-miRNAs (each containing one intron), positive effect of splicing on generation of mature miRNAs was observed and the interaction between active 5' splice sites (5'ss); (found downstream of the stem-loop structure of miRNA) and U1 snRNP was reported to be very important for stimulation of miRNA biogenesis [8, 143]. Current researches on *Arabidopsis* intronic miR402 production exhibited different results. *Arabidopsis*, miR402 is elicited by heat stress and it is encoded within the first intron of a protein-coding gene At1g77230. Inactivation of the 5'ss of the intron hosting miR402 resulted in enhanced accumulation of the intronic mature miR402. On the other hand mutation of the first-intronic 5'ss when the miR402 stemloop structure was mobilized into the first exon caused reduced miRNA level [78]. Therefore, in these studies, the importance of functional 5'ss in controlling the efficiency of biogenesis of *Arabidopsis* miRNAs generated from intron-containing genes was observed. Importance of the position of the miRNA hairpin in context to the 5'ss on maturation of miRNA was also observed. Hence, in case of plant miRNAs formed from intron containing genes, a new mechanism for controlling their biogenesis and target mRNA level is exhibited. Proteins also assist mRNA splicing and maturation of miRNA. Nuclear cap-binding complex (CBC) is a protein involved in pri-miRNA processing. It is responsible for the appropriate splicing of the first intron by binding to the cap of mRNAs [47, 74, 84, 127]. CBC is a dimeric protein formed of two different protein subunits (CBP80 and CBP20). The CBP80 protein subunit of *Arabidopsis* is also known as Abscisic acid (ABA) Hypersensitive1 (ABH1); because it was isolated from mutants that are hypersensitive to the phytohormone ABA. Role of *Arabidopsis* CBC in promoting efficient intron splicing was shown in studies where increased levels of unspliced transcripts were observed in loss-of-function mutations in *ABH1/CBP80* and *CBP20* thus indicating towards defective intron splicing in *cbp80/abh1* and *cbp20* muatnts [84]. Additionally, increased levels of pri-miRNAs and reduced levels of mature miRNA were observed in both the *cbp80/abh1* and *cbp20* mutants [47, 74, 84]. Hence above-mentioned reports have suggested dual roles of the *Arabidopsis* CBC in pre-mRNA splicing and processing of pri-miRNA. SE interacts with CBP20, suggesting that CBP20/80 may be a part of the processing complex [51, 109]. The pleiotropic abnormalities including the "serrated leaves" are exhibited by *cbp80/abh1* and *cbp20* mutants that are similar to the phenotype observed in the *serrate (se)* mutant [49]. This indicates towards a possible role of SE also in pre-mRNA splicing. Analysis of intron-containing pri-miRNAs suggests that the pri-miRNA processing function of CBC and SE are possibly independent from their role in pre-mRNA splicing [84]. SE is possibly involved in the interactions between the CBC and the spliceosome, in case of pre-mRNA splicing, and also between the CBC and the pri-miRNA processing machinery [84]. Colocalization and interaction between SE and many U1 snRNP proteins (small nuclear ribonucleoproteins which are part of spliceosome complex) was observed which may affect maturation of miRNA [78]. Besides CBP80/20, CDC5, which has important role in miRNA biogenesis, enhances splicing of some mRNAs, but role of CDC5 in pri-miRNA splicing is not clear [175]. Another protein AtGRP7, which is a hnRNP-like glycine-rich RNA-binding protein also plays an important role in pri-miRNAs processing in addition to regulating pre-mRNA splicing in *Arabidopsis* [79].

9.3.2.3   Feedback Regulation of miRNA Biogenesis and Processing: DCL1 and AGO1

The biogenesis of miRNA is controlled in a feedback process. The DCL1 and AGO1 are vital players in miRNA biogenesis pathway. A negative feedback regulatory mechanism operates in control of *DCL1* and *AGO1* gene expression by miRNAs. MiR162 cleaves the target *DCL1* mRNA [178]. There are experimental observations supporting these studies where levels of *DCL1* mRNA are increased in the mutants of miRNA processing genes having decreased levels of miR162. Spatio-temporal changes in miR162 expression pattern may affect the mature miRNA production levels in different tissues or at different developmental stages. miRNA also regulates *AGO1* gene. There is a complementary site for miR168 binding in the *AGO1* mRNA, and like *DCL1*, miR 168 targets and subsequently cleaves the *AGO1* mRNA [159]. siRNAs generated from *AGO1* mediate the silencing of *AGO1* itself [103]. Another *Arabidopsis* AGO protein, AGO2 is regulated by miR403 [2], but role of AGO2 in miRNA pathway is not clearly worked out.

9.3.3 MODIFICATION AND MATURATION OF MIRNAS: STABILITY OF MIRNAS

Due to processing of pri-miRNA by the DCL1 complex, a small RNA duplex made up of a miRNA strand and another miRNA* strand is generated. This double stranded miRNA/miRNA* complex has a 2-nt overhang at the 3'-end of each small RNA. Addition of a methyl group (methylation) to the 2'-hydroxyl group of the 3'-terminal nucleotide in each of the small RNA strands of the miRNA/miRNA* duplex is done in plants by HUA ENHANCER1 (HEN1), the Mg^{2+} requiring small RNA methyltransferase [57, 185]. HEN1 possess two dsRNA-binding domains and a signal for nuclear localization. Methylation by HEN1 a phenomenon, which possibly happens before AGO1 loading. It is not clear whether methylation takes place in cytoplasm or nucleus due to the reports of presence of HEN1 in both cytoplasm and nucleus [89]. The methylation of the miRNA/miRNA* duplex protects miRNAs against the 3'-end uridylation activity and subsequent degradation [89]. In the *hen1* mutants, lower level of mature miRNAs are observed, which points towards the importance of HEN1-mediated 2'-O-methylation in plant miRNA biogenesis and stability. On the other hand uridylation destabilizes miRNAs in plants. In *Arabidopsis*, HEN1 SUPPRESSOR1 (HESO1), a uridyl transferase, adds uridines (uridylation) to most of the miRNAs [133, 195]. In mutants *heso1* levels of miRNAs are increased whereas the levels of miRNAs in *hen1* are decreased on overexpression of *HESO1*. These reports suggest the fact that uridylation causes degradation of miRNAs in higher plants [133]. HESO1 interacts with AGO1 and also uridylates miRNAs attached with AGO1 *in vitro* [135]. Furthermore on mutation in AGO1, uridylation of miRNAs is impaired in *hen1* [135, 189]. These observations suggest that place of miRNA uridylation is AGO1 complex and they also show the critical role of methylation, which protects miRNAs from AGO1-associated HESO1 action and subsequently maintains the function of the AGO1-miRNA complex.

9.3.3.1   Turnover of miRNAs

The biogenesis and subsequently, degradation of miRNAs is important for homeostasis of miRNA levels and function. In *Arabidopsis*, small RNA degrading nuclease 1, 2, and 3 (SDN1, SDN2, and SDN3) which belong to family of 3′-to-5′ exoribonucleases, participate in turnover of mature miRNAs. Consistent with their role in miRNA turnover, inactivation of SDN proteins results in increased miRNA levels and abnormal development of plant. Possibly, SDN1 and HESO1 act together to regulate the degradation of 2′-*O* -methylated miRNAs in *Arabidopsis* which is supported by the observations that SDN1 can degrade 2′-*O*-methylated miRNAs, but not 3′ uridylated miRNAs [130]. The miRISC association with mRNA stops miRNA degradation. If miRNA does not find its target, it can be detached from miRISC and become prone to XRN-2 activity and later on destroyed, rendering AGO proteins to be loaded with new miRNAs. This is supported by the reports of requirement of free 5′ end of mature miRNAs by 5′ to 3′ exonuclease XRN-2 for their degradation in *Caenorhabditis elegans*. Free 5′ end of mature miRNAs is accessible in case of release of miRNAs from the miRISC [22]. In another study in *Chlamydomonas reinhardtii*, non-functional exosome constituents RRP6, resulted in increased abundance of miRNAs, indicating towards the point that miRNAs can be degraded from 3′-to-5′ by exosome [60]. Therefore, the, turnover of miRNAs may be an important step in the regulation of miRNA function.

9.3.4   REGULATION OF ACTION OF MIRNAS

9.3.4.1   Loading miRNA Strand into AGO Containing RNA-Induced Silencing Complex

Ultimately, preferential loading of the miRNA strand into RNA-induced silencing complex (RISC), which is an AGO containing effector complex, completes the miRNA maturation. The miRNA strand of the double stranded miRNA/miRNA* complex is selectively loaded into an AGO complex and this strand subsequently guides the sequence-specific interaction with target mRNAs, while miRNA* strand is released from the AGO assembly and degraded. The four domains of AGO are: the N-terminal domain, the PAZ domain, the middle (MID) domain and the PIWI domain [155, 157]. The PAZ domain attaches to the 3' end of single-stranded RNAs and the PIWI domain cleaves target within miRNA binding site, at a position opposite to the 10th and 11th nucleotides of miRNAs [138]. The PIWI domain has a structure like RNase H [20, 120]. Ten AGO proteins are found in *Arabidopsis* of which AGO1 is the most important [69, 157]. AGO proteins, e.g., AGO7 and AGO10 attach with miR390 and miR165/166, respectively [113]. In *AGO1* gene mutants, levels of some miRNAs were reduced while miRNA target genes were ectopically expressed [73, 140, 159]. Studies where abnormal polarity of lateral organ and leaf and defects in flower phenotype like that in *dcl1* mutants, can be seen in the *AGO1* hypomorphic alleles; and additionally embryonically lethality in null *AGO1* alleles, suggest that AGO proteins are linked with miRNA processing [72, 159]. Various strategies are adopted to control activity and levels of AGO1. In a feedback regulatory mechanism, AGO1, which is linked with miRNA processing, is itself targeted by miR168 [137, 159]. High expression of osmotically responsive genes 1 (HOS1), which is involved in response of plants to abiotic stress, for example, cold, has been reported to control miR168a/b levels in *Arabidopsis* by controlling transcription of *MIR168b*. MiR168a/b, on the other hand regulates level of *AGO1* gene in *Arabidopsis* [163]. A F-box protein FBW2 negatively regulates AGO1. Observations that AGO1 protein levels are reduced on overexpression of FBW2 while in mutants *fbw2*, AGO1 protein levels are enhanced, support the above-mentioned function of FBW2 [34].

Preventing miRNAs from associating with AGO1 is also a mechanism for their control. For example, the function of miR165/miR166 is limited and thus the proper development of shoot apical meristem (SAM) is ensured if AGO10 binds miR165/miR166 in SAM which prevents the formation of the AGO1-miR165/166 complex [198]. In a different process, autophagy pathway, the AGO1-miRNA assembly is disturbed and AGO1 is degraded [31].

Many things such as protein factors, structure of double stranded miRNA/miRNA* complex and 5′ nucleotides affect the loading of miR-NAs into AGO1 [33, 111, 113]. Somewhat like animals, one of the factors affecting preferential loading of the small RNA strand from a miRNA/miRNA* duplex into the RISC complex in plants, is differential thermo-stability of the two ends of the functional small RNA duplexes. Usually, the small RNA strand with its 5′-end found at the less stable end of the duplex preferentially functions as guide strand and becomes part of RISC [33, 69]. Type of last nucleotide at the 5′ terminal end of small RNA deter-mines which type of small RNA (miRNA) will be selectively loaded into some of the 10 AGO proteins of *Arabidopsis*. In this context, AGO1 has a preference of small RNAs having a 5′-terminal uridine (5′U); AGO2 and AGO4 have a preference for small RNAs bearing a 5′-terminal adenosine (5′A), and AGO5 binds selectively to small RNAs having a 5′-terminal cytosine (5′C), respectively [19, 113, 121]. The selective loading mecha-nism of the small RNA strand into RISC complex has not been worked out in detail yet but it plays an important role in miRNA action, for example replacing the nucleotide at the 5′ terminal end from uridine-to-adenosine of artificially engineered miRNAs caused the change in small RNA load-ing form AGO1- to-AGO2 which therefore inhibited their silencing activ-ity [111].

Convergence of location of accumulation of many *Arabidopsis* miR-NAs in cytoplasm and location of cleavage of their target mRNAs by many plant miRNAs in cytoplasm [69, 118] seems to indicate towards possibility that plant miRNAs function only in the cytoplasm but it is not totally clear yet. A protein HASTY found in the *Arabidopsis* which is a homolog of exportin5 protein found in metazoans, is possibly associated with transport of miRNAs from nucleus to cytoplasm [10, 118]. The mor-phological characters of *hasty* mutants, that is, pleiotropic abnormalities in development of plant and decreased accumulation of most miRNAs are similar to that observed in the *dcl1* or *ago1* mutants. But, whether miRNA/miRNA* duplexes or the miRNA-RISC complexes are transported by the HASTY protein, is not clearly worked out. Reports that proteins HEN1 and AGO1 are involved in various small RNA silencing pathways, is supported by their localization in both the cell nucleus and cytoplasm [38, 158]. Co-localization of HEN1, AGO1 and HYL1 in the subnuclear D-bodies [38], points towards the possibility of interlinking of steps mediated by these proteins, for example, dicing, modification of ends and formation of miRNA-containing silencing complexes (miRISCs) in the nucleus. HYL1 reportedly has important role in deciding the preferential loading of strand during the formation of miRNA-AGO1 complex. Another protein CPL1 along with HYl1 facilitates the miRNA strand selection [33, 105]. Interaction of HYL1 and AGO1 in the nuclear D-bodies [38] linked with presence of single stranded, mature miRNAs in the nucleus [118], possibly suggests that loading of some of the plant miRNAs into AGO1 and formation of functional miRISCs occur in the nucleus. Additionally some proteins which asoociate with AGO1 and help them in their activities, are identified such as squint (SQN); a cyclophilin 40 (CPY40 protein) and HSP90 protein [146]. Heat shock protein 90 (HSP90) and CyP40 attach with AGO1 thus helping in the miRNA loading in a cell free system [35, 61, 62]. Further, upon ATP-hydrolysis by HSP90; HSP90-CYP40 complex and miRNA* detach from the AGO1-miRNA complex, possibly due to the HSP90 mediated conformational changes of AGO1 [61, 62]. Majority of miRNA*s are destroyed after detachment from AGO1 [33, 83]. But some miRNA*s can be incorporated into other AGOs.

9.3.4.2   Target Cleavage

Plant miRNAs recognize their substrate if there is high degree of sequence complementarity between target mRNA and miRNA [6, 95, 102]. Recently it has been shown that besides complementarity between miRNA and target, the context of miRNA binding site and expression levels perhaps also have role in the recognition of target [95]. In plants, predominant process for miRNA-mediated down-regulation of gene

expression is cleavage of target [5], which is possibly supported by observations of the inability of AGO1 having mutations in the catalytic region
to reverse the abnormalities in *ago1* mutant [16]. A 5' fragment with a 3'
hydroxyl group and a 3' fragment with a 5' phosphate group are formed
due to target RNA cleavage by AGO proteins [97]. Post-target cleavage
by AGO1, exonucleases without the requirement for 3' deadenylation or
5' decapping comes into action and cause decay of the target mRNAs.
3' RNA fragments are possibly degraded by XRN4 (a cytoplasmic 5'-3'
exoribonuclease) in plants [149]. In the algae *C. reinhardtii*, 3'-to-5' degradation of 5' fragments, mediated by the exosome, is initiated when 3'
end of 5' fragments is adenylated by the nucleotidyl transferase MUT68
[59]. In animals and higher plants, 5' fragments are uridylated at 3' end
[59]. In *Arabidopsis*, HESO1 mainly uridylates 5' fragments. On decrease
in levels of HESO1, the abundance of 5' fragments is increased, which
shows the critical role of uridylation in stimulation of the degradation
of 5' fragments [135]. A process different from 3'-to-5' degradation is
involved in degradation of uridylated 5' fragments. In 5' fragments, 5'-to-
3' degradation occurs, possibly through a XRN4 independent mechanism,
which is supported by observations in which, in *xrn4* mutants, the accumulation of 5' fragments is increased [135]. In studies of the nonprotein-
coding gene IPS1 (induced by phosphate starvation 1) from *Arabidopsis*,
a control mechanism of miRNA action was reported [39]. The IPS1 RNA
has an uncleavable sequence with sequence complementarity to miRNA
399. Consequently, the IPS1 RNA is not cleaved following miR399 pairing but it sequesters miR-399-loaded RISC thus causing inhibition of
miR399 activity and resulting in accumulation of target *PHO2* mRNA
[39]. Whether this "target mimicry" is somewhat common mechanism in
plants to control the action of other miRNAs, is not clear. But this strategy may be potentially important for sequence-specific inhibition of plant
miRNAs *in vivo* [166].

9.3.4.3   Translational Inhibition

Although target cleavage was reported to be predominant mechanism for
regulation of gene expression in plants but some studies indicated that

some plant miRNAs employ translational inhibition to down-regulate gene expression [14, 90, 142]. There are many studies that elaborate the mechanisms of miRNA-mediated translational inhibition. One such mechanism is inhibition of initiation or elongation steps of translation of target RNAs by miRNAs in plants. It is executed by blocking the recruitment or movement of ribosomes by AGO1-miRNA after attaching to the 5′ untranslated region (UTR) or the open reading frame of target RNAs [63]. Processing bodies (P-bodies), which are found in cytoplasm, are the sites for storage of many mRNA degrading enzymes [94]. In plants, P-bodies are possibly involved in translational inhibition mediated by miRNAs and may store translationally-repressed target mRNAs, which is based upon the observations of localization of a part of AGO1 in the P-bodies and involvement of the P-body constituents varicose (VCS) and SUO (a GW-repeat possessing protein) in miRNA-mediated translational inhibition [14, 183]. In case of plants, the clear relationship between target RNA decay and translational repression has not been established [14, 114].

Some miRNAs employ both translational inhibition and target cleavage to repress gene expression. Therefore, studies were required to gain knowledge about how it is determined that which mechanism, that is, translation inhibition or target cleavage will be used to repress gene expression. It has been observed that many factors decide whether cells will employ translational repression or target cleavage as mechanism of gene repression. Giving clues about the mechanism of down-regulation of gene expression employed, are the reports of dominant translational inhibition in case of presence of binding sites of miRNAs at the 5′ coding region and increase in translational repression in male germ cells of plants by miRNAs [46, 90]. In some types of cells, both translational inhibition and target cleavage phenomenon are observed. The reason for this may be the preferential interaction of AGO1 with protein factors. Another reason may be that different subcellular compartments have different specificities for translational inhibition or target cleavage and therefore AGO1-miRNA-targets may be selected and transported into designated subcellular compartment thus deciding which mechanism will be employed to repress gene expression. There are some reports that possibly support above mentioned observations; for example association of AGO1 with the endoplasmic reticulum (ER); [90] and abnormalities in miRNA-mediated translational repression, but not in target transcript cleavage on absence of altered meristem program1 (*AMP1*), an integral ER membrane protein, which indicate that ER may be the site for translational inhibition [90]. There are some other proteins which are involved in translational inhibition by miRNAs such as enzyme katanin (KTN), 3-hydroxy-3-methylglutaryl CoA reductase (HMG1) and the sterol C-8 isomerase hydra1 (HYD1). KTN1 is needed for the proper cortical microtubule assembly [151]. Disruption of KTN1 inhibits miRNA-mediated translational repression, pointing towards the role of microtubule in translational inhibition. This observation is in line with the role of microtubules in ER organization and P-body dynamics, both of which have been associated with translational inhibition [14]. HMG1 is indispensable for the biosynthesis of isoprenoids, which are involved in numerous metabolic pathways such as membrane sterols and many phytohormones, while HYD1 is needed for the biosynthesis of sterols. Involvement of HMG1 and HYD1 in sterol biosynthesis on one hand and in miRNA mediated translational repression on the other hand, possibly indicates towards the role of sterol in miRNA activity [14, 15].

9.3.4.4   DNA Methylation

Inhibition of gene expression at the transcriptional levels by DNA methylation may be other mechanism for control of gene expression by miRNAs. For example, some AGO4-24-nt miRNAs complex directs methylation of DNA at the *miRNA* and target loci and thus may repress gene expression at transcriptional level [173]. AGO4-24-nt siRNAs complex cause cytosine methylation through RNA-directed-DNA methylation [12]. Possibly mechanism of DNA methylation by miRNAs is like that of siRNAs.

9.3.5   *REGULATION OF EXPRESSION OF MIRNAS*

9.3.5.1   Transcriptional Control of miRNA Expression: Transcription Factors

Transcription factors (TFs) encoded by target mRNAs are also involved in the regulation of miRNA expression in plants [69]. Putative promoters of *miRNA* genes contain sequence motifs present in the promoters of protein coding genes besides core promoter elements such as the TATA box [179]. Many conserved *miRNA* gene promoters in *Arabidopsis* show excess representation of binding sites for the transcription factors; Auxin Response Factors (ARFs), (induces gene repression on lack of phytohormone auxin); LEAFY (LFY), (Important floral genes become functional upon their stimulation by gibberellic acid; GA) and AtMYC2 (a basic helix–loop–helix [bHLH] type transcription factor); (enhances sensitivity to abscisic acid; ABA and thus helps in coping with drought conditions) as compared to the promoters of protein-coding genes [109]. These studies also suggest a relationship between miRNA transcription and plant hormones involving transcription factors [109]. Additionally, some of these transcription factor families are themselves controlled by miRNAs, thus pointing towards the presence of complex transcriptional feedback loops between miRNAs, transcription factors and hormones [109]. The experimental evidences for these come through the discovery of ARF-binding motifs in *miRNA* promoters and targeting of mRNAs for many ARF family members by miR160, miR167, and miR390 [69]. Interaction between two important miRNAs, miR156, and miR172 gives clue about the transcriptional regulation of miRNA expression in plants. miR156 and miR172 follow a complementary temporal expression scheme in *Arabidopsis*; while miR156 expression is reduced during the transition from the juvenile to the adult phase, the expression of miR172 in opposition to that of miR156 increases [171]. The miR156 interacts with mRNAs of several SQUAMOSA Promoter Binding Protein-Like (SPL) TFs while miR172 controls members of the APETALA2 (AP2)-like family of TFs to regulate developmental timing, vegetative to reproductive phase transition and floral development [27, 69, 142, 172]. AP2 also represses expression of *MIR172* loci through transcriptional repressor Leunig (LEG) and Seuss (SU) [50]. SPL9 and SPL10 TFs that are targeted by miR156, increase the expression of miR172 by activating its transcription [171]. SPL9 and SPL10 possibly also promote the expression of miR156 by activation of transcription, therefore, forming a negative feedback loop that regulates their own expression [171]. This SPL-mediated regulation of miR156 and miR172 helps in regulating developmental timing [165, 171]. Another transcription factor FUSCA3 may be a positive regulator of *MIR156A* and *MIR156C* [164]. Environmental signals, such as shortage in supply of nutrients, cold, or light stimulate many plant miRNAs [67, 69, 145, 197]. Copper is one of the important nutrients and part of many proteins and other important biomolecules. Shortage in copper supply has been observed to induce expression of *Arabidopsis* miR398 [152, 180]. The TF SPL7 activates miR398 transcription under low copper conditions [181] by its interaction with numerous GTAC motifs found in the promoters of both *MIR398b* and *MIR398c* [181]. SPL7 possibly has an important function in miRNA-mediated regulation of copper levels [181]. MiR398 acts on cytosolic CSD1 and a chloroplastic CSD2, two intimately related Cu/Zn superoxide dismutases, as well as on COX5b-1, a subunit of the mitochondrial cytochrome-c oxidase [69, 152]. MiR398-mediated down-regulation of its target gene expression under limited copper conditions possibly helps plants in managing copper economy through transferring copper to essential proteins involved in important life-processes like photosynthesis [180]. Like miR398, other three *Arabidopsis* miRNAs— miR397, miR408, and miR857—induced under low cooper conditions, help in proper response to abiotic stress conditions of copper deficiency [1].

Some *MIRs* exhibit peculiar spatio-temporal expression schemes [13, 65, 156]. The TF scarecrow activates the expression of *MIR165/166* in the root endodermis, which is important for the patterning of the cell types in root xylem [17].

9.3.5.2   Control of miRNA Expression at Post-transcriptional Level

The miRNA expression can be affected by the spatio-temporal expression pattern of proteins involved in pri-miRNA processing and miRNA maturation such as SE and AGO. These are supported by the studies in *Arabidopsis* where SE has been observed to exhibit specific spatial expression patterns during development of embryos [126]. Specific spatial expression schemes for SE were also observed in developing *Arabidopsis* leaves and floral organs [126]. Therefore, these studies may indicate towards the possibility that miRNA expression is affected by SE expression schemes. In a different research, the specific spatial expression pattern of AGO7 in *Arabidopsis* has been shown to play important role in *TAS3* ta-siRNA biogenesis regulated by miR390 [37, 43]. The localized expression of AGO7, coupled with that of *TAS3* helps in deciding the activity region of miR390, thereby contributing towards *TAS3* ta-siRNA-mediated leaf patterning during leaf development [25, 113].

9.4 CONCLUSIONS AND FUTURE PROSPECTS

In this review, studies on miRNA biogenesis, mechanism of action and its regulation in plants have been summarized. Knowledge of many factors such as protein components and structure of pri-miRNAs and miRNAs have increased our knowledge of miRNA biogenesis and function. miRNA pathway and other biological processes may be interlinked. This needs to be worked out in detail, which will subsequently fine-tune our knowledge of control of various biological processes. Roles of HASTY and DCL1 has to be elaborately inquired. Studies are also required for determining exact subcellular localization of the miRNA-RISC complex. The knowledge about additional components/proteins involved in miRNA processing and maturation needs to be enhanced. The interactions between miRNAs need to be worked out in fine detail. Other areas of miRNA action, the mechanisms of which need to be worked out in detail are mechanism of miRNA-mediated translation inhibition and criteria used by cells in selecting translation inhibition or target cleavage as their mode of repression of gene activity. Because miRNAs control numerous processes vital to the development of the plant as well as plant's response to various stress conditions such as drought, cold, viral defense, etc., further knowledge obtained on various aspects of miRNA biogenesis and its regulation as well mechanism of action of miRNAs, may be potentially beneficial in increasing agricultural productivity by positively altering plant's response to various stress conditions such as prolonged drought, cold, salinity, encountered in some agricultural areas and modifying other agricultural traits through miRNA technology.

ACKNOWLEDGMENTS

Thanks are due to friends and colleagues for their valuable advice and help during preparation of manuscript.

KEYWORDS

- biogenesis
- control
- MicroRNA
- mRNA cleavage
- plants
- stability

REFERENCES

1. Abdel-Ghany, S. E., & Pilon, M. (2008). MicroRNA-mediated systemic down-regulation of copper protein expression in response to low copper availability in *Arabidopsis*. *J. Biol. Chem. 283*, 15932–15945.
2. Allen, E., Xie, Z., Gustafson, A. M., & Carrington, J. C. (2005). MicroRNA-directed phasing during trans-acting siRNA biogenesis in plants. *Cell 121*, 207–221.
3. Arteaga-Vazquez, M., Caballero-Perez, J., & Vielle-Calzada, J. P. (2006). A family of microRNAs present in plants and animals. *Plant Cell 18*, 3355–3369.
4. Aukerman, M. J., & Sakai, H. (2003). Regulation of flowering time and floral organ identity by a microRNA and its APETALA2-like target genes. *Plant Cell 15*, 2730–2741.
5. Bartel, D. P. (2004). MicroRNAs: genomics, biogenesis, mechanism, and function. *Cell 116*, 281–297.
6. Baumberger, N., & Baulcombe, D. C. (2005). *Arabidopsis* ARGONAUTE1 is an RNA Slicer that selectively recruits microRNAs and short interfering RNAs. *Proc. Natl. Acad. Sci. U.S.A., 102*, 11928–11933.
7. Ben Chaabane, S., Liu, R., Chinnusamy, V., Kwon, Y., Park, J. H., Kim, S. Y., Zhu, J. K., Yang, S. W., & Lee, B. H. (2013). STA1, an *Arabidopsis* pre-mRNA processing factor-6 homolog, is a new player involved in miRNA biogenesis. *Nucleic Acids Res. 41*, 1984–1997.
8. Bielewicz, D., Kalak, M., Kalyna, M., Windels, D., Barta, A., Vazquez, F., Szweykowska-Kulinska, Z., & Jarmolowski, A. (2013). Introns of plant pri-miRNAs enhance miRNA biogenesis. *EMBO Rep. 14*, 622–628.
9. Bohmert, K., Camus, I., Bellini, C., Bouchez, D., Caboche, M., & Benning, C. (1998). *AGO1* defines a novel locus of *Arabidopsis* controlling leaf development. *EMBO J. 17*, 170–180.
10. Bollman, K. M., Aukerman, M. J., Park, M. Y., Hunter, C., Berardini, T. Z., & Poethig, R. S. (2003). HASTY, the *Arabidopsis* ortholog of exportin5/MSN5, regulates phase change and morphogenesis. *Development 130*, 1493–1504.
11. Bologna, N. G., Mateos, J. L., Bresso, E. G., & Palatnik, J. F. (2009). A loop-to-base processing mechanism underlies the biogenesis of plant microRNAs miR319 and miR159. *EMBO J. 28*, 3646–3656.

12. Bologna, N. G., & Voinnet, O. (2014). The diversity, biogenesis, and activities of endogenous silencing small RNAs in *Arabidopsis*. *Annu. Rev. Plant Biol. 65*, 473–503.

13. Breakfield, N. W., Corcoran, D. L., Petricka, J. J., Shen, J., Sae-Seaw, J., Rubio-Somoza, I., Weigel, D., Ohler, U., & Benfey, P. N. (2012). High resolution experimental and computational profiling of tissue specific known and novel miRNAs in *Arabidopsis*. *Genome Res. 22*, 163–176.

14. Brodersen, P., Sakvarelidze-Achard, L., Bruun-Rasmussen, M., Dunoyer, P., Yamamoto, Y. Y., Sieburth, L., & Voinnet, O. (2008). Widespread translational inhibition by plant miRNAs and siRNAs. *Science 320*, 1185–1190.

15. Brodersen, P., Sakvarelidze-Achard, L., Schaller, H., Khafif, M., Schott, G., Bendahmane, A., & Voinnet, O. (2012). Isoprenoid biosynthesis is required for miRNA function and affects membrane association of argonaute1 in *Arabidopsis*. *Proc. Natl. Acad. Sci. U.S.A., 109*, 1778–1783.

16. Carbonell, A., Fahlgren, N., Garcia-Ruiz, H., Gilbert, K. B., Montgomery, T. A., Nguyen, T., Cuperus, J. T., & Carrington, J. C. (2012). Functional analysis of three *Arabidopsis* argonautes using slicer defective mutants. *Plant Cell 24*, 3613–3629.

17. Carlsbecker, A., Lee, J. Y., Roberts, C. J., Dettmer, J., Lehesranta, S., Zhou, J., Lindgren, O., Moreno-Risueno, M. A., Vaten, A., Thitamadee, S., Campilho, A., Sebastian, J., Bowman, J. L., Helariutta, Y., & Benfey, P. N. (2010). Cell signalling by microRNA165/6 directs gene dose-dependent root cell fate. *Nature 465*, 316–321.

18. Carrington, J. C., & Ambros, V. (2003). Role of microRNAs in plant and animal development. *Science 301*, 336–338.

19. Cenik, E. S., & Zamore, P. D. (2011). Argonaute proteins. *Curr. Biol. 21*(12), R446-R449.

20. Cerutti, L., Mian, N., & Bateman, A. (2000). Domains in gene silencing and cell differentiation proteins: the novel PAZ domain and redefinition of the PIWI domain. *Trends Biochem. Sci. 25*, 481–482.

21. Chapman, E. J., & Carrington, J. C. (2007). Specialization and evolution of endogenous small RNA pathways. *Nat. Rev. Genet. 8*, 884–896.

22. Chatterjee, S., & Grosshans, H. (2009). Active turnover modulates mature microRNA activity in *Caenorhabditis elegans*. *Nature 461*, 546–549.

23. Chen, T., Cui, P., & Xiong, L. (2015). The RNA-binding protein HOS5 and serine/arginine-rich proteins RS40 and RS41 participate in miRNA biogenesis in *Arabidopsis*. *Nucleic Acids Res., 43*(17), 8283–8298.

24. Chen, X. (2004). A microRNA as a translational repressor of APETALA2 in *Arabidopsis* flower development. *Science 303*, 2022–2025.

25. Chitwood, D. H., Nogueira, F. T., Howell, M. D., Montgomery, T. A., Carrington, J. C., & Timmermans, M. C. (2009). Pattern formation via small RNA mobility. *Genes Dev. 23*, 549–554.

26. Chuck, G., Candela, H., & Hake, S. (2009). Big impacts by small RNAs in plant development. Curr. Opin. *Plant Biol. 12*, 81–86.

27. Chuck, G., Cigan, A. M., Saeteurn, K., & Hake, S. (2007). The heterochronic maize mutant Corngrass1 results from overexpression of a tandem microRNA. *Nat. Genet. 39*, 544–549.

28. Collart, M. A., & Panasenko, O. O. (2012). The Ccr4–not complex. *Gene 492*(1), 42–53.

29. Coruh, C., Shahid, S., & Axtell, M. J. (2014). Seeing the forest for the trees: annotating small RNA producing genes in plants. *Curr. Opin. Plant Biol. 18C*, 87–95.

30. Cuperus, J. T., Montgomery, T. A., Fahlgren, N., Burke, R. T., Townsend, T., Sullivan, C. M., & Carrington, J. C. (2010). Identification of MIR390a precursor processing-defective mutants in *Arabidopsis* by direct genome sequencing. *Proc. Natl. Acad. Sci. U.S.A., 107*, 466–471.

31. Derrien, B., Baumberger, N., Schepetilnikov, M., Viotti, C., De Cillia, J., Ziegler-Graff, V., Isono, E., Schumacher, K., & Genschik, P. (2012). Degradation of the antiviral component argonaute1 by the autophagy pathway. *Proc. Natl. Acad. Sci. U.S.A., 109*, 15942–15946.

32. Dong, Z., Han, M. H., & Fedoroff, N. (2008). The RNA-binding proteins HYL1 and SE promote accurate *in vitro* processing of pri-miRNA by DCL1. *Proc. Natl. Acad. Sci. U.S.A., 105*, 9970–9975.

33. Eamens, A. L., Smith, N. A., Curtin, S. J., Wang, M. B., & Waterhouse, P. M. (2009). The *Arabidopsis thaliana* double-stranded RNA binding protein DRB1 directs guide strand selection from microRNA duplexes. *RNA 15*(12), 2219–2235.

34. Earley, K., Smith, M., Weber, R., Gregory, B., & Poethig, R. (2010). An endogenous F-box protein regulates argonaute-1 in *Arabidopsis thaliana*. *Silence 1*, 15.

35. Earley, K. W., & Poethig, R. S. (2011). Binding of the cyclophilin-40 ortholog SQUINT to Hsp90 protein is required for SQUINT function in *Arabidopsis*. *J. Biol. Chem. 286*, 38184–38189.

36. Engelsberger, W. R., & Schulze, W. X. (2012). Nitrate and ammonium lead to distinct global dynamic phosphorylation patterns when resupplied to nitrogen-starved *Arabidopsis* seedlings. *Plant, J. 69*, 978–995.

37. Fahlgren, N., Montgomery, T. A., Howell, M. D., Allen, E., Dvorak, S. K., Alexander, A. L., & Carrington, J. C. (2006). Regulation of *Auxin Response Factor-3* by *TAS3* ta-siRNA affects developmental timing and patterning in *Arabidopsis*. *Curr. Biol. 16*, 939–944.

38. Fang, Y., & Spector, D. L. (2007). Identification of nuclear dicing bodies containing proteins for microRNA biogenesis in living *Arabidopsis* plants. *Curr. Biol. 17*, 818–823.

39. Franco-Zorrilla, J. M., Valli, A., Todesco, M., Mateos, I., Puga, M. I., Rubio-Somoza, I., Leyva, A., Weigel, D., Garcia, J. A., & Paz-Ares, J. (2007). Target mimicry provides a new mechanism for regulation of microRNA activity. *Nat. Genet. 39*, 1033–1037.

40. Fujioka, Y., Utsumi, M., Ohba, Y., & Watanabe, Y. (2007). Location of a possible miRNA processing site in SmD3/SmB nuclear bodies in *Arabidopsis*. *Plant Cell Physiol. 48*, 1243–1253.

41. Gandikota, M., Birkenbihl, R. P., Hohmann, S., Cardon, G. H., Saedler, H., & Huijser, P. (2007). The miRNA156/157 recognition element in the 3' UTR of the *Arabidopsis* SBP box gene *SPL3* prevents early flowering by translational inhibition in seedlings. *Plant, J. 49*, 683–693.

42. Garcia, D. (2008). A miRacle in plant development: role of microRNAs in cell differentiation and patterning. Semin. *Cell Dev. Biol. 19*, 586–595.

43. Garcia, D., Collier, S. A., Byrne, M. E., & Martienssen, R. A. (2006). Specification of leaf polarity in *Arabidopsis* via the trans-acting siRNA pathway. *Curr. Biol. 16*, 933–938.

44. Ghildiyal, M., & Zamore, P. D. (2009). Small silencing RNAs: an expanding universe. *Nat. Rev. Genet. 10*, 94–108.

45. Golden, T. A., Schauer, S. E., Land, J. D., Pien, S., Mushegian, A. R., Grossniklaus, U., Meinke, D. W., & Ray, A. (2002). Short Integuments1/Suspensor1/Carpel Factory, a Dicer homolog, is a maternal effect gene required for embryo development in *Arabidopsis*. *Plant Physiol. 130*, 808–822.

46. Grant-Downton, R., Kourmpetli, S., Hafidh, S., Khatab, H., Le Trionnaire, G., Dickinson, H., & Twell, D. (2013). Artificial microRNAs reveal cell-specific differences in small RNA activity in pollen. *Curr. Biol. 23*, R599–R601.

47. Gregory, B. D., O'Malley, R. C., Lister, R., Urich, M. A., Tonti-Filippini, J., Chen, H., Millar, A. H., & Ecker, J. R. (2008). A link between RNA metabolism and silencing affecting *Arabidopsis* development. *Dev. Cell. 14*, 854–866.

48. Griffiths-Jones, S., Saini, H. K., Dongen, S. V., & Enright, A. J. (2008). miRBase: tools for microRNA genomics. *Nucleic Acids Res. 36*, D154–158.

49. Grigg, S. P., Canales, C., Hay, A., & Tsiantis, M. (2005). SERRATE coordinates shoot meristem function and leaf axial patterning in *Arabidopsis*. *Nature 437*, 1022–1026.

50. Grigorova, B., Mara, C., Hollender, C., Sijacic, P., Chen, X., & Liu, Z. (2011). Leunig and Seuss co-repressors regulate miR172 expression in *Arabidopsis* flowers. *Development 138*, 2451–2456.

51. Gruber, J. J., Zatechka, D. S., Sabin, L. R., Yong, J., Lum, J. J., Kong, M., Zong, W. X., Zhang, Z., Lau, C. K., Rawlings, J., Cherry, S., Ihle, J. N., Dreyfuss, G., & Thompson, C. B. (2009). Ars2 links the nuclear capbinding complex to RNA interference and cell proliferation. *Cell 138*, 328–339.

52. Ha, M., & Kim, V. N. (2014). Regulation of microRNA biogenesis. *Nat. Rev. Mol. Cell. Biol. 15*, 509–524.

53. Hajheidari, M., Farrona, S., Huettel, B., Koncz, Z., & Koncz, C. (2012). CDKF-1 and CDKD protein kinases regulate phosphorylation of serine residues in the C-terminal domain of *Arabidopsis* RNA polymeraseII. *Plant Cell 24*(4), 1626–1642.

54. Haley, B, & Zamore, P. D. (2004). Kinetic analysis of the RNAi enzyme complex. *Nat. Struct. Mol. Biol. 11*(7), 599–606.

55. Hiraguri, A., Itoh, R., Kondo, N., Nomura, Y., Aizawa, D., Murai, Y., Koiwa, H., Seki, M., Shinozaki, K., & Fukuhara, T. (2005). Specific interactions between Dicer-like proteins and HYL1/DRB-family dsRNAbinding proteins in *Arabidopsis thaliana*. *Plant Mol. Biol. 57*, 173–188.

56. Hirsch, J., Lefort, V., Vankersschaver, M., Boualem, A., Lucas, A., Thermes, C., d'Aubenton-Carafa, Y., & Crespi, M. (2006). Characterization of 43 non-protein-coding mRNA genes in *Arabidopsis*, including the MIR162a-derived transcripts. *Plant Physiol. 140*, 1192–1204.

57. Huang, Y., Ji, L., Huang, Q., Vassylyev, D. G., Chen, X., & Ma, J. B. (2009). Structural insights into mechanisms of the small RNA methyltransferase HEN1. *Nature 461*, 823–827.

58. Humphreys, D. T., Westman, B. J., Martin, D. I., & Preiss, T. (2005). MicroR-NAs control translation initiation by inhibiting eukaryotic initiation factor 4E/cap and poly(A) tail function. *Proc. Natl. Acad. Sci. U.S.A., 102*, 16961–16966.

59. Ibrahim, F., Rohr, J., Jeong, W. J., Hesson, J., & Cerutti, H. (2006). Untemplated oligoadenylation promotes degradation of RISCcleaved transcripts. *Science 314*(5807), 1893.

60. Ibrahim, F., Rymarquis, L. A., Kim, E. J., Becker, J., Balassa, E., Green, P. J., & Cerutti, H. (2010). Uridylation of mature miRNAs and siRNAs by the MUT68 nucleotidyltransferase promotes their degradation in *Chlamydomonas. Proc. Natl. Acad. Sci. U.S.A., 107*, 3906–3911.

61. Iki, T., Yoshikawa, M., Meshi, T., & Ishikawa, M. (2012). Cyclophilin40 facilitates HSP90-mediated RISC assembly in plants. *EMBO J. 31*, 267–278.

62. Iki, T., Yoshikawa, M., Nishikiori, M., Jaudal, M. C., Matsumoto-Yokoyama, E., Mitsuhara, I., Meshi, T., & Ishikawa, M. (2010). *In vitro* assembly of plant RNA-induced silencing complexes facilitated by molecular chaperone HSP90. *Mol. Cell 39*, 282–291.

63. Iwakawa, H. O., & Tomari, Y. (2013). Molecular insights into microRNA mediated translational repression in plants. *Mol. Cell 52*, 591–601.

64. Iwata, Y., Takahashi, M., Fedoroff, N. V., & Hamdan, S. M. (2013). Dissecting the interactions of SERRATE with RNA and DICERLIKE1 in *Arabidopsis* microRNA precursor processing. *Nucleic Acids Res. 41*, 9129–9140.

65. Jeong, D. H., Park, S., Zhai, J., Gurazada, S. G., De Paoli, E., Meyers, B. C., & Green, P. J. (2011). Massive analysis of rice small RNAs: mechanistic implications of regulated microRNAs and variants for differential target RNA cleavage. *Plant Cell 23*, 4185–4207.

66. Jia, F., & Rock, C. D. (2013). MIR846 and MIR842 comprise a cistronic miRNA pair that is regulated by abscisic acid by alternative splicing in roots of *Arabidopsis. Plant Mol. Biol. 81*, 447–460.

67. Jia, X., Ren, L., Chen, Q. J., Li, R., & Tang, G. (2009). UV-B-responsive microR-NAs in *Populus tremula. J. Plant Physiol. 166*, 2046–2057.

68. Jones-Rhoades, M. W., & Bartel, D. P. (2004). Computational identification of plant micro-RNAs and their targets, including a stress-induced miRNA. *Mol. Cell 14*, 787–799.

69. Jones-Rhoades, M. W., Bartel, D. P., & Bartel, B. (2006). MicroRNAs and their regulatory roles in plants. Annu. Rev. *Plant Biol. 57*, 19–53.

70. Jung, J. H., Seo, P. J., & Park, C. M. (2009). MicroRNA biogenesis and function in higher plants. *Plant Biotechnol. Rep. 3*, 111–126.

71. Jung, J. H., Seo, Y. H., Seo, P. J., Reyes, J. L., Yun, J., Chua, N. H., & Park, C. M. (2007). The GIGANTEA-regulated microRNA172 mediates photoperiodic flowering independent of CONSTANS in *Arabidopsis. Plant Cell 19*, 2736–2748.

72. Kidner, C. A., & Martienssen, R. A. (2004). Spatially restricted microRNA directs leaf polarity through ARGONAUTE1. *Nature 428*, 81–84.

73. Kidner, C. A., & Martienssen, R. A. (2005). The role of ARGONAUTE1 (AGO1) in meristem formation and identity. *Dev. Biol. 280*, 504–517.

74. Kim, S., Yang, J. Y., Xu, J., Jang, I. C., Prigge, M. J., & Chua, N. H. (2008). Two cap-binding proteins CBP20 and CBP80 are involved in processing primary MicroR-NAs. *Plant Cell Physiol. 49*, 1634–1644.

75. Kim, W., Benhamed, M., Servet, C., Latrasse, D., Zhang, W., Delarue, M., & Zhou, D. X. (2009). Histone acetyltransferase GCN5 interferes with the miRNA pathway in *Arabidopsis. Cell Res. 19*, 899–909.

76. Kim, Y. J., Zheng, B., Yu, Y., Won, S. Y., Mo, B., & Chen, X. (2011). The role of Mediator in small and long noncoding RNA production in *Arabidopsis thaliana. EMBO J. 30*, 814–822.

77. Kiriakidou, M., Tan, G. S., Lamprinaki, S., De Planell-Saguer, M., Nelson, P. T., & Mourelatos, Z. (2007). An mRNA m7G cap binding-like motif within human Ago2 represses translation. *Cell 129* (6), 1141–1151.

78. Knop, K., Stepien, A., Barciszewska-Pacak, M., Taube, M., Bielewicz, D., Michalak, M., Borst, J. W., Jarmolowski, A., & Szweykowska-Kulinska, Z. (2017). Active 5' splice sites regulate the biogenesis efficiency of *Arabidopsis* microRNAs derived from intron-containing genes. *Nucleic Acids Res. 45*(5), 2757–2775. https://doi.org/10. 1093/nar/gkw895.

79. Köster, T., Meyer, K., Weinholdt, C., Smith, L. M., Lummer, M., Speth, C., Grosse, I., Weigel, D., & Staiger, D. (2014). Regulation of pri-miRNA processing by the hnRNP-like protein AtGRP7 in *Arabidopsis. Nucleic Acids Res. 42(15)*, 9925–9936.

80. Kruszka, K., Pacak, A., Swida-Barteczka, A., Stefaniak, A. K., Kaja, E., Sierocka, I., Karlowski, W., Jarmolowski, A., & Szweykowska-Kulinska, Z. (2013). Developmentally regulated expression and complex processing of barley pri-microRNAs. *BMC Genom. 14*, 34.

81. Kurihara, Y., Takashi, Y., & Watanabe, Y. (2006). The interaction between DCL1 and HYL1 is important for efficient and precise processing of pri-miRNA in plant microRNA biogenesis. *RNA 12*, 206–212.

82. Kurihara, Y., & Watanabe, Y. (2004). *Arabidopsis* micro-RNA biogenesis through Dicer-like1 protein functions. *Proc. Natl. Acad. Sci. U.S.A., 101*, 12753–12758.

83. Kwak, P. B., & Tomari, Y. (2012). The N domain of Argonaute drives duplex unwinding during RISC assembly. *Nat. Struct. Mol. Biol. 19*, 145–151.

84. Laubinger, S., Sachsenberg, T., Zeller, G., Busch, W., Lohmann, J. U., Rätsch, G., & Weigel, D. (2008). Dual roles of the nuclear cap-binding complex and SERRATE in pre-mRNA splicing and microRNA processing in *Arabidopsis thaliana. Proc. Natl. Acad. Sci. U.S.A., 105*, 8795–8800.

85. Laubinger, S., Zeller, G., Henz, S. R., Buechel, S., Sachsenberg, T., Wang, J. W., Ratsch, G., & Weigel, D. (2010). Global effects of the small RNA biogenesis machinery on the *Arabidopsis thaliana* transcriptome. *Proc. Natl. Acad. Sci. U.S.A., 107*, 17466–17473.

86. Lee, R. C., Feinbaum, R. L., & Ambros, V. (1993). The, *C. elegans* heterochronic gene lin-4 encodes small RNAs with antisense complementarity to lin-14. *Cell 75*, 843–854.

87. Lewis, B. P., Burge, C. B., & Bartel, D. P. (2005). Conserved seed pairing, often flanked by adenosines, indicates that thousands of human genes are microRNA targets. *Cell 120*(1), 15–20.

88. Li, J., Reichel, M., & Millar, A. A. (2014). Determinants beyond both complementarity and cleavage govern microR159 efficacy in *Arabidopsis. PLoS Genet. 10*, e1004232.

89. Li, J., Yang, Z., Yu, B., Liu, J., & Chen, X. (2005). Methylation protects miR-NAs and siRNAs from a 3'-end uridylation activity in *Arabidopsis*. *Curr. Biol. 15*, 1501–1507.

90. Li, S., Liu, L., Zhuang, X., Yu, Y., Liu, X., Cui, X., Ji, L., Pan, Z., Cao, X., Mo, B., Zhang, F., Raikhel, N., Jiang, L., & Chen, X. (2013). MicroRNAs inhibit the translation of target mRNAs on the endoplasmic reticulum in *Arabidopsis*. *Cell 153*, 562–574.

91. Lin, S., Henriques, R., Wu, H., Niu, Q., Yeh, S., & Chua, N. (2007a). Strategies and mechanisms of plant virus resistance. *Plant Biotechnol. Rep., 1*, 125–134.

92. Lin, Z., Yin, K., Zhu, D., Chen, Z., Gu, H., & Qu, L. J. (2007b). AtCDC5 regulates the G2 to M transition of the cell cycle and is critical for the function of *Arabidopsis* shoot apical meristem. *Cell Res. 17*, 815–828.

93. Liu, C., Axtell, M. J., & Fedoroff, N. V. (2012). The helicase and RNaseIIIa domains of *Arabidopsis* Dicer-Like1 modulate catalytic parameters during microRNA biogenesis. *Plant Physiol. 159*, 748–758.

94. Liu, J., Valencia-Sanchez, M. A., Hannon, G. J., & Parker, R. (2005). MicroRNA-dependent localization of targeted mRNAs to mammalian *P-bodies*. *Nat. Cell. Biol. 7*, 719–723.

95. Liu, Q., Wang, F., & Axtell, M. J. (2014). Analysis of complementarity requirements for plant MicroRNA targeting using a *Nicotiana benthamiana* quantitative transient assay. *Plant Cell 26*, 741–753.

96. Llave, C., Kasschau, K. D., Rector, M. A., & Carrington, J. C. (2002a). Endogenous and silencing-associated small RNAs in plants. *Plant Cell; 14*, 1605–1619.

97. Llave, C., Xie, Z., Kasschau, K. D., & Carrington, J. C. (2002b). Cleavage of Scarecrow-like mRNA targets directed by a class of *Arabidopsis* miRNA. *Science; 297*, 2053–2056.

98. Lobbes, D., Rallapalli, G., Schmidt, D. D., Martin, C., & Clarke, J. (2006). SERRATE: a new player on the plant microRNA scene. *EMBO Rep. 7*, 1052–1058.

99. Lu, C., & Fedoroff, N. (2000). A mutation in the *Arabidopsis* HYL1 gene encoding a dsRNA binding protein affects responses to abscisic acid, auxin, and cytokinin. *Plant Cell 12*, 2351–2366.

100. Machida, S., Chen, H. Y., & Adam Yuan, Y. (2011). Molecular insights into miRNA processing by *Arabidopsis thaliana* Serrate. *Nucleic Acids Res. 39*, 7828–7836.

101. Machida, S., & Yuan, Y. A. (2013). Crystal structure of *Arabidopsis thaliana* Dawdle forkhead-associated domain reveals a conserved phospho-threonine recognition cleft for dicer-like-1 binding. *Mol. Plant 6*, 1290–1300.

102. Mallory, A. C., Reinhart, B. J., Jones-Rhoades, M. W., Tang, G., Zamore, P. D., Barton, M. K., & Bartel, D. P. (2004). MicroRNA control of PHABULOSA in leaf development: importance of pairing to the microRNA 5' region. *EMBO J. 23*(16), 3356–3364.

103. Mallory, A. C., & Vaucheret, H. (2009). ARGONAUTE 1 homeostasis invokes the coordinate action of the microRNA and siRNA pathways. *EMBO Rep. 10*, 521–526.

104. Mallory, A. C., & Vaucheret, H. (2006). Functions of microRNAs and related small RNAs in plants. *Nat. Genet., 38*(Suppl):S31–S36.

105. Manavella, P. A., Hagmann, J., Ott, F., Laubinger, S., Franz, M., Macek, B., & Weigel, D. (2012). Fast-forward genetics identifies plant CPL phosphatases as regulators of miRNA processing factor HYL1. *Cell 151*, 859–870.

106. Margis, R., Fusaro, A. F., Smith, N. A., Curtin, S. J., Watson, J. M., Finnegan, E. J., & Waterhouse, P. M. (2006). The evolution and diversification of Dicers in plants. *FEBS Lett. 580*, 2442–2450.

107. Mateos, J. L., Bologna, N. G., Chorostecki, U., & Palatnik, J. F. (2010). Identification of microRNA processing determinants by random mutagenesis of *Arabidopsis* MIR172a precursor. *Curr. Biol. 20*, 49–54.

108. Matranga, C., Tomari, Y., Shin, C., Bartel, D. P., & Zamore, P. D. (2005). Passenger-strand cleavage facilitates assembly of siRNA into Ago2-containing RNAi enzyme complexes. *Cell 123*(4), 607–620.

109. Megraw, M., Baev, V., Rusinov, V., Jensen, S. T., Kalantidis, K., & Hatzigeorgiou, A. G. (2006). MicroRNA promoter element discovery in *Arabidopsis*. *RNA 12*, 1612–1619.

110. Merchan, F., Boualem, A., Crespi, M., & Frugier, F. (2009). Plant polycistronic precursors containing non-homologous microRNAs target transcripts encoding functionally related proteins. *Genome Biol. 10*, R136.

111. Mi, S., Cai, T., Hu, Y., Chen, Y., Hodges, E., Ni, F., Wu, L. Li, S., Zhou, H., Long, C., Chen, S., Hannon, G. J., & Qi, Y. (2008). Sorting of small RNAs into *Arabidopsis* argonaute complexes is directed by the 5' terminal nucleotide. *Cell 133*(1), 116–127.

112. Michlewski, G., Guil, S., Semple, C. A., & Ca´ceres, J. F. (2008). Posttranscriptional regulation of miRNAs harboring conserved terminal loops. *Mol. Cell 32*, 383–393.

113. Montgomery, T. A., Howell, M. D., Cuperus, J. T., Li, D., Hansen, J. E., Alexander, A. L., Chapman, E. J., Fahlgren, N., Allen, E., & Carrington, J. C. (2008). Specificity of argonaute-7-miR390 interaction and dual functionality in TAS3 trans-acting siRNA formation. *Cell 133*, 128–141.

114. Motomura, K., Le, Q. T., Kumakura, N., Fukaya, T., Takeda, A., & Watanabe, Y. (2012). The role of decapping proteins in the miRNA accumulation in *Arabidopsis thaliana*. RNA *Biol. 9*, 644–652.

115. Nogueira, F. T. S., Chitwood, D. H., Madi, S., Ohtsu, K., Schnable, P. S., Scanlon, M. J., & Timmermans, M. C. P. (2009). Regulation of small RNA accumulation in the maize shoot apex. *PLoS Genet. 5*(1), e1000320. doi: 10.1371/journal.pgen.1000320.

116. Nozawa, M., Miura, S., & Nei, M. (2012). Origins and evolution of microRNA genes in plant species. *Genome Biol. Evol. 4*(3), 230–239.

117. Palma, K., Zhao, Q., Cheng, Y. T., Bi, D., Monaghan, J., Cheng, W., Zhang, Y., & Li, X. (2007). Regulation of plant innate immunity by three proteins in a complex conserved across the plant and animal kingdoms. *Genes Dev. 21*, 1484–1493.

118. Park, M. Y., Wu, G., Gonzalez-Sulser, A., Vaucheret, H., & Poethig, R. S. (2005). Nuclear processing and export of microRNAs in *Arabidopsis*. *Proc. Natl. Acad. Sci. U.S.A., 102*, 3691–3696.

119. Park, W., Li, J., Song, R., Messing, J., & Chen, X. (2002). Carpel Factory, a Dicer homolog, and HEN1, a novel protein, act in microRNA metabolism in *Arabidopsis thaliana*. *Curr. Biol. 12*, 1484–1495.

120. Parker, J. S., Roe, S. M., & Barford, D. (2004). Crystal structure of a PIWI protein suggests mechanisms for siRNA recognition and slicer activity. *EMBO J. 23*, 4727–4737.

121. Pashkovskiy, P. P., & Ryazansky, S. S. (2013). Biogenesis, evolution, and functions of plant microRNAs. *Biochem. (Moscow) 78*(6), 627–637.

122. Peragine, A., Yoshikawa, M., Wu, G., Albrecht, H. L., & Poethig, R. S. (2004). *SGS3* and *SGS2/SDE1/RDR6* are required for juvenile development and the production of trans-acting siRNAs in *Arabidopsis. Genes Dev. 18*, 2368–2379.

123. Phillips, J. R., Dalmay, T., & Bartels, D. (2007). The role of small RNAs in abiotic stress. *FEBS Lett. 581*, 3592–3597.

124. Piriyapongsa, J., & Jordan, I. K. (2008). Dual coding of siRNAs and miRNAs by plant transposable elements. *RNA 14*, 814–821.

125. Pouch-Pelissier, M. N., Pelissier, T., Elmayan, T., Vaucheret, H., Boko, D., Jantsch, M. F., & Deragon, J. M. (2008). SINE RNA induces severe developmental defects in *Arabidopsis thaliana* and interacts with HYL1 (DRB1), a key member of the DCL1 complex. PLoS Genet. 4, e1000096. Doi: 10.1371/journal.pgen.1000096.

126. Prigge, M. J., & Wagner, D. R. (2001). The *Arabidopsis* SERRATE gene encodes a zinc finger protein required for normal shoot development. *Plant Cell 13*, 1263–1279.

127. Raczynska, K. D., Simpson, C. G., Ciesiolka, A., Szewc, L., Lewandowska, D., McNicol, J., Szweykowska-Kulinska, Z., Brown, J. W., & Jarmolowski, A. (2010). Involvement of the nuclear cap-binding protein complex in alternative splicing in *Arabidopsis thaliana. Nucleic Acids Res. 38*, 265–278.

128. Raczynska, K. D., Stepien, A., Kierzkowski, D., Kalak, M., Bajczyk, M., McNicol, J., Simpson, C. G., Szweykowska-Kulinska, Z., Brown, J. W., & Jarmolowski, A. (2014). The SERRATE protein is involved in alternative splicing in *Arabidopsis thaliana. Nucleic Acids Res. 42*, 1224–1244.

129. Rajagopalan, R., Vaucheret, H., Trejo, J., & Bartel, D. P. (2006). A diverse and evolutionarily fluid set of microRNAs in *Arabidopsis thaliana. Genes Dev. 20*, 3407–3425.

130. Ramachandran, V., & Chen, X. (2008). Degradation of microRNAs by a family of exoribonucleases in *Arabidopsis. Science 321*, 1490–1492.

131. Reinhart, B. J., Slack, F. J., Basson, M., Pasquinelli, A. E., Bettinger, J. C., Rougvie, A. E., Horvitz, H. R., & Ruvkun, G. (2000). The 21-nucleotide let-7 RNA regulates developmental timing in *Caenorhabditis elegans. Nature 403*, 901–906.

132. Reinhart, B. J., Weinstein, E. G., Rhoades, M. W., Bartel, B., & Bartel, D. P. (2002) MicroRNAs in plants. *Genes Dev. 16*, 1616–1626.

133. Ren, G., Chen, X., & Yu, B. (2012a). Uridylation of miRNAs by hen1 suppressor1 in *Arabidopsis. Curr. Biol. 22*, 695–700.

134. Ren, G., Xie, M., Dou, Y., Zhang, S., Zhang, C., Yu, B. (2012b). Regulation of miRNA abundance by RNA binding protein TOUGH in *Arabidopsis. Proc. Natl. Acad. Sci. U.S.A., 109*, 12817–12821.

135. Ren, G., Xie, M., Zhang, S., Vinovskis, C., Chen, X., & Yu, B. (2014). Methylation protects microRNAs from an AGO1-associated activity that uridylates 5' RNA fragments generated by AGO1 cleavage. *Proc. Natl. Acad. Sci. U.S.A., 111*(17), 6365–6370.

136. Ren, G., & Yu, B. (2012). Critical roles of RNA-binding proteins in miRNA biogenesis in *Arabidopsis*. *RNA Biol. 9*, 1424–1428.

137. Rhoades, M. W., Reinhart, B. J., Lim, L. P., Burge, C. B., Bartel, B., & Bartel, D. P. (2002). Prediction of plant microRNA targets. *Cell 110*, 513–520.

138. Rivas, F. V., Tolia, N. H., Song, J. J., Aragon, J. P., Liu, J., Hannon, G. J., & Joshua-Tor, L. (2005). Purified Argonaute2 and an siRNA form recombinant human RISC. *Nat. Struct. Mol. Biol. 12*, 340–349.

139. Rogers, K., & Chen, X. (2013). Biogenesis, turnover, and mode of action of plant microRNAs. *Plant Cell 25*, 2383–2399.

140. Ronemus, M., Vaughn, M. W., & Martienssen, R. A. (2006). MicroRNA-targeted and small interfering RNA-mediated mRNA degradation is regulated by Argonaute, Dicer, and RNA-Dependent-RNA Polymerase in *Arabidopsis*. *The Plant Cell 18*, 1559–1574.

141. Schauer, S. E., Jacobsen, S. E., Meinke, D. W., & Ray, A. (2002). DICERLIKE1, blind men and elephants in *Arabidopsis* development. *Trends Plant Sci. 7*, 487–491.

142. Schwab, R., Palatnik, J. F., Riester, M., Schommer, C., Schmid, M., & Weigel, D. (2005). Specific effects of microRNAs on the plant transcriptome. *Dev. Cell 8*, 517–527.

143. Schwab, R., Speth, C., Laubinger, S., & Voinnet, O. (2013). Enhanced microRNA accumulation through stem loop-adjacent introns. *EMBO Rep. 14*, 615–621.

144. Shen, B., & Goodman, H. M. (2004). Uridine addition after microRNA directed cleavage. *Science 306*, 997.

145. Sire, C., Moreno, A. B., Garcia-Chapa, M., Lopez-Moya, J. J., & San Segundo, B. (2009). Diurnal oscillation in the accumulation of *Arabidopsis* microRNAs, miR167, miR168, miR171 and miR398. *FEBS Lett. 583*, 1039–1044.

146. Smith, M. R., Willmann, M. R., Wu, G., Berardini, T. Z., Moller, B., Weijers, D., & Poethig, R. S. (2009). Cyclophilin40 is required for microRNA activity in *Arabidopsis*. *Proc. Natl. Acad. Sci. U.S.A., 106*, 5424–5429.

147. Song, L., Axtell, M. J., & Fedoroff, N. V. (2010). RNA secondary structural determinants of miRNA precursor processing in *Arabidopsis*. *Curr. Biol. 20*, 37–41.

148. Song, L., Han, M. H., Lesicka, J., & Fedoroff, N. (2007). *Arabidopsis* primary microRNA processing proteins HYL1 and DCL1 define a nuclear body distinct from the Cajal body. *Proc. Natl. Acad. Sci. U.S.A., 104*, 5437–5442.

149. Souret, F. F., Kastenmayer, J. P., & Green, P. J. (2004). AtXRN4 degrades mRNA in *Arabidopsis* and its substrates include selected miRNA targets. *Mol. Cell 15*, 173–183.

150. Speth, C., Willing, E. M., Rausch, S., Schneeberger, K., & Laubinger, S. (2013). RACK1 scaffold proteins influence miRNA abundance in *Arabidopsis*. *Plant, J. 76*, 433–445.

151. Stoppin-Mellet, V., Gaillard, J., & Vantard, M. . (2006). Katanin's severing activity favors bundling of cortical microtubules in plants. *Plant, J 46*, 1009–1017.

152. Sunkar, R., Kapoor, A., & Zhu, J. K. (2006). Posttranscriptional induction of two Cu/Zn superoxide dismutase genes in *Arabidopsis* is mediated by down-regulation of miR398 and important for oxidative stress tolerance. *Plant Cell 18*, 2051–2065.

153. Szarzynska, B., Sobkowiak, L., Pant, B. D., Balazadeh, S., Scheible, W. R., Mueller-Roeber, B., Jarmolowski, A., & Szweykowska-Kulinska, Z. (2009). Gene

structures and processing of *Arabidopsis thaliana* HYL1-dependent pri-miRNAs. *Nucleic Acids Res. 37*, 3083–3093.

154. Tang, G., Reinhart, B. J., Bartel, D. P., & Zamore, P. D. (2003). A biochemical framework for RNA silencing in plants. *Genes Dev. 17*, 49–63.

155. Tolia, N. H., & Joshua-Tor, L. (2007). Slicer and the argonautes. *Nat. Chem. Biol. 3*, 36–43.

156. Valoczi, A., Varallyay, E., Kauppinen, S., Burgyan, J., & Havelda, Z. (2006). Spatio-temporal accumulation of microRNAs is highly coordinated in developing plant tissues. *Plant, J. 47*, 140–151.

157. Vaucheret, H. (2008). Plant Argonautes. *Trends Plant Sci. 13*, 350–358.

158. Vaucheret, H. (2006). Post-transcriptional small RNA pathways in plants: mechanisms and regulations. *Genes Dev. 20*, 759–771.

159. Vaucheret, H., Vazquez, F., Crete, P., & Bartel, D. P. (2004). The action of ARGONAUTE1 in the miRNA pathway and its regulation by the miRNA pathway are crucial for plant development. *Genes Dev. 18*, 1187–1197.

160. Vazquez, F., Gasciolli, V., Crete, P., & Vaucheret, H. (2004a). The nuclear dsRNA binding protein HYL1 is required for microRNA accumulation and plant development, but not posttranscriptional transgene silencing. *Curr. Biol. 14*, 346–351.

161. Vazquez, F., Vaucheret, H., Rajagopalan, R., Lepers, C., Gasciolli, V., Mallory, A. C., Hilbert, J. L., Bartel, D. P., & Crete, P. (2004b). Endogenous trans-acting siRNAs regulate the accumulation of *Arabidopsis* mRNAs. *Mol. Cell 16*(1), 69–79.

162. Voinnet, O. (2009). Origin, biogenesis, and activity of plant microRNAs. *Cell 136*, 669–687.

163. Wang, B., Duan, C. G, Wang, X., Hou, Y. J., Yan, J., Gao, C., Kim, J. H., Zhang, H., & Zhu, J. K. (2015). HOS1 regulates Argonaute1 by promoting the transcription of the microRNA gene *MIR168b* in *Arabidopsis*. *Plant, J. 81*(6), 861–870.

164. Wang, F., & Perry, S. E. (2013). Identification of direct targets of FUSCA3, a key regulator of *Arabidopsis* seed development. *Plant Physiol. 161*, 1251–1264.

165. Wang, J. W., Czech, B., & Weigel, D. (2009). MiR156-regulated SPL transcription factors define an endogenous flowering pathway in *Arabidopsis thaliana*. *Cell 138*, 738–749.

166. Wang, J. W., Schwab, R., Czech, B., Mica, E., & Weigel, D. (2008). Dual effects of miR156-targeted SPL genes and CYP78A5/KLUH on plastochron length and organ size in *Arabidopsis thaliana*. *Plant Cell 20*, 1231–1243.

167. Wang, L., Song, X., Gu, L., Li, X., Cao, S., Chu, C., Cui, X., Chen, X., & Cao, X. (2013). NOT2 proteins promote polymeraseII-dependent transcription and interact with multiple microRNA biogenesis factors in *Arabidopsis*. *Plant Cell 25*, 715–727.

168. Werner, S., Wollmann, H., Schneeberger, K., & Weigel, D. (2010). Structure determinants for accurate processing of miR172a in *Arabidopsis thaliana*. *Curr. Biol. 20*, 42–48.

169. Willmann, M. R., & Poethig, R. S. (2005). Time to grow up: the temporal role of smallRNAs in plants. *Curr. Opin. Plant Biol. 8*, 548–552.

170. Wu, F., Yu, L., Cao, W., Mao, Y., Liu, Z., & He, Y. (2007). The N-terminal double-stranded RNA binding domains of *Arabidopsis* Hyponastic Leaves-1 are sufficient for pre-microRNA processing. *Plant Cell 19*, 914–925.

171. Wu, G., Park, M. Y., Conway, S. R., Wang, J. W., Weigel, D., & Poethig, R. S. (2009). The sequential action of miR156 and miR172 regulates developmental timing in *Arabidopsis*. *Cell 138*(4), 750–759.

172. Wu, G., & Poethig, R. S. (2006). Temporal regulation of shoot development in *Arabidopsis thaliana* by miR156 and its target SPL3. *Development 133*, 3539–3547.

173. Wu, L., Zhou, H., Zhang, Q., Zhang, J., Ni, F., Liu, C., & Qi, Y. (2010). DNA methylation mediated by a microRNA pathway. *Mol. Cell 38*(3), 465–475.

174. Wu, X., Shi, Y., Li, J., Xu, L., Fang, Y., Li, X., & Qi, Y. (2013). A role for the RNA binding protein MOS2 in microRNA maturation in *Arabidopsis*. *Cell Res. 23*, 645–657.

175. Xie, M., Zhang, S., & Yu, B. (2015). microRNA biogenesis, degradation and activity in plants. *Cell. Mol. Life Sci. 72*(1), 87–99.

176. Xie, Z., Allen, E., Fahlgren, N., Calamar, A., Givan, S. A., & Carrington, J. C. (2005a). Expression of *Arabidopsis* miRNA genes. *Plant Physiol. 138*, 2145–2154.

177. Xie, Z., Allen, E., Wilken, A., & Carrington, J. C. (2005b). DICER-LIKE4 functions in transacting small interfering RNA biogenesis and vegetative phase change in *Arabidopsis thaliana*. *Proc. Natl. Acad. Sci. U.S.A., 102*, 12984–12989.

178. Xie, Z., Kasschau, K. D., & Carrington, J. C. (2003). Negative feedback regulation of Dicer-Like1 in *Arabidopsis* by microRNA-guided mRNA degradation. *Curr. Biol. 13*, 784–789.

179. Xie, Z., Khanna, K., & Ruan, S. (2010). Expression of microRNAs and its regulation in plants. Semin. *Cell Dev. Biol. 21*, 790–797.

180. Yamasaki, H., Abdel-Ghany, S. E., Cohu, C. M., Kobayashi, Y., Shikanai, T., & Pilon, M. (2007). Regulation of copper homeostasis by micro-RNA in *Arabidopsis*. *J. Biol. Chem. 282*, 16369–16378.

181. Yamasaki, H., Hayashi, M., Fukazawa, M., Kobayashi, Y., & Shikanai, T. (2009). SQUAMOSA Promoter Binding Protein-Like7 is a central regulator for copper homeostasis in *Arabidopsis*. *Plant Cell 21*, 347–361.

182. Yang, L., Liu, Z. Q., Lu, F., Dong, A. W., & Huang, H. (2006a). SERRATE is a novel nuclear regulator in primary microRNA processing in *Arabidopsis*. *Plant, J.; 47*, 841–850.

183. Yang, L., Wu, G., & Poethig, R. S. (2012). Mutations in the GW-repeat protein SUO reveal a developmental function for microRNA mediated translational repression in *Arabidopsis*. *Proc. Natl. Acad. Sci. U.S.A., 109*, 315–320.

184. Yang, S. W., Chen, H. Y., Yang, J., Machida, S., Chua, N. H., & Yuan, Y. A. (2010). Structure of *Arabidopsis* HYPONASTIC LEAVES1 and its molecular implications for miRNA processing. *Structure 18*, 594–605.

185. Yang, Z., Ebright, Y. W., Yu, B., & Chen, X. (2006b). HEN1 recognizes 21–24 nt small RNA duplexes and deposits a methyl group onto the 2' OH of the 3' terminal nucleotide. *Nucleic Acids Res. 34*(2), 667–675.

186. Yu, B., Bi, L., Zheng, B., Ji, L., Chevalier, D., Agarwal, M., Ramachandran, V., Li, W., Lagrange, T., Walker, J. C., & Chen, X. (2008). The FHA domain proteins DAWDLE in *Arabidopsis* and SNIP1 in humans act in small RNA biogenesis. *Proc. Natl. Acad. Sci. U.S.A., 105*, 10073–10078.

187. Yuan, Y. R., Pei, Y., Ma, J. B., Kuryavyi, V., Zhadina, M., Meister, G., Chen, H. Y., Dauter, Z., Tuschl, T., & Patel, D. J. (2005). Crystal structure of, *A. aeolicus* argo-

naute, a site-specific DNA-guided endoribonuclease, provides insights into RISC-mediated mRNA cleavage. *Mol. Cell 19*(3), 405–419.

188. Yumul, R. E., Kim, Y. J., Liu, X., Wang, R., Ding, J., Xiao, L., & Chen, X. (2013). Powerdress and diversified expression of the MIR172 gene family bolster the floral stem cell network. *PLoS Genet. 9*, e1003218.

189. Zhai, J., Zhao, Y., Simon, S. A., Huang, S., Petsch, K., Arikit, S., Pillay, M., Ji, L., Xie, M., Cao, X., Yu, B., Timmermans, M., Yang, B., Chen, X., & Meyers, B. C. (2013). Plant microRNAs display differential 3' truncation and tailing modifications that are ARGONAUTE1 dependent and conserved across species. *Plant Cell 25*, 2417–2428.

190. Zhan, X., Wang, B., Li, H., Liu, R., Kalia, R. K., Zhu, J. K., & Chinnusamy, V. (2012). *Arabidopsis* proline rich protein important for development and abiotic stress tolerance is involved in microRNA biogenesis. *Proc. Natl. Acad. Sci. U.S.A., 109*, 18198–18203.

191. Zhang, B. H., Pan, X. P., Wang, Q. L., Cobb, G. P., & Anderson, T. A. (2005). Identification and characterization of new plant microRNAs using EST analysis. *Cell Res. 15*, 336–360.

192. Zhang, S., Liu, Y., & Yu, B. (2014). PRL1, an RNA-binding protein, positively regulates the accumulation of miRNAs and siRNAs in Arabidopsis. *PLoS Genet. 10*(12), e1004841. doi: 10.1371/journal.pgen.1004841.

193. Zhang, S., Xie, M., Ren, G., & Yu, B. (2013). CDC5, a DNA binding protein, positively regulates posttranscriptional processing and/or transcription of primary microRNA transcripts. *Proc. Natl. Acad. Sci. U.S.A., 110*, 17588–17593.

194. Zhao, X., Zhang, H., & Li, L. (2013). Identification and analysis of the proximal promoters of microRNA genes in *Arabidopsis. Genomics 101*, 187–194.

195. Zhao, Y., Yu, Y., Zhai, J., Ramachandran, V., Dinh, T. T., Meyers, B. C., Mo, B., & Chen, X. (2012). The *Arabidopsis* nucleotidyl transferase HESO1 uridylates unmethylated small RNAs to trigger their degradation. *Curr. Biol. 22*, 689–694.

196. Zhou, X., Ruan, J., Wang, G., & Zhang, W. (2007). Characterization and identification of MicroRNA core promoters in four model species. *PLoS Comput. Biol. 3*(3), e37. doi: 10.1371/journal.pcbi.0030037.

197. Zhou, X., Wang, G., Sutoh, K., Zhu, J. K., & Zhang, W. (2008). Identification of cold-inducible microRNAs in plants by transcriptome analysis. *Biochim. Biophys. Acta. 1779*(11), 780–788.

198. Zhu, H., Hu, F., Wang, R., Zhou, X., Sze, S. H., Liou, L. W., Barefoot, A., Dickman, M., & Zhang, X. (2011). *Arabidopsis* Argonaute-10 specifically sequesters miR166/165 to regulate shoot apical meristem development. *Cell 145*, 242–256.

199. Zhu, H., Zhou, Y., Castillo-Gonzalez, C., Lu, A., Ge, C., Zhao, Y. T., Duan, L., Li, Z., Axtell, M. J., Wang, X. J., & Zhang, X. (2013). Bidirectional processing of pri-miRNAs with branched terminal loops by *Arabidopsis* Dicer-like-1. *Nat. Struct. Mol. Biol. 20*, 1106–1115.

CHAPTER 10

MINERAL NUTRITION IN PLANTS AND ITS MANAGEMENT IN SOIL

ANEG SINGH and A. K. TIWARI

U.P. Council of Sugarcane Research, Shahjahanpur – 242001 (U.P.), E-mail: ajju1985@gmail.com

CONTENTS

10.1   INTRODUCTION

The growth and development of plants are determined by numerous factors of soils, fertilizers, water, and climate. Some of these factors are under the control of human being, but some are not. For example, rainfall, temperature, light, air, etc., cannot be controlled by the human. These factors influence the supply of nutrients in soils. Translocation of nutrients within a plant body is an ever-continuing process. In this regard, there is a considerable difference in the mobility of different nutrients. Nitrogen is very mobile in plant in comparison to phosphorus, potassium and magnesium. Nitrogen (a high mobility nutrient) deficient plants have yellow colored lower leaves and green upper leaves but nutrients with weak mobility such as phosphorus, potassium and magnesium deficient plant produce symptoms on newer leaves and growing points.

10.2   ESSENTIAL PLANT NUTRIENTS AND THEIR FUNCTIONS

For past centuries, people knew that substances such as manure, ashes, bone meal, etc., had a stimulating effect on plant growth and this effect was found to result from the essential elements contained in the materials. The plant nutrients are necessary for the growth of green plants. In the absence of any one of these essential elements, a plant can, however, be corrected by the addition of that particular element. Two criteria commonly used in establishing the essentiality of a plant nutrient are: (i) its necessity for the plant to complete its life cycle, and (ii) its direct involvement in the nutrition of the plant apart from the possible effect in correcting some unfavorable condition in the soil or culture medium. The elements generally required by plants are divided in to three groups based on the amount that plants required. A list of the macro, secondary, and micro nutrients are as follows.

- *Macronutrients*: C, H, O, N, P, K
- *Secondary nutrients:* Ca, Mg, S
- *Micronutrients:* Zn, Fe, Cu, Mn, Bo, Mo, Co, Cl, Na, Va, Si, Se, Ni

Carbon is obtained from CO_2 of the air, oxygen from air and water, hydrogen from water, nitrogen from air and/or soil and all other nutrients from the soil. Soil is thus most important source or medium for plant growth. N, P, and K are known as primary plant nutrients and they are required in more than 500 ppm concentration in plant body while Ca, Mg, and S are secondary nutrients required in less than 150 ppm. Other nutrients, as micronutrients, are required in less than 50 ppm but they are as important as the major elements in plant nutrition for the normal growth and proper development of any crop.

10.3 NITROGEN

Nitrogen is the constituent of all proteins, chlorophyll, coenzymes, and nucleic acids. Air is the primary source of nitrogen for plant nutrition and only leguminous crops can directly use this free nitrogen with help of symbiotic bacteria of the genus *Rhizobium*. Nitrogen encourage the vegetative development of plant by imparting a healthy green color to the leaf. It also regulates to some extent, the efficient utilization of P and K.

10.4 PHOSPHORUS

Phosphorus influences the vigor of plant and improves the quality of crops. It is essential component of the organic compound often called the energy currency of the living cells, A.T.P. Phosphorus usually contributes only 0.2 and 0.4 percent of the dry matter of plant. The first precipitated calcium phosphate may slowly change from relatively soluble tri-calcium phosphate, $Ca_3(PO_4)_2$ to apatite forms. In highly weathered tropical soils, some phosphorus may be coated with iron and aluminum oxides and protected from solution [23]. Phosphorus deficiency is characterized by stunted plants that have about equally affected root and top growth.

10.5 POTASSIUM

Potassium is especially important in helping the plants to adopt under environmental stress and also plays many complex role in plant nutrition. Potassium is immobilized, but it does not accumulate in the soil organic fractions. Soils with the lowest available potassium are low textured light soils of India [1]. A deficiency of potassium shows the symptoms of leaf scorch in most crops. Sugarcane plants with symptoms of yellowing of the tips and margins of the lower leaves indicate Potassium deficiency and a need for potassium [1, 4].

10.6 SULPHUR

Sulphur is an indispensable element for carbohydrate metabolism and crop production. It is an important constituent of plant protein, amino acid and co-enzymes A. Most of the sulphur is absorbed from soils as sulphate (SO_4^-) but some is absorbed through leaves as SO_2. The crop response to sulphur application are increasingly reported from different parts of India [1, 3, 4, 16]. Sulphur is mobile in plant and deficiency symptoms are similar to those observed under nitrogen deficiency, i.e., plant are stunted with leaves light green to yellow in color.

10.7 CALCIUM AND MAGNESIUM

There are many similarities between the behavior of calcium magnesium and potassium in soils. Calcium is a cell wall component and it plays role in the structure and permeability of membranes. Magnesium is a constituent of chlorophyll and acts as enzyme activator. Calcium and magnesium as are absorbed as cations Ca^{++} and Mg^{++}. A deficiency of calcium is characterized by a malformation and disintegration of the terminal portion of the plant. Calcium deficiency for plants has occasionally been observed on very acidic soil with low calcium saturation.

10.8 IRON AND MANGANESE

Iron is an important element playing role in chlorophyll synthesis and in enzymes for electron transfer and it is weathered from minerals as divalent (Fe^{++}) cations in soil solution. Manganese controls several oxidation-reduction system, Formation of O_2 in photosynthesis and it realized in weathering as Mn^{++} which are absorbed by plant and adsorbed on cation exchange site. Both play an important role in enzyme systems that regulate various metabolic activities. Deficiency symptoms for iron are striking and are commonly seen on plants growing on calcareous or alkaline soils. Fe deficient plants have a light-yellow leaf color, which are more evident on younger leaves. In some very acidic soils, there are toxic concentration of iron and manganese due to dilution of soil suspension. Iron content in soils of various sugar mill zones ranges from 1.8 to 32.6 ppm which showed wide range and irregular trend. The available manganese content at most of the places varied from 1.2 to 30.3 ppm which indicated that 19% of soil were below the critical limit of 3.0 ppm in soils.

10.9 CHLORINE AND COBALT

Cobalt is essential for symbiotic nitrogen fixation by *Rhizobium* and it is required in very small amount by some crops. Chlorine activates system for production of O_2 in photosynthesis and it is also required by plants in very small amount. Generally, potassium fertilizers contain chlorine and it is amended in Indian soil for the correction of chlorine deficiency.

10.10 ZINC AND COPPER

Zinc helps in production of growth hormones especially in dole-3-acetic acid and it is a constituent of several enzymes like carbonic anhydrase, tryptophan synthetase, etc. Several enzymes regulate various metabolic activities. Copper is a constituent of several enzymes and acts as a catalyst for respiration. Copper and zinc are released in weathering as Cu^{++} and Zn^{++} absorbed by plant and adsorbed on cation exchange site [10]. Copper is immobile in plants and deficiency symptoms are highly variable from crop to crop. In some plants, older leaf tips become necrotic in small patches. Zinc deficiency is common in high calcareous soil and is characterized by spotting of the lower leaves. It is generally caused due to high phosphorus fertilization.

10.11   BORON

Boron plays an important role in sugar translocation, maintenance of the ratio of K and Ca, carbohydrate metabolism and occurs in soil in the mineral tourmaline. The borate ion is absorbed by plants and accumulated in soil organic matter. Many physiological diseases of plants such as top rot, pokkah boeng disease, cracked stems are associated with the deficiency of boron [19]. Maximum availability of boron has been observed in the neutral soil of India.

10.12   MOLYBDENUM

It plays an important role in activities of nitrate reductase enzymes. Molybdenum accumulates in soil organic matter and it is adsorbed as an anion by the clay fraction. This element is needed for nitrogen fixation in legumes, when it is deficient legumes show symptoms of nitrogen deficiency. Molybdenum causes other disturbances in plants such as cupping of leaves and reduced rate of expansion in leaf margin.

10.13   AVAILABLE LAND AND SOIL RESOURCES

- On a global basis, the main limitation for using the world soil resources for agricultural production are drought 28%, mineral stress 23%, shallow depth 22%, water excess 10%, and permafrost 6%.
- Only 11% of the world soils are without serious limitation.
- These situations suggest either improvement in soil management on existing cultivated land or expansion of cultivated area on new lands to meet the increasing demand of food.

Most of the cultivated soils in India are alluvial belonging to the order Entisol. Soil reactions varied from almost neutral to saline. Most of the

Indian soils are deficient in organic carbon (<5.0 g kg⁻¹), poor in available nitrogen (< 280 kg ha⁻¹) and low in available phosphorus (<24.0 kg ha⁻¹). Both light and heavy textured soils are sufficient in potash. The available sulphur for all the soils are approachable to critical limit [20]. Micronutrient cations are deficient in soil especially in rice and sugarcane growing soils [15]. The physico-chemical properties of soils in U.P. are given in Table 10.1.

10.14   CHANGES IN ORGANIC MATTER DUE TO CULTIVATION

Even on non erosive land that is brought under cultivation, rapid losses of organic matter usually occur. It has been found that as a result of cultivation over a period of 60 years, soil in a non eroded condition lost over one third of their organic matter (Figure 10.1). The losses were much greater during the earlier than the later periods. The organic matter losses account up to about 25% during the first 20 years, about 10% during the second 20 years and only about 7% during the third 20 years.

10.15   SOIL FERTILITY DEPLETION

Nutrient levels are decreasing continuously in soil due to intensive cultivation and continuous application of primary nutrient as a chemical

TABLE 10.1   Physico-Chemical Properties of Soils in U.P.

Characteristics	Range	Average	Norms of normal soil
pH	6.3–8.9	7.4	7.5
EC (dSm⁻¹)	0.03–1.30	0.24	>0.5
Org. C (%)	0.03–0.95	0.38	<0.5
P (kg/ha)	1.18–77.77	9.5	<40.0
K (kg/ha)	84–300	135.0	<175.0
Zn (mg/kg)	0.02–3.8	0.58	<1.0
Fe (mg /kg)	1.8–32.6	10.5	<10.0
Cu (mg/kg)	0.1–0.9	0.34	<3.0
Mn (mg/kg)	1.2–30.3	9.8	<5.0

Source: [6, 7, 12, 13].

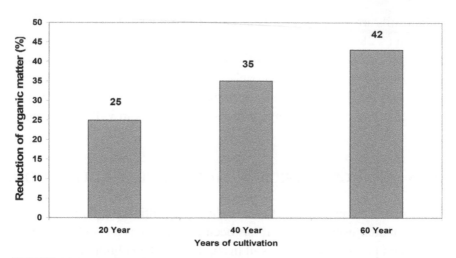

FIGURE 10.1   Decline of soil organic matter with 60 years cultivation.

fertilizer which usually deplete the soil organic matter resulting into inherent loss of organic carbon, soil nitrogen, phosphorus, potassium and sulphur. Most of the growers use only nitrogenous fertilizers, resulting in more losses on nutrient due to leaching, volatilization, erosion and large scale shifts towards organic free materials in the fertilizer product. All the factors in a given area are prone to imbalanced nutrient in the soil which is likely to be exhausted in fertility levels in a shorter period. Nutrient status of soils in western U.P. during 1953–54 and 1999–2000 is presented in Table 10.2.

10.16   SITE SELECTION AND SOIL SAMPLING

Collection and testing of soil samples have three main objectives.
1. Fertilizer recommendation.
2. Reclamation of salt affected soil.
3. Plantation of crop.

For fertilizer recommendation, the soil samples should be taken in such a way that it represent the whole field. For this purpose slope of field, agronomical practices, topography are kept in view. From 1 acre field at 10–15 places "V" type of trenches of 23 cm depth should be dug

TABLE 10.2   Physico-Chemical Properties of Soils in Western U.P. (1953–54 and 2004–2005)

Characteristics	1953–54	2004–2005
pH	7.6	7.2
Org. C (%)	0.61	0.34
N (%)	0.053	0.028
P (%)	0.120	0.067
K (%)	0.720	0.50

Source: [24, 25].

out. The 1.5 to 2.0 cm layer is crapped from the trench and the soil is collected in bucket. The roots and other unwanted matter are removed during the collection of soil. This collected soil is given a shape of sphere which is divided into four parts. Two opposite parts are removed and again the above process is repeated till the soil sample remains up to 0.5 kg. This soil is taken in clean muslin cloth bags. A tag demarcating the field identity like name, village, tehsil, district, and grown crop is attached to the muslin cloth bags. This soil sample will be called as an ideal soil sample, which is representative of whole field. For the analysis of salt affected soil and for plantation crop the soil sample should be taken from 90 cm and 180 cm depth, respectively and soil sample is collected depth wise (30 cm in each).

10.17   INSTRUCTION FOR SENDING SOIL SAMPLES

1. Dry the soil sample in the open air.
2. Mix the sample completely and take out a sub sample of approximately 250 gm.
3. Remove any insect and vegetative matter that are obviously present in this sample.
4. Place the sub sample in a container that is leak proof and put on the identification label sticking it to the container.
5. Place the samples inside the mailing container addressed to the concerned laboratory for analysis.

10.18   MANAGEMENT STRATEGIES FOR IMPROVEMENT IN SOIL

10.18.1   MANAGEMENT STRATEGIES FOR CONTROLLING SOIL EROSION

1. The contouring alone is effective in controlling erosion during storms of low or moderate intensity, but it provides little protection against the occasional severe storms that cause break over of the contoured rows. Strip cropping along with contouring, provided more protection and terraces are an effective way to reduce slop length.
2. Deep ploughing should be employed to control wind erosion where sandy surface soils are underlain by Bt horizons that contain 20 to 40% clay. 1 cm of sub soil should be ploughed up every 2 cm of surface soil thickness to control wind erosion. The kind of ploughing increases the clay content of surface soil by 5 to 12% in some areas.
3. Tree plantation for wind breaks. Row of trees are often planted across the area at right angles to the prevailing wind. A number of shrubs also make good wind breaks. These plantations also check the direct falling of rain drops on the ground and thereby help in soil erosion control. Wind erosion is controlled based on one or more of the following:
    i)   Trap soil particles with rough surface (tillage) and on the use of crop residues and strip cropping.
    ii)   Deep ploughing to increase clay content of surface soil.
    iii) Protect surface soil with complete vegetation cover.

10.18.2   WATER MANAGEMENT STRATEGIES FOR SOILS

1. In the critical and semi critical waterlogged areas, canal irrigation should be stopped for at least three successive years and 100% irrigation requirement should be met with private shallow tube wells. These shallow tubes well acts as vertical drainage and or most effective in lowering down ground water level to the desired depth

for reclamation of waterlogged area, and for removal of excess salt from the soil profiles.

2. In the area having a water level of 3.0 to 5.0 m, 30–50% canal irrigation and 50–70% ground water irrigation should be given depending on the hydrogeology of the area.

3. Ground water is deeper than 10 m, intensive canal irrigation (>50%) should be given in the absence of canal water, rain water harvesting structure should be intensified.

4. In the area of excessive seepage, and high water logging adjacent to the canal network.

5. In the critical and semi critical areas, for its sustainable reclamation and to avoid reversion in sodic soils, cropping system should be intensified supported with ground water irrigation through wells or shallow tube wells.

6. to assess the environmental impact and judicious management of water resources in the problem areas, periodical monitoring of surface and ground water is essentially required.

10.18.3   NUTRIENT MANAGEMENT STRATEGIES FOR THE SOILS

1. Nitrogen use efficiency is poor due to leaching, volatalization, denitrification, and run off losses under water logged soil. Among the different nitrogen sources, ammonical form is more efficient in water logged soil and application of urea coated with neem product is also beneficial for the water logged soils. The experiments conducted at SRI, Shahjahanpur revealed that application of nitrogen @ 125 kg/ha through coated urea gave cane yield at par with recommended dose of nitrogen @ 150 kg/ha in sugarcane (Table 10.3). Thus, with the application of Neem product about 25 kg N/ha can be saved (Table 10.4).

2. The sodic soils were required higher dose of nitrogenous fertilizers by about 20–25% as compared to the normal soils [5].

3. Foliar application of N is much economical and beneficial in water logged and saline soils [9, 14].

TABLE 10.3   Fertilizer Recommendation for Sugarcane Plant and Ratoon Crop

	Nitrogen	Phosphorus	Potash	Sulphur
Autumn plant	200	80	60	40
Spring plant	180	80	60	40
Ratoon	250	30	20	-

*1/3 N will be applied through organic manure.

**Phosphorus will be applied through single super phosphate for the correction of sulphur deficiency.

TABLE 10.4   Effect of Urea Coating Agent on the Yield of Sugarcane

S. No.	Treatments	Yield MT/ha
1.	Nitrogen alone (100 kg/ha)	61.40
2.	Nitrogen alone (125 kg/ha)	65.25
3.	Nitrogen alone (150 kg/ha)	68.76
4.	Nitrogen through coated urea (100 kg/ha)	64.81
5.	Nitrogen through coated urea (125 kg/ha)	69.18
6.	Nitrogen through coated urea (150 kg/ha)	71.19

Source: [16–18].

4. In saline soil, top dressing of N should always be done after every irrigation. Foliar application of N along with soil application is also beneficial for sugarcane cultivation [3, 5, 7].

5. Amongst the nitrogenous fertilizers, Ammonium sulphate has given better results in rice in sodic soils compared to urea and calcium ammonium nitrate.

6. Single super phosphate has been found best source of phosphorus fertilizer as it also contains sulphur [2, 24, 25, 30].

7. The deficiency of zinc is a common feature in the rice and sugarcane soils, therefore, basal application of zinc sulphate @ 30 kg/ha each to rice and sugarcane [9, 11, 14, 27].

8. Utilization of phosphorus and zinc are increased about 10.93% and 7.79% by the application of fertilizer along with SPMC @5q/ha before planting [26, 28, 29].

9. Integrated nutrient supply system (INSS) through advanced application of organic manure, green manuring, biofertilizer can be conveniently adopted. This results in greater nitrogen economy and higher nitrogen use efficiency [8]. Use of following organics will help to bridge up the increasing gap between demand and supply of fertilizer as:

10.18.3.1 Legumes

The legumes in sugarcane farming are grown either as intercrops or as green manuring. Green manuring legumes preceding sugarcane give a benefit of 19–43% increase in the yield of spring sugarcane and contribute 41–85 kg ha^{-1} nitrogen through biological nitrogen fixation.

10.18.3.2 Crop Residues

Addition of green leaves such as potato foliage, sugarcane trash, etc. from outside has the similar effect to that of green manuring crops incorporated into the soil. Potato leaves contain 2.42 to 2.56% nitrogen and 46 kg N/ha can be saved.

The cane trash obtained at 10–15% of the cane produced contains 0.42% N, 0.15% P, and 0.57% K besides 25.7, 20.45, 236.4, and 16.8 ppm Zn, Fe, Mn, and Cu, respectively. At the rate of 10% of the cane produced, at least 38 million tones of trash will be available during 2020 AD. This much trash will have the potential to supply 1.6 and 2.2 lakh tonnes of N and K respectively besides 57.0, 77.7, 8.98, 1.0, and 0.68 thousand tonnes of P, Fe, Mn, Zn, and Cu, respectively.

10.18.3.3 Factory Waste

On an average, our country produces about 6–7 million tones of pressmud (sugar mill waste) annually. Their direct use in the soil is harmful. The conventional method of composting yields low quality of compost and that takes a very long period (8–10 months) [21, 22].

10.18.3.4 Biofertilizers

Biofertilizers use for sustainable agriculture is very needful in present scenario. They save at least 25% inorganic fertilizers and increase availability of plant nutrients by improving the efficiency of inorganic fertilizers. The nutrient retention capacity of soil is also increased by their use. About 10 kg of each (N – fixing and PSB) bio-fertilizer is sufficient for one hectare.

10.19 BENEFITS OF ORGANIC FARMING

- Green manuring of legumes contributes about 50–84 Kg N/ha to sugarcane crop.
- The use of organic manure directly supplies macro and micronutrients and indirectly improves the physico-chemical, biological properties of soil.
- Organic manure reduces the leaching losses of nitrogen.
- The decomposition of organic matters liberates CO_2 leading to a lowering of soil pH and solubilization of P, K, and other nutrients.
- Use of organics opens the tilth of clayey soils and thereby increases the aeration.
- Organics and green manuring are specially beneficial in the WHC and reclamation of salt affected soils.

KEYWORDS

- **fertilizer recommendation**
- **nutrient management**
- **organic farming**
- **soil**
- **sugarcane**
- **Uttar Pradesh**

REFERENCES

1. Singh, A., Singh V., & Mehta, V. S. (1998). Effect of nitrogen and sulphur on yield and nutrient by rape seed. *J. Ind. Soc. Soil Sci. 36*(1), 182–184.
2. Singh, A., Singh, V., & Mehta, V. S. (1991). Sulphur status in alluvial soils, *J. Agril and Sci. Res. 33,* 109–110.
3. Singh, A., Dua, S. P., & Singh, G. P. (1995). Effect of sulphur on yield and quality of sugarcane. *Indian Sugar, 11*(1), 237–238.
4. Singh, A., Rao, S. P., Gupta, A. K., & Kushwaha, H. P. (1998). Effect of phosphorus and sulphur application on yield, quality and nutrient content in sugarcane. *Indian Sugar 48*(2), 33–37.
5. Singh, A., Srivastava, R. N., Singh, D. N., & Singh, G. P. (1999). Sulphur status of soils growing sugarcane. *Annals Pl. and Soil Res. 1*(1), 69–70.
6. Singh, A., Lal, K., & Singh, S. B. (2001). DTPA-extractable Fe, Mn, Zn and Cu in sugarcane growing soils. *Indian J. Sugarcane Technology, 16*(2), 48–51.
7. Singh, A., Srivastava, R. P., Lal, K., & Singh, S. B. (2001). Critical limit of potassium in sugarcane plant and soil. *Indian J. Sugarcane Technology, 16*(1), 104–108.
8. Singh, A., Gupta, A. K., Srivastava, R. N., Lal, K., & Singh, S. B. (2001). Nutrient status of sugarcane growing soil in central Uttar Pradesh, India. *Sugar Tech. 3*(3), 117–119.
9. Singh, A., Gupta, A. K., Srivastava, R. N., Lal, K., & Singh, S. B. (2002). Response of zinc and manganese to sugarcane. *Sugar Tech. 4*(2), 74–76.
10. Singh, A., Singh, A., Lal, K., & Singh, S. B. (2003). Effect of zinc and sulphur on sugarcane wilt, yield and juice quality. *Annals of Pl. and Soil Res. 5*(1), 81–83.
11. Singh, A., Srivastava, R. N., & Singh, S. B. (2003). Effect of nutrient combination on sugarcane productivity. *Sugar Tech, 5*(4), 311–313.
12. Singh, A., Srivastava, R. P., Lal, K., & Singh, S. B. (2001). Critical limit of potassium in sugarcane plant and soil. *Indian J. Sugarcane Technology, 16*(1), 104–108.
13. Singh, A., Gupta, A. K., Srivastava, R. N., Lal, K., & Singh, S. B. (2001). Nutrient status of sugarcane growing soil in central Uttar Pradesh. *Sugar Tech, 3*(3), 117–119.
14. Singh, A., Gupta, A. K., Srivastava, R. N., Lal, K., & Singh, S. B. (2002). Response of zinc and manganese to sugarcane. *Sugar Tech, 4*(2), 74–76.
15. Singh, A., Kumar, A., Gupta, A. K., & Singh, S. B. (2004). Status of micronutrient cations in soils. *Annals of Plant and Soil Res. 6*(2), 218–220.
16. Singh, A., Gupta, A. K., & Singh, S. B. (2005). Nitrogen use efficiency under urea coating aids in entisol. *Indian J. Sugarcane Technology, 20*(1), 40–44.
17. Singh, S. B., Singh, A., & Srivastava, R. P. (2005). Micro nutrient status in alluvial soils of Uttar Pradesh. *STAI, 25,* 55–60.
18. Singh, A., Srivastava, R. P., Singh, S. P., & Singh, S. B. (2005). Role of iron and manganese on sugarcane productivity, *STAI, 25,* 49–54.
19. Singh, A., Singh, A., & Singh, S. B. (2006). Deterioration in sugarcane due to pokkah being disease. *Sugar Tech, 8*(3), 187–190.
20. Singh, A., Srivastava, R. N., & Singh, S. B. (2007). Effect of sources of sulphur on yield and quality of sugarcane. *Sugar Tech, 9*(1), 98–100.

21. Chauhan, N., Singh, M. P., Singh, A., & Singh, S. B. (2007). Decomposition of press-mud by various cellulolytic fungi in vivo. *Sugar Tech, 9*(2), 227–229.
22. Chauhan, N., Singh, M. P., Singh, A., & Singh, S. B. (2008). Effect of biocompost application on sugarcane crop. *Sugar Tech, 10*(2), 174–176.
23. Singh, A., Srivastava, R. N., Gupta, A. K., & Sharma, M. L. (2008). Effect of sulphur and iron nutrition on the yield, juice quality of sugarcane. *Indian J. Agri. Science, 78*(10), 887–891.
24. Sharma, M. L., Singh, A., Gupta, A. K., & Srivastava, R. N. (2010). Nutrient status of sugarcane growing soils in Uttar Pradesh. *Indian J. Sugarcane Technology, 25*(1&2), 20–24.
25. Singh, A., Srivastava, R. N., Singh, S. P., Srivastava, R. P., & Sharma, M. L. (2010). Effect of effluent of sugar mills on growth, yield and juice quality of sugarcane. *Indian J. Sugarcane Technology, 25*(1&2), 63–65.
26. Singh, A., Srivastava, R. N., Kumar, A., & Sharma, M. L. (2011). Effect of SPMC treated dose of phosphorus and zinc fertilizer on the yield and quality of sugarcane, June 2011. *Indian Sugar LXI, 3,* 31–36.
27. Singh, A., Kumar, R., & Sharma, B. L. Response of SPMC treated Zn and Cu on yield and quality of sugarcane. *Agrica 5*(2), 90–94.
28. Yadav, S. P., Kumar, A., Yadav, S., Singh, A. K., Tiwari, A. K., & Ram, B. Yield, quality and nutrient uptake in autumn sugarcane as influenced by phosphorus levels and inoculation of phosphate solubilizing bacteria in legume based intercropping systems. *Agrica 5*(1), 63–65.
29. Yadav, S. P., Singh, S. C., Singh, A. K., Tiwari, A. K., Yadav, S. K., & Sharma, B. L. (2016). Yield and quality of sugarcane influenced by various plant nutrients. *Agrica 5*(1), 47–50.
30. Yadav, S. P., Singh, S. C., Yadav, S., Tiwari, A. K., & Sharma, B. L. (2016). Studies on yield, quality and availability of N and P of soil in autumn cane as influenced by P levels and P. S. B. inoculation in legume inter-cropping systems. *Agrica 5*(2), 98–101.

CHAPTER 11

SUPERMOLECULAR ORGANIZATION OF PHOTOSYNTHETIC COMPLEXES IN ARABIDOPSIS

MOHAMMED SABAR,[1,5] ALEXANDRE MARÉCHAL,[1] DAVID JOLY,[2] SRIDHARAN GOVINDACHARY,[2] ERIC BONNEIL,[4] PIERRE THIBAUT,[3,4] ROBERT CARPENTIER,[2] and NORMAND BRISSON[1]

[1]*Department of Biochemistry, University of Montreal, PO Box 6128, Station Centre-Ville, Montreal, Quebec, H3C 3J7, Canada*

[2]*Research Groupe in Plant Biology, University of Quebec in Trois-Rivieres, PO Box 500, Trois-Rivieres, Quebec, G9A 5H7, Canada*

[3]*Department of Chemistry, University of Montreal, PO Box 6128, Station Centre-Ville, Montreal, Quebec, H3C 3J7, Canada*

[4]*Institute for Research in Immunology and Cancer, University of Montreal, PO Box 6128, Station Centre-Ville, Montreal, Quebec, H3C 3J7, Canada*

[5]*McGill University, Department of Biology, Stewart Building 1205 Dr. Penfield, Montreal, Quebec Canada H3A 1B1, E-mail: mohammed.sabar@mcgill.ca*

CONTENTS

ABSTRACT

The light collecting and electron transfer complexes are dynamically distributed within the thylakoid membrane of *Arabidopsis thaliana* and assemble into supramolecular structures. Beside the basic structure of PSI and PSII bound to LHCI and LHCII, respectively, we found that these supercomplexes may interact together to form a stable or transient supracomplex (PSI-LHCI-PSII-LHCII). This supracomplex is preferentially formed in an *AtWhy1* overexpressing mutant affected in chloroplast biogenesis and may represent an intermediate for energy redistribution and compensation. Spectroscopic data indeed reveals that the distribution of absorbed energy is favoured towards photosystem I in the mutant as evidenced from the absence of a transition from state 2 to state 1. Besides, the absorbed light energy is also efficiently dissipated via non-photochemical quenching as photosystem II is down-regulated, a key process that operates when plants have to deal with extreme stress conditions. Several intermediates of this supracomplex containing only the core complexes of PSI and PSII with or without the PSI antenna (LHCI) have been observed. Furthermore we show that large populations of PSI are bound to LHCI and LHCII, supercomplexes known to be formed under energy state transition. The distribution between these forms varies in abundance, suggesting their oligomerization.

11.1   INTRODUCTION

The proteome of the chloroplast has been extensively investigated during the last few years, mostly by high-throughput mass spectrometry coupled to traditional two-dimensional electrophoresis (IEF). In these studies, several alternative approaches for efficient recovery of non redundant proteins have been used. These studies targetted the soluble proteins comprised in the free space of the chloroplast and the hydrophobic proteins enclosed in the thylakoid membranes and the envelope [1–5]. This led to the resolution of native complexes present in the photosynthetic machinery of a wide range of oxygen evolving organisms including Synechocystis, Chlamydomonas [6, 7] and higher plants such as tobacco, spinach and Arabidopsis [8–12]. In many reports, particular attention has been paid to the dynamic fate of the photosynthetic complexes in order to depict their interactions and reorganization into supercomplexes under external cues or after mutations [7, 13, 14].

Identification of interactions between macromolecules represents an important step to understand their actual *in vivo* function. For example, stable or transient interactions between macromolecules and their assembly into super molecular organizations often lead to multiple functionalities. This provides enzymatic advantages such as substrate channelling, which in turn may raise the efficiency, specificity and speed of metabolic pathways. A super organization of macromolecules into supracomplexes, either by their oligomerization or by interaction with other complexes, could also lead to several functions being induced and completed simultaneously. Knowledge of the composition and structural organization of supercomplexes could thus provide deeper understanding of metabolic pathways and cellular processes [13, 15].

Interactions between complexes and their reorganization into supercomplexes have been extensively studied in the mitochondrial respiratory chain of animals, plants and fungi [16–22]. These studies, which have been made possible in large part by the emergence and optimization of native gel systems such as the blue native polyacrylamide gel electrophoresis (BN-PAGE) [23–25], have contributed to enlarge the current view of the organization of the respiratory chain. In the chloroplast, little is known about the physical associations of the photosystem I (PSI), photosystem II (PSII) and their respective light harvesting complexes (LHCI and LHCII) into supercomplexes in the photosynthetic membrane organization. Dynamic fates of their reorganization under intrinsic or extrinsic cues that affect chloroplast biogenesis are only starting to unfold [7, 26, 27]. In this work, we revisited the native proteome of the photosynthetic machinery from freshly isolated chloroplasts of full photosynthesis-adapted plants of *Arabidopsis thaliana*.

Using BN-PAGE coupled to nano liquid chromatography mass spectrometry (nanoLC-MS), we found an arrangement of the photosynthetic machinery that is similar to that described previously [9–12]. However, we also obtained evidences of new organizations of the classical photosynthetic super complexes into more complex systems. This includes the identification of several populations of the supercomplex PSI-LHCI-LHCII involved in energy state transition. In addition, taking advantage of an Arabidopsis mutant line altered in chloroplast biogenesis, an induction of a supracomplex comprising PSI, LHCI, LHCII, and PSII has been identified. The Arabidopsis mutant is an overexpressor of the plastid targeted single-stranded DNA binding Whirly protein encoded by the At*Why1* gene [28, 29, 30, 31]. This supracomplex, isolated as a single band, is more probably favoured upon chloroplast decay caused by the overexpression of the Whirly protein.

11.2   METHODS

Unless otherwise mentioned, all chemicals were purchased from BioShop Canada Inc. (Burlington, Ontario, Canada). Primers, vectors and restriction enzymes were from Invitrogen Canada Inc. (Burlington, Ontario, Canada). 6-aminohexanoic acid, n-dodecyl-β-D-maltoside and Coomassie Brilliant Blue G-250 were from Sigma-Aldrich (Oakville, Ontario, Canada).

11.2.1   PLANT MATERIAL

The Arabidopsis seeds, wild type and mutant lines, of Columbia (Col-0) ecotype were sown on soil (Agromix 20, Fafard Inc., Agawam, Massachusetts, USA) in individual pots and kept for three days at 4°C. The plants were grown in the growth chamber under 16 hours light and 8 hours dark (long day conditions) under normal light intensity (150 µmol photons/m²/s) at 22°C.

11.2.2   PRODUCTION OF PLANTS CONSTITUTIVELY OVEREXPRESSING PLASTIDIAL WHIRLIES

Six copies of the c-myc epitope were amplified by PCR from the pCR-blunt-II-TOPO-myc vector (kind gift of Dr. Jeff L. Dangl) using the following primers: 5'-CCCAAGCTTGCCCTTCCGGTCGACAAAGCTATG-3' and 5'-CCGCTCGAGTCATCGATTTCGAA CCCGGGGTAC-3'. The amplicon was then digested with HindIII and XhoI and cloned into a pBS-SK(+) vector. The full length *AtWhy1* was amplified from cDNA using the following primers: 5'-CGGGATCCATGTCGCAACTCTTATCGACTCCT-3' and 5'-AACTGCAGTCT ATTCCATTCATAGTCTCCTCC-3'. The amplicon was subsequently digested with BamHI and XhoI restriction enzymes and cloned in frame with the c-myc epitopes in the pBS-SK(+)-6-c-myc vector. The tagged *AtWhy1* was then reamplified using the following primers: 5'-TCTAGAAGGCCTATGTCGCAACTCTTATCGACTCCT-3'and 5'-GGATCCACTAGTTC ATCGATTTCGAACCCGGGGTAC cloned into the XhoI and BamHI sites of a pGREENII022935S vector [32]. The vector was cotransformed with a pSOUP vector into a GV3101 pMP90 *Agrobacterium tumefaciens* strain. Plant transformation was carried out using the floral dip approach as described [33]. Transformed plants were selected on soil using the BASTA resistance conferred by the pGREENII0229 vector. The phenotype presented here appeared in the second generation of plants (T2) in homozygous individuals of lines expressing a high level of the transgene. Plants of the third and subsequent generations derived from one of these lines were used for the experiments presented here.

11.2.3 ISOLATION OF CHLOROPLASTS

The chloroplasts were isolated from 6 weeks old Arabidopsis plants. The timing for the chloroplast isolation was respected i.e. the extraction and purification procedures were always performed at mid-day, therefore after a 6 hours light period, representing full photosynthesis activity *in planta*. The leaves were harvested and ground in ice-cold grinding buffer containing 50 mM HEPES pH 7.3, 0.33 M sucrose, 1 mM $MnCl_2$, 1 mM $MgSO_4$, 2 mM EDTA, 5 mM 6-aminohexanoic acid, 10 mM β-mercaptoethanol and 0.1% (w/v) BSA using a Waring blender. All further procedures were carried out at 4°C. The homogenate was filtered through Miracloth (Calbiochem, Mississauga, Ontario, Canada) and centrifuged for 10 min at 1000 *g*. The pellet was resuspended in grinding buffer without reductant and BSA (wash buffer) and the chloroplasts were further purified on a step gradient of 40–75% (v/v) Percoll (Amersham-Pharmacia Biotech, Baie-d'Urfée, Québec, Canada) in wash buffer. The gradient was centrifuged at 6000 g for 30 min and the chloroplasts were collected at the interface of the gradient. The chloroplasts were washed twice in wash buffer followed by centrifugation at 1000 g for 10 min. The pellet was resuspended at 4°C in membrane solubilization buffer containing 750 mM 6-aminohexanoic acid, 0.5 mM EDTA and 50 mM Bis-Tris-HCl, pH 7.0. Protein content was determined using the Bradford reagent (Bio-Rad Laboratories, Inc., Mississauga, Ontario, Canada).

11.2.4   FRACTIONATION OF CHLOROPHYLL-PROTEIN COMPLEXES BY BN-PAGE

n-dodecyl-β-D-maltoside was added to the purified chloroplasts at a final concentration of 1% (w/v). The suspension was immediately centrifuged at 25,000 g for 10 min. The solubilized complexes were subsequently analyzed by 1D Blue-native PAGE without prior storage at −80°C. The supernatant enriched with complexes was supplemented with a 5% (w/v) stock solution of Coomassie Brilliant blue G-250 (Fluka-Sigma) in membrane solubilization buffer to a final ratio of 1:3 (w:w) to detergent, and subjected to BN-PAGE according to Schägger and von Jagow, (1991), excepted that the separating gel consisted of a linear gradient of 5–13% (w/v) acrylamide and a stacking gel of 4.5% (w/v) acrylamide. The cathode buffer, kept at 4°C, was freshly prepared and contained 50 mM Tricine, 15 mM Bis-Tris pH 7.0, 0.02% (w/v) Coomassie blue G-250, and 0.02% (w/v) n-dodecyl-β-D-maltoside. Electrophoresis was carried out at 4°C and set at 140 V overnight for standard medium gels. After electrophoresis, the gels were fixed in 12% (w/v) trichloro-acetic acid for at least 1 h and stained with Coomassie according to Ref. [23].

11.2.5   IDENTIFICATION OF THE CHLOROPLAST PHOTOSYNTHETIC SUPERCOMPLEXES

Coomassie stained bands were carefully excised under a laminar flow hood to minimize keratin contamination. Bands were subsequently reduced with dithiothreitol and alkylated with iodoacetamide before digestion with trypsin. The corresponding tryptic peptides were analyzed by nano liquid chromatography mass spectrometry (nanoLC-MS) using a NanoAcquity UPLC system interfaced with a Q-TOF Premier spectrometer via a nano-electrospray interface (Waters, Milford, MA). LC separations were performed using custom made C_{18} pre-column (5 mm x 300 mm i.d. Jupiter 3 mm, C_{18}) and an analytical column (10 cm x 150 mm i.d., Jupiter 3 mm C_{18}). Sample injection was 5 uL, and tryptic digests were first loaded on the pre-column at a flow rate of 4 uL/min and subsequently eluted onto the analytical column using a gradient from 10% to 60% aqueous acetonitrile (0.2% formic acid) over 56 min with a flow rate of 0.6 uL/min. External calibration of the instrument was made using a Glu-Fib B (Sigma) solution of 83 fmol/L. Data-dependent acquisition of MS-MS spectra was obtained for up to three precursor ions per survey spectrum using argon as a target gas, with collision energies ranging from 20 to 45 eV (laboratory frame of reference). Fragment ions formed in the RF-only quadrupole were recorded by a time-of-flight mass analyzer. MS/MS spectra acquired from LC-MS/MS analyses were processed using Mascot Distiller (version 2.1.1.0, Matrix Science) to reduce spectral redundancy and to correctly identify precursor m/z from survey scans. Data base searches were performed against a non-redundant NCBI database (3310354 entries) (version 3.24, released 20060303) using Mascot version 2.1 (Matrix Science, London, UK) and selecting *Arabidopsis thaliana* species. Parent ion and fragment ion mass tolerances were both set at +/– 0.1 and 0.4 Da, respectively.

11.2.6   CYTOLOGY

Hand-made thin sections of fresh leaves from 4 weeks-old plants were examined in sterile water. The chloroplasts morphological fate was examined by the visualization of their chlorophyll autofluorescence at 575–630 nm using a laser scanning confocal microscope (Olympus FV300, America, Melville, NY). Excitation wavelength for the chlorophyll was at 543 nm.

Fully expanded young leaves were used for the determination of chloroplast density per leaf area. Chloroplasts were rapidly isolated with equal volumes of grinding buffer from equal leaf area using a cork borer. The number of chloroplasts was estimated using an improved Neubauer cell counting chamber (Weber Scientific International, Cambridge, UK). In this experiment, only non-altered chloroplasts were taken into consideration.

11.2.7   WESTERN BLOT ANALYSIS

Proteins separated by SDS-PAGE were electro-blotted onto nitrocellulose membrane (Amersham Hybond ECL, Baie-d'Urfée, Québec, Canada) using a semi-dry transfer apparatus (Bio-Rad laboratories Ltd., Mississauga, Ontario, Canada). The membranes were blocked with 5% milk (w/v), incubated with primary antibody for about 2 hours in Tris-buffered saline (TBS) supplemented with 0.05% (v/v) Tween-20. After washing, the membranes were incubated with an anti-rabbit IgG peroxidase-conjugated secondary antibody (Jackson Immunoresearch Laboratories, Westgrove, PA, USA) for the detection of the Rubisco large subunit or with an anti-mouse IgG peroxidase-conjugated secondary antibody (Jackson Immunoresearch Laboratories, Westgrove, PA, USA) for the detection of the c-myc epitope. Immunolabelling was detected by chemiluminescence (Pierce, Brockville, Ontario, Canada) and exposure to Kodak Bio-Max ML films (PerkinElmer, Woodbridge, Ontario., Canada).

11.2.8   CHLOROPHYLL FLUORESCENCE

Chlorophyll fluorescence from detached leaves was measured with a pulse amplitude modulated fluorometer (PAM 101–103; Walz, Germany) as described previously [34]. In all measurements, plants were dark adapted for at least 3 hours before detecting F_o with a weak measuring light modulated at 1.6 kHz that does not initiate charge separation in PSII. The maximal level of fluorescence (F_{mD}) was determined with the application of 800-ms width white pulse of about 5600 μmol m^{-2}s^{-1} photon flux density (Schott KL 1500; Walz, Germany) which was sufficient to saturate fluorescence yield. State 2 was induced by exposure of a leaf to a beam of white light from KL 1500 filtered through a Schott BG39 (with 80% transmittance around 500 nm and no transmittance beyond 600 nm). The intensity of this blue light was about 100–120 μmol m^{-2}s^{-1} as measured with LI-250 light meter (Li-Cor, Lincoln, Nebraska, USA). A beam of white light passed through a Schott RG-9 filter (far-red light, PFD 120–130 μmol m^{-2}s^{-1}) was used to induce state 1 transition. Under steady state level of fluorescence, a saturating white pulse was applied to measure the yields of F_{m1} and F_{m2}, respectively. The relative change in fluorescence, F_r was determined from $F_r = [(F_{I'} - F_I) - (F_{II'} - F_{II})]/F_{I'} - F_I$ [34], wherein F_I and F_{II} correspond to the fluorescence in the presence of far-red light in states 1 and 2, respectively. The parameters, $F_{I'}$ and $F_{II'}$ are the fluorescence observed in the absence of far-red light in states 1 and 2, respectively.

11.2.9   LOW TEMPERATURE FLUORESCENCE EMISSION SPECTRA

Low temperature (77 K) spectra of fluorescence emission were recorded with a Perkin-Elmer LS55 spectrofluorometer equipped with a red-sensitive photomultiplier R928 as described previously [35, 36]. Chlorophyll fluorescence was excited at 436 nm and emission spectra were corrected according to the photomultiplier sensitivity using the correction factor spectrum provided by Perkin-Elmer. Thylakoids were suspended in a medium containing 30 mM Hepes NaOH (pH 7.6), 100 mM sorbitol, 5 mM MgCl$_2$, 10 mM NaCl, 20 mM KCl and 60% glycerol (v/v) to a final concentration of 5 μg Chl/mL. State 2 transition was induced by incubating a thylakoid suspension under blue illumination for 15 min in the presence of 10 mM NaF and 0.4 mM ATP. State 1 transition was achieved by illuminating a thylakoid suspension with far red light for 15 min.

11.3   RESULTS

11.3.1   PHOTOSYTHETIC COMPLEXES NATIVE-PROTEOMICS

Eighteen bands containing photosynthetic complexes were reproducibly observed after separation of the chloroplast proteins by BN-PAGE (Figure 11.1). These protein complexes were highly resolved and delimited as clearly visible green bands with little background in the gel, suggesting that the complexes had been well preserved throughout the procedure. The green bands were excised from the gel and subjected to in-gel

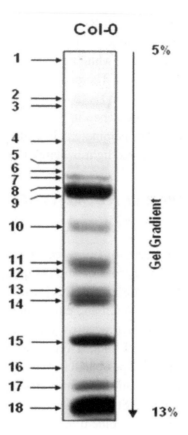

FIGURE 11.1 Chloroplast photosynthetic machinery from 6 weeks old Arabidopsis wild type Col-0 resolved by BN-PAGE. The gel was loaded with proteins from the solubilized complexes after detergent treatment of the chloroplasts. Native complexes or supercomplexes correspond to the bands numbered 1 to 18.

trypsin digestion. The resulting peptide fragments were further analyzed by nano liquid chromatography coupled to electrospray ionization tandem mass spectrometry (nanoLC-MS/MS). Each band contained many protein subunits with high identity score and most of them matched a predicted Arabidopsis protein sequence having a known function (Table 11.1). No protein belonging to mitochondria or other cellular organelles was identified in these experiments, indicating the high purity of the isolated complexes. Almost all of the proteins identified are of chloroplast

TABLE 11.1 Identification of Components of Single Complexes Separated by Blue Native Polyacrylamide Gel Electrophoresis (BN-PAGE) from Isolated Arabidopsis Chloroplasts Using Nano Liquid Chromatography Tandem Mass Spectrometry (nanoLC-MS/MS)*

Gel sample	Gene and Protein description	Organism	Uniprot Accession	MW (Da)	Matched Peptides	Score	Pi	Sequence Length
Photosystem I								
1–8	PsaD1; PSI Reaction Center SU II	S. oliracea	gi|15235503	22584	12–30	247–977	9.78	208
1–7; 9; 14	PsaF; Reaction center su III	A. thaliana	gi|5734518	24158	3–15	89–509	9.58	221
1–5; 7; 8	PsaA; P700 Apoprotein A1	A. thaliana	gi|5881694	83178	7–16	147–535	6.60	750
1–8; 14	PsaB; Apoprotein A2	A. thaliana	gi|5881693	82423	3–14	224–570	6.89	734
1–9	PsaG; Reaction Center SU VI	A. thaliana	gi|15222757	17075	2–8	50–339	9.57	160
1–9	Putative PsaE; Reaction Center SU IV 1	A. thaliana	gi|15225331	15180	3–8	119–341	9.94	145
1–4; 6–9	PsaE; Reaction Center SU IV	A. thaliana	gi|11692884	11703	3–10	127–302	9.79	110
1–8	PsaC; PSI 9kD protein (UMF)	N. tabacum	gi|2924280	9032	3–10	137–351	6.67	81
1–9	PsaL; Reaction Center SU XI	A. thaliana	gi|5738542	23142	1–9	44–341	9.85	219
1–3; 7; 8	PsaH1; Reaction Center SU VI 1	A. thaliana	gi|15233291	15207	3–9	76–229	9.95	145

TABLE 11.1 (Continued)

Gel sample	Gene and Protein description	Organism	Uniprot	MW	Matched	Score	Pi	Sequence
2–8	PsaK; SU X precursor	A. thaliana	gi|5738540	13232	1–3	53–135	10.52	130
4–6; 9	PsaH2; Reaction Center SU VI	N. sylvestris	gi|15218186	15264	1–3	62–85	9.90	145
1; 8	PsaF; Reaction Cente SU II 1	A. thaliana	gi|15221681	24158	19	166–605	9.58	221
4; 14	PSA D, PSI Reaction Center SU II precursor 1	A. thaliana	gi|21595031	22599	3	113–169	9.78	208
1	PSI L	A. thaliana	gi|15235490	23037	2	128	9.85	219
4	photosystem I subunit PSI-L	A. thaliana	gi|6006283	13069	1	67	9.17	124
LHCI								
1; 2; 4–8; 16	LHCA3*1; Type III Chlorophyll a/b-binding Protein	A. thaliana	gi|79320443	23731	2–17	87–460	5.62	218
1–8;	LHCA1; Type I Chlorophyll a/b binding Protein	A. thaliana	gi|15233115	25979	3–10	117–435	6.21	241
1–5; 7; 8	LHCA4; Chlorophylla/b-binding Protein 4 (CAB4)	A. thaliana	gi|30692874	27716	4–17	169–494	6.22	251
1; 3	LHCA3.1; Type III Chlorophyll a/b-binding Protein	A. thaliana	gi|15219941	29163	6	106–159,09	8.61	273
6	LHCI, Type III (CAB4)	A. thaliana	gi|11762172	16718	3	111	5.76	148

8	LHCA2; chlorophylla/b-binding Protein	A. thaliana	gi|15233120	27768	4	118	6.90	257
16	LHCA1; Type I Chlorophyll a/b binding Protein 1	A. thaliana	gi|13265501	25966	2	63	6.52	241
1	Lhca5 protein	A. thaliana	gi|4741942	27800	2	106	6.66	256
Photosystem II								
1; 4; 9–11; 13; 14	PsbaA; D1 (Reaction Center)	A. thaliana	gi|515374	38911	1–14	179–287	5.12	353
4; 8; 9; 13; 14	PsbE; cytb559	N. tabacum	gi|11848	9391	1–3	53–105	4.83	83
1; 4; 8–14	PsbD; D2 (Reaction Center Protein)	A. thaliana	gi|5881689	39522	2–12	65–405	5.46	353
1; 4; 8–15; 17	PsbB; CP47	A. thaliana	gi|5881720	56001	5–31	581–1120	6.40	508
1; 4; 8; 9; 11; 12; 17; 18	PsbC, CP43	A. thaliana	gi|5881690	51835	2–22	123–716	6.71	473
9; 11; 17	CP43	A. thaliana	gi|27435858	51835	10–15	442–593	7.18	400
1; 4; 9; 11; 14	PsbH; 8–10 kD phosphoprotein	S. montanum	gi|5881723	7697	2–4	101–136	6.02	73
4; 11; 14	PsbL; L SU	A. thaliana	gi|37721762	4467	1	88–96	4.53	38
14	(LHB1 B1; Chlorophylle binding protein 1	A. thaliana	gi|13265505	28800	3	82	6.60	268
17	CP22	A. thaliana	gi|6006279	16371	2	95	4.59	155

TABLE 11.1 (Continued)

Gel sample	Gene and Protein description	Organism	Uniprot	MW	Matched	Score	Pi	Sequence
1; 4	photosystem II G protein	A. thaliana	gi|5881698	25350	3	48–171	9.17	225
4	CP 29	A. thaliana	gi|15081739	31192	4	219–311	5.76	290
1, 18	33 kDa oxygen-evolving protein	A. thaliana	gi|22571	35114	3	222	5.68	332
4	PSBO-1 (OXYGEN-EVOLVING ENHANCER 33)	A. thaliana	gi|15240013	35121	8	391	5.55	332
1,4, 11	PSBO-2/PSBO2; oxygen evolving	A. thaliana	gi|15230324	34998	2	312	5.92	331
18	PsbQ (16 kDa oxygen-evolving complex)	A. thaliana	gi|4583542	22991	1	45	9.64	216
LHCII								
1; 2; 4–7; 13; 16;17	LHCB4; CP29	A. thaliana	gi|15081739	31192	3–11	104–327	5.76	290
1; 2; 4–7; 15; 16,18	LHCB2; Chlorophyll A-B binding Protein Type II	A. thaliana	gi|4741944	28631	3–7	217–714	5.48	265
1; 4; 6; 13; 16, 17, 18	LHCB6; Chlorophyll A-B binding Protein CP24	A. thaliana	gi|4741960	27505	1–7	46–339	6.75	258
1; 2; 4; 7; 13; 16; 17; 18	LHCB4.2 Chlorophyll A-B binding Protein	A. thaliana	gi|15231990	31174	2–12	57–433	5.85	287

1; 2; 4–7; 13; 15; 16; 18	LHCII Type - CAB2/ CAB3 (LHCP AB 180)	A. thaliana	gi|16374	24979	3–7	84–470	5.12	233
1; 4; 16; 17; 18	LHCB5; CP26	A. thaliana	gi|15235029	30138	3–11	115–769	6.00	280
4; 13; 15; 16; 18	LHB1B1; Chlorophyll A-B binding Protein Type I	A. thaliana	gi|18403549	28152	6–7	195–360	5.15	266
13; 15; 16; 18	LHCB3; Chlorophyll a/bbinding Protein type III	A. thaliana	gi|15239602	28688	3	113–195	4.96	265
1	Lhcb2 protein	A. thaliana	gi|4741948	28645	2	223	5.48	265
1; 4	Chlorophyll A-B binding protein	A. thaliana	gi|11762172	16719	2	73–129	5.76	148
18	LHCB4.3; chlorophyll binding	A. thaliana	gi|15225630	30193	2	146	5.23	276
18	CP29	A. thaliana	gi|15241005	31120	6	42–623	5.76	290
4	Catalase 3	A. thaliana	gi|2347178	56661	3	89	8.31	492
9;13	Catalase 2	A. thaliana	gi|1246399	56881	4–9	126–334	6.63	492
14	Identical cat2,similar to cat3	A. thaliana	gi|16215	56849	1–3	86	6.75	492

[1] Putative.

*The presence or absence of a protein in a given complex is indicated by the number of the band (1–18) seen on the gel (Figure 11.1) where this protein is found.

origin and are part of various complexes of the photophosphorylation machinery.

Table 11.2 indicates which complex corresponds to each band in the gel shown in Figure 11.1. The pattern of BN-PAGE complexes is generally conserved as compared to that found in other species [6–12]. However in our conditions, where the chloroplasts are isolated under full photosynthesis period, there are marked variations in the make up of these complexes. The results show that several of them are organized into supercomplexes, which may be generated either by their own oligomerization or by interaction with other complexes. To simplify the analysis, data from Table 11.1 were extracted and plotted as the number of subunits found in all bands for

TABLE 11.2　Identification of Supercomplexes Separated by BN-PAGE from Isolated Arabidopsis Chloroplasts Using nanoLC-MS/MS

BN-Gel bands	Supercomplexes
1	PSI/LHCI/LHCII
2	PSI/LHCI/LHCII
3	PSI/LHCI
4	PSI/LHCI/PSII/LHCII/Catalase
5	PSI/LHCI/LHCII
6	PSI/LHCI/LHCII
7	PSI/LHCI/LHCII/ATPase
8	PSI/LHCI/PSII/ATPase
9	PSI/PSII/ATPase/Catalase
10	PSII/CF1/Catalase
11	CytB6f/PSII/ATPase
12	CytB6f/PSII/ATPase
13	CytB6f/PSII/LHCII/ATPase
14	PSI/PSII/ATPase
15	CytB6f/PSII/LHCII/ATPase
16	CytB6f/LHCII/LHCI/ATPase/FstH protease Complex
17	PSII/LHCII/ATPase/FstH protease complex
18	LHCII/PSII/ATPase

each of PSI, PSII, LHCI, and LHCII (Figure 11.2). Under our growth conditions, photosystems are organized in different clusters where they vary in abundance. Results show that PSI was always associated with other components of the chloroplast photosynthetic machinery. Similar results were observed for PSII, although in some cases, it accumulated without the light harvesting complexes (Figure 11.2G). Two classical photosystem supercomplexes were identified. They contain basically the reaction centers subunits associated with the light harvesting complexes (PSI-LHCI and PSII-LHCII; Figure 11.2A and 2B, respectively). The basic PSI-LHCI supercomplex, which has been identified by others [8, 11, 12], was present only as a single band (band 3 in Figure 11.1). PSII-LHCII, the other basic super complex, was present in many other bands (bands 13, 15, 17 and 18 in Figure 11.1). These different forms of the super complex PSII-LHCII may represent the various forms of PSII oligomerization reported in the previous work [37]. Figure 11.2C reveals another supercomplex that migrated in the gel at different positions but contained similar components.

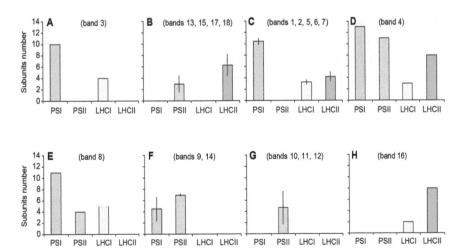

FIGURE 11.2 Composition of the photophosphorylation machinery supercomplexes found in Arabidopsis chloroplasts (A–H). Number of protein components of a specific photosynthetic complex found in each supercomplex are added together and plotted against the corresponding complex. Error bars represent the variability of the number of subunits of the complexes represented at different positions in the gel.

It is present in bands 1, 2, 5, 6 and 7 in Figure 11.1, which may represent different oligomerization forms. Subunits for PSI, LHCI and LHCII are abundant in this super complex, which suggests that it could represent the theoretical super complex that was shown to accumulate during state transition (state 2) when LHCII migrates from PSII to PSI in a reversible manner [38]. Figure 11.2 also revealed another new super complex containing the components required for light absorption, electron transfer and water splitting (Figure 11.2D, band 4 in the gel). This complex contains PSI and PSII, the light harvesting complexes LHCI and LHCII as well as catalase (see Table 11.1). However this super complex was barely detectable in the Col-0 ecotype under normal photosynthetic conditions, suggesting its instability or low abundance in the thylakoid membranes. Several intermediates of this super complex were also isolated. The first intermediate lacks LHCII (Figure 11.2E, band 8 in the gel) and is composed only of subunits of PSI, PSII and LHCI. This intermediate is very abundant and may represent the major form of the photosynthetic machinery at the protein level (Figure 11.1, band 8). The second and third intermediates contained only PSI and PSII (Figure 11.2F) or only PSII reaction center subunits (Figure 11.2G). These forms of PSII may represent intermediates of the photoinhibition-repair cycle, which consists of permanent disassembling, repair and activation of damaged PSII centers (see [39] for a review). Similar to the association of PSI and PSII, light harvesting complexes were also found to be associated together as a super complex (Figure 11.2H).

11.3.2   EFFECT OF THE ALTERATION OF CHLOROPLAST BIOGENESIS ON THE ORGANIZATION OF PHOTOSYNTHETIC COMPLEXES

While studying the function of the chloroplast targeted single-stranded DNA binding protein AtWhy1 [28–30], we isolated a chloroplast over expresser (OEX1) mutant line (Figure 11.3A) with a variegated pale-green phenotype reflecting chloroplast anomalies. This line provided a tool to verify whether chloroplast biogenesis failure would be linked to any quantitative or qualitative variations in the super complexes observed in Arabidopsis wild type Col-0.

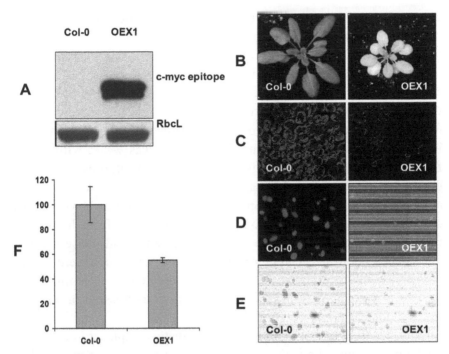

FIGURE 11.3   (a) Western blot analysis of the expression of the AtWhy1-c-myc construct. The antibody raised against the c-myc epitope shows the constitutive overexpression of AtWhy1 at the protein level. Equal protein loading was controlled by the immunodetection of the expression of the stromal large subunit of Rubisco (RbcL)., **B-F.** Morphological and cytological characterization of the Arabidopsis *AtWhy1* overexpressor mutant (OEX1) compared with wild-type Arabidopsis (Col-0). Photographs of five week-old plants grown in soil under normal conditions as described in methods (B). Confocal microscopy scanning of sections from the first leaf pair of 5–6 weeks-old plants. The chloroplasts are visualized as red spots (chlorophyll auto-fluorescence) within the cells (C). Epifluorescence (D) and phase-contrast (E) microscopy of isolated chloroplasts visualized at a magnification of 100X., **F.** Comparison of chloroplast content in Col-0 and OEX1 leaves. The chloroplasts were isolated from the same amount of starting fresh leaf material.

Under normal light and temperature growth conditions (150 μmol photons m^{-2} s^{-1}, 16 hours light photoperiod at 22°C), the *AtWhy1* over-expressing plants grew slower than the wild-type Col-0 and showed leaf pigmentation defects (variegated appearance) and pale green-phenotype (Figure 11.3B). Germination of seeds was not affected in the mutant and visual inspection did not reveal any drastic change in their production and yield as compared with wild-type (not shown). This suggests that the alteration of the phenotype was mostly restricted to green tissues and was linked to chloroplast biogenesis during plant development. To further characterize the vegetative phenotype, leaf sections and isolated chloroplasts were analysed by confocal microscopy. In wild-type plants, the chlorophyll red auto-fluorescence revealed near homogeneous and well shaped chloroplasts, whereas in OEX1, the intensity of the chlorophyll auto-fluorescence was weak, with only a few chloroplasts showing bright chlorophyll auto-fluorescence (Figure 11.3C). These results indicate that either the chloroplast number is altered or the amount of chlorophyll antenna within the photosynthetic complexes is reduced. Isolation of the chloroplasts from leaf tissue followed by their examination by epifluorescence and phase-contrast microscopy confirmed that chloroplast number and shape were altered in OEX1 (Figure 11.3D-F). These results indicate that chloroplast biogenesis or development is affected by the over expression of AtWhy1 in Arabidopsis.

The pattern of migration of the photosynthetic complexes in Col-0 and OEX1 was compared by 1D BN-PAGE by loading protein samples either on the basis of the same amount of total proteins released after chloroplast solubilisation with the detergent, n-dodecyl-β-D-maltoside (Figure 11.4A), or on the basis of total chloroplast proteins extracted from equal mass fresh weight of leaf tissue (Figure 11.4B). On the basis of equal protein load, the overall band pattern was similar in both OEX1 and Col-0, except for band 4, which was more abundant in OEX1. Sequencing of this band from OEX1 confirmed that it corresponds to the supracomplex PSI-LHCI-PSII-LHCII described above (see Figure 11.2D). The difference in abundance of complexes between Col-0 and OEX1 was more evident when gels were loaded on the basis of equal initial leaf fresh weight (Figure 11.4B). Under these conditions bands 5, 6, 7, 11, 12, 13, 14, 16 and 18 were the most reduced in intensity, while band 15 was barely affected. These results are in agreement with the cytological analysis of OEX1 showing a net defect in the biogenesis of the photosynthetic apparatus and suggest that the complex present in band 4 (PSI-LHCI-PSII-LHCII supracomplex), reflects a rearrangement of the photosynthetic apparatus in this line.

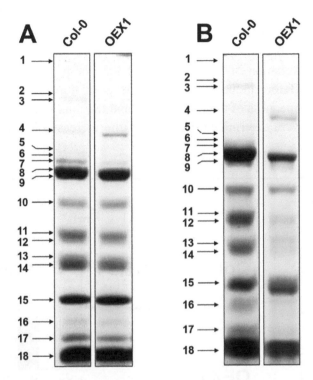

FIGURE 11.4 Chloroplast photosynthetic machinery from 6 week-old Col-0 and OEX1 plants resolved by BN-PAGE. The gels were loaded either on the basis of the same amount of proteins from the solubilized complexes after detergent treatment of chloroplasts **(A)** or on the basis of total proteins extracted from the same amount of starting leaf material for chloroplast isolation **(B)**. Native complexes or supercomplexes are visualized as bands numbered from 1 to 18.

11.3.3 CHANGES IN THE DISTRIBUTION OF EXCITATION ENERGY BETWEEN PHOTOSYSTEMS FOLLOWING ALTERED CHLOROPLAST BIOGENESIS

The distribution of absorbed light energy between PSII and PSI is efficiently balanced to maximize the photosynthetic electron transport. In this process, LHCII, a mobile fraction of the peripheral antenna complex, plays a key role. Thus it was of interest to understand how a new rearrangement of the photosynthetic apparatus, caused by the formation of the supra-complex PSI-LHCI-PSII-LHCII in OEX1 would influence the excitation energy distribution between the photosystems. Figure 11.5 shows the evolution of chlorophyll fluorescence from dark-adapted leaves of Col-0 and the OEX1 line after induction of state 1 or state 2 transition upon preferential excitation of PSI with far-red light or PSII with blue light, respectively. Before induction, the basal level of fluorescence (F_o), that indicates the openness of the PSII centers and the maximum fluorescence signal (F_{mD}) were determined. Because of the increased F_o level, F_v, the

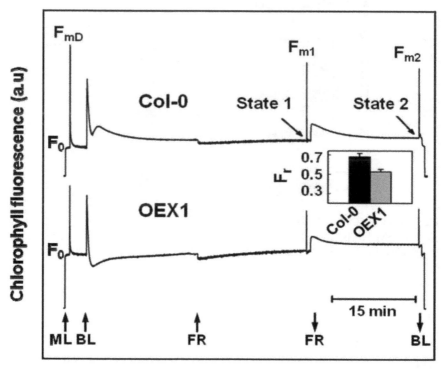

FIGURE 11.5 State transitions in Arabidopsis Col-0 and OEX1. Blue light (BL) and blue light together with far-red light (FR) were used to induce transitions to state 2 and state 1, respectively. Fo, the basal level of chlorophyll fluorescence was measured a few seconds after the weak modulated measuring light (ML) was switched on. FmD, the maximal fluorescence, was induced by an 800-ms width white saturating pulse in dark-adapted leaf. Fm1 and Fm2 are the yields of maximal fluorescence with the saturating white pulse in state 1 and state 2, respectively. Upward and downward arrows indicate the switching on and termination of ML, BL and FR, respectively. Inset: Fr is the change in fluorescence upon a re-distribution of the absorbed light energy between the photosystems during state transitions.

variable fluorescence (=F_{md} – F_o), markedly decreased in the OEX1 plants (Figure 11.5). Once the fluorescence was relaxed to its basal level, leaves were exposed to blue light so that PSII centers could be excited preferentially. Col-0 leaves showed the expected kinetics of fluorescence rise and decay that occurs before the onset of steady state conditions. However, in leaves of OEX1 plants the yield of fluorescence observed shortly after its rise to the maximum declined well below the observed F_o (Figure 11.5). In order to induce transition from state 2 to state 1, PSI centers were then preferentially excited with far-red light. The maximal fluorescence yield in state 1 (F_{m1}) was determined with a saturating white pulse. Then, the far-red light was switched off and the yield of maximal fluorescence in state 2 (F_{m2}) induced by blue light was measured. The yields of F_{m1} and F_{m2} obtained with Col-0 and the pattern of the fluorescence yield, relative to the sequence of light events, are consistent with earlier observations made with Arabidopsis (Figure 11.5). Also, in all Col-0 leaf samples tested, F_{m1} was always higher than F_{m2}. By contrast, the yields of F_{m1} and F_{m2} were almost identical in the OEX1 line (Figure 11.5). These results indicate that the energy distribution between the photosystems is markedly altered in the OEX1 line. This was confirmed by measuring the relative change in fluorescence (F_r), which represents the redistribution of excitation energy between photosystems during state transitions; see Materials and Methods section). Figure 11.5 (inset) indicates that the state transitions involved in balancing the light energy distribution between PSII and PSI centers are modified to the extent of about 25% in OEX1 plants.

Since the yields of maximal fluorescence measured with a white saturating pulse remained the same under state 1 or state 2 in the OEX1 leaves, it can be assumed that the mode of transition from state 2 to state 1 is being altered. This was verified by measuring fluorescence emission from PSII and PSI in isolated thylakoids at low-temperature (77K). Before detecting the emission of fluorescence, the induction of state 1 and state 2 was carried out as described above. The low temperature fluorescence emission spectra presented in Figure 11.6 were normalized at 685 nm, which corresponds to the fluorescence that originates from PSII complexes. In Col-0, transition from state 2 to state 1 induced by far-red light is illustrated by the profound decrease in the fluorescence emission at 725–735 nm, attributed to fluorescence that emanates from the PSI centers (Figure 11.6). This

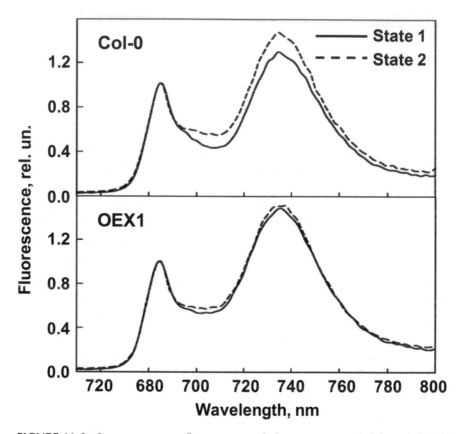

FIGURE 11.6 Low temperature fluorescence emission spectra recorded from thylakoid membranes isolated from Arabidopsis Col-0 (upper panel) or OEX1 (lower panel) and kept in state 1 (solid line) or state 2 (dashed line).

clearly indicates that the excitation energy is re-distributed from the PSI to the PSII centers when the samples are subjected to state 1 transition. Under the same conditions, the fluorescence emission corresponding to PSI centers was not significantly changed in the thylakoids from OEX1. This indicates that the mutant plants are unable to adapt to state 1, as their photosynthetic membranes remain in state 2 under the light conditions employed, and which probably reflects the new organization of the photosynthetic membranes as a consequence of the PSI-LHCI-PSII-LHCII supracomplex integration.

11.4   DISCUSSION

In higher plants, light energy conversion into chemical energy is mediated by four large complexes. These multiprotein complexes, PSI, PSII, Cytochrome b_6f and CF_oCF_1 ATP-synthase, are embedded in the thylakoid membrane where the biophysical reactions of photosynthesis take place. Evidences show that both PSI and PSII are structured into large supercomplexes with variable amounts of the membrane-bound light harvesting complexes LHCI and LHCII respectively. Further, these supercomplexes have a tendency to associate into megacomplexes (for a review see Ref. [14]. However, despite the wealth of information available on these multiproteins complexes, their supramolecular organization in the membrane remains unclear. In the present study, we show a special organization of the light conversion machinery in higher plants resolved by means of BN-PAGE, a technique for membrane complexes isolation. The findings suggest that the occurrence of photosynthetic complexes in the thylakoid membrane is dynamic and that they are distributed as supramolecular structures. We found several types of these supercomplexes, including the basic PSI-LHCI and PSII-LHCII complexes which may interact together to form a stable or transient supracomplex (PSI-LHCI-PSII-LHCII). This supracomplex is abundant in an AtWhy1 overexpressing mutant affected in chloroplast biogenesis. Minor intermediates of this supracomplex were found to contain only the core complexes of PSI and PSII together with or without PSI antenna (LHCI). Furthermore, spectroscopic data indeed revealed that the distribution of absorbed energy is favoured towards photosystem I in the mutant as evidenced from the absence of a transition from state 2 to state 1. Finally, the accumulation of a large population of PSI-LHCI-LHCII supercomplexes observed in this work is known to occur under energy state transition during photosynthesis acclimation [38].

11.4.1   POSSIBLE PHYSICAL INTERACTION BETWEEN PSI-LHCI AND PSII-LHCII SUPERCOMPLEXES

The continuous disassembly and reassembly of PSII involves dynamic relocation of its reaction centers to different membrane domains. PSII can be found in different forms in terms of its subunits, oligomerization and complex formation with other components, making it very heterogeneous *in vivo* [37]. A striking finding from the present study is that PSII can also be associated physically with PSI to form a supracomplex (PSI-LHCI-PSII-LHCII). Such an association is likely to optimize light absorption and its utilization *in vivo* by favouring efficient energy redistribution between the two photosystems. It has long been assumed that PSI and PSII are physically separated in the thylakoid membranes. It was recently shown, however, that PSI and PSII fractions could be localized to the same thylakoidal domains [37], although a direct physical association between the two complexes was neither reported nor discussed in detail previously. Ciambella and coworkers [12] have noticed the existence of two high molecular weight supercomplexes comprising PSI and PSII proteins which were highly expressed in dicots (tomato) but less expressed in monocots (barley). In *Prochloron didemni* (a class of cyanobacteria) have separated on sucrose density gradient, three thylakoid membrane fractions solubilized with -dodecyl maltoside. Two of these fractions contained photosynthetic supercomplexes associated or not with chlorophyll-binding proteins (Pcb). Beside these fractions which contained either PSI or PSII-Pcb supercomplexes, a third fraction contained a mixture of PSI and PSII. In the present study the PSI-LHCI-PSII-LHCII supracomplex accumulated to higher levels in the OEX1 line than in Col-0 and this correlated with a global decline in all other complexes, specifically those containing subunits of PSII. This suggests that a re-arrangement of PSII into a new organization is favoured in response to chloroplast alteration in this mutant. A parallel could be made with the dynamic reorganizations of photosynthetic complexes under environmental stress that have been reported in cyanobacteria, where it was shown that iron deficiency causes the accumulation of a giant chlorophyll-protein supercomplex [26]. In *Chlamydomonas reinhardtii*, alteration of culture media was shown to lead to dynamic responses of the photosynthetic system. Differences in PSI organization were shown to occur in cells grown under photoautotrophic or photomixotrophic (photoautotrophic condition supplemented with acetate as a second source of carbon) conditions [7]. These features shed new light on the association of the reaction centers and the light-harvesting complexes. Altogether, these evidences may account for the flexibility and physiological adaptation of the photosynthetic apparatus in response to the genetic background and external cues.

11.4.2   PSI-LHCI-LHCII SUPERCOMPLEX

In addition to the supracomplex PSI-LHCI-PSII-LHCII, a large population of the supercomplex PSI-LHCI-LHCII has been isolated from Col-0 chloroplasts by BN-PAGE. Theoretically, the physical association between the basic PSI-LHCI supercomplex and the LHCII complex occurs under state 2 of energy transition between the two photosystems. It is postulated that in this particular situation, LHCII dissociates from PSII, migrates laterally, and transfers excitation energy to PSI. The reverse process likely takes place when plants are shifted to state 1, which is thought to be the mechanism for redistribution of excitation energy between PSI and PSII [38].

Several attempts have been made to isolate the PSI–LHCI-LHCII supercomplex in higher plants [47, 48]. Using 2-dimension BN-PAGE, [11] noticed the co-migration of subunits of PSI, LHCI and LHCII, which suggested the existence of such a supercomplex. However, the complex has not yet been purified from higher plants. The difficulties encountered in isolating this super complex may be linked to a possible loose binding of the mobile LHCII protein(s) to the PSI core under state 2 of energy transition and also to the small amount of energy redistribution occurring in higher plants [27]. In Chlamydomonas, where the amount of energy redistribution is much higher, Takahashi and coworkers [27] have succeeded in purifying the PSI–LHCI-LHCII super complex using sucrose gradient separation of digitonin-solubilized membranes. Our results now provide new biochemical evidences for the existence of such a super complex in higher plants. Four populations of this supercomplex were resolved by BN-PAGE, which may account for their oligomerization. The highest molecular weight species harbours all subunits of LHCII that are associated with PSI in Chlamydomonas. These include a major monomeric type II LHCII protein which could have the same role as lhcbM5 (type II) in Chlamydomonas and for which no homologue has been found in higher plants [27]. Furthermore, our data suggest that two minor monomeric LHCII proteins, CP29 and CP24, which have long been considered to associate solely with PSII, may also shuttle between PSI and PSII in higher plants.

11.4.3 POSSIBLE PHYSIOLOGICAL DYNAMICS OF THE PHOTOSYNTHETIC SUPERCOMPLEXES

Figure 11.7 summarizes the occurrence of the different supercomplexes found in this work. Our findings suggest a dynamic interaction between the photosynthetic supercomplexes and their organization into supracomplexes. This may occur *in planta* either permanently or transiently under some particular physiological cues such as energy redistribution through

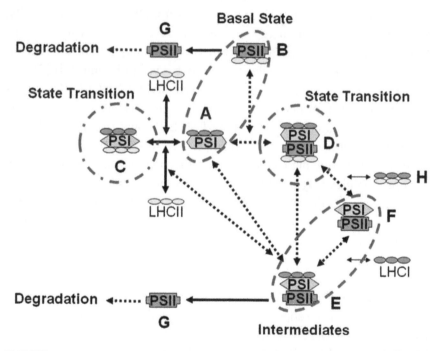

FIGURE 11.7 Model for the dynamic interactions between photosynthetic supercomplexes and their organization into supracomplexes. The basic supercomplexes, PSI-LHCI and PSII-LHCII may interact together to give rise to a supracomplex (PSI-LHCI-PSII-LHCII). The formation of this supracomplex may be considered as a transitory adaptation device for the maximum absorption and conversion of light. This supracomplex may split and give rise to intermediates which in turn may serve as precursors for the reconstitution of this supracomplex in a reversible manner. On the other hand, the free LHCII may integrate the supercomplex PSI-LHCI to give rise to another supracomplex PSI-LHCI-LHCII which is predicted to accumulate under state 2 of energy transition during photosynthesis.

the thylakoid membranes. The basic supercomplexes are formed by PSI-LHCI and PSII-LHCII (A-B). This corresponds to state 1 conditions in Col-0 induced by far-red light as shown in Figures 11.5 and 11.6. This model also incorporates the classical form of state 2 transition of photosynthesis as shown in Col-0 (Figures 11.5 and 11.6). The transition is represented by the movement of LHCII, originating from PSII-LHCII, to PSI-LHCI, giving rise to the PSI-LHCI-LHCII supercomplex (C). Then, these supercomplexes may interact together to give rise to a supracomplex under some specific conditions which are limiting photosynthetic efficiency and may represent another form of state transition (D). The formation of this supracomplex may be considered as a compensatory device for the maximum absorption and conversion of light as it is more favoured in the OEX1 mutant deficient in chloroplast biogenesis. In this respect, it should be emphasized that the excitation energy distribution between the photosystems is greatly altered in OEX1 as detected from the fluorescence spectroscopic data (Figures 11.5 and 11.6). Further, these data revealed that a transition from state 2 to state 1 is prevented in this case. Clearly, the above illustrates the origin of fluorescence from membranes which were predominantly re-organized with PSI-LHCI-PSII-LHCII and PSI-LHCI-LHCII supercomplexes, since the distribution of the excitation energy is more favoured towards PSI in order to down-regulate PSII. This supracomplex may split and give rise to its intermediates which in turn may serve as precursors to reconstitute the supracomplex in a reversible manner (E, F). The finding of different PSII complex species devoid of any form of chlorophyll (G) may represent degradation forms. Unlike other photosynthetic complexes, PSII operates at high oxidizing potentials and can be easily damaged, giving rise to photo-inhibition-repair cycle where damaged PSII centers are continuously disassembled, repaired and finally activated.

11.5 CONCLUSION

In biological systems, the isolation of integrated enzyme clusters, which may represent the *in vivo* state, depends largely on the utilization of experimental designs and technologies that are minimally disruptive. Using BN-PAGE as a mild technique for complexes isolation, our results indicate that the chloroplast photosynthetic complexes interact together to form supracomplexes such as PSI-LHCI-PSII-LHCII. This molecular organization into supercomplexes, compared to an organization where complexes are separated, could have several advantages, including the promotion of substrate channelling, catalysis improvement and sequestration of intermediates, thus permitting rapid transfer reactions through intramolecular functional groups. As for mitochondria, where the occurrence of supramolecular organization of the respiratory chain has been confirmed by several means including electron microscopy, the next step of this work will include the visualization of the supramolecular organization of the photosynthetic machinery, and confirmation of their existence in other species. Furthermore, as the PSI-LHCI-PSII-LHCII is more favoured in a chloroplast altered mutant (OEX1), the occurrence and dynamic fate of this supracomplex could be analysed quantitatively and qualitatively under conditions where alterations of the chloroplast functions are induced, such as abiotic stress.

ACKNOWLEDGMENTS

This research was supported by grants from the Natural Science and Engineering Research Council of Canada (NSERC). We thank Amina Tazi for references entry.

KEYWORDS

- arabidopsis
- BN-PAGE
- chloroplast
- gene over-expression
- light harvesting chlorophyll
- photosynthesis adaptation
- photosystems I and II
- protein oligomerization
- supercomplexes
- whirly

REFERENCES

1. Peltier, J. B., Friso, G., Kalume, D. E., Roepstorff, P., Nilsson, F., Adamska I, & van Wijk, K. J., (2000). Proteomics of the chloroplasts: systemic identification and targeting analysis of lumenal and peripheral thylakoid proteins. *Plant Cell. 12*, 319–341.
2. Schubert, M., Petersson, U. A., Haas, B. J., Funk, C., Schröder, W. P., & Kieselbach, T., (2002). Proteome map of the chloroplast lumen of Arabidopsis thaliana. *J Biol Chem. 277*, 8354–8365.
3. Friso, G., Giacomelli, L., Ytterberg, A. J., Peltier, J. B., Rudella, A., Sun, Q., & van Wijk, K. J., (2004). In-depth analysis of the thylakoid membrane proteome of Arabidopsis thaliana chloroplasts: new proteins, new functions, and a plastid proteome database. *Plant Cell. 16*, 478–499.
4. Ferro, M., Salvi, D., Riviere-Rolland, H., Vermat, T., Seigneurin-Berny, D., Grunwald, D., Garin, J., Joyard, J., & Rolland, N. (2002). Integral membrane proteins of the chloroplast envelope: identification and subcellular localization of new transporters. *Proc Natl Acad Sci USA. 99*, 11487–11492.
5. Froehlich, J. E., Wilkerson, C. G., Ray, W. K., McAndrew, R. S., Osteryoung, K. W., Gage, D. A., & Phinney, B. S. (2003). Proteomic study of the Arabidopsis thaliana chloroplastic envelope membrane utilizing alternatives to traditional two-dimensional electrophoresis. *J Proteome Res. 2*, 413–425.
6. Herranen, M., Battchikova, N., Zhang, P., Graf, A., Sirpiö, S., Paakkarinen, V., & Aro, E. M. (2004). Towards functional proteomics of membrane protein complexes in Synechocystis sp. PCC 6803. *Plant Physiol. 134*, 470–481.
7. Rexroth, S., Meyer zu Tittingdorf, J. M., Krause, F., Dencher, N. A., & Seelert, H. (2003). Thylakoid membrane at altered metabolic state: challenging the forgotten realms of the proteome. *Electrophoresis. 24*, 2814–2823.
8. Kügler, M., Jänsch, L., Kruft, V., Schmitz, U. K., & Braun, H. P. (1997). Analysis of the chloroplast protein complexes by blue-native polyacrylamide gel electrophoresis (BN-PAGE). *Photosynthesis Res. 53*, 35–44.
9. Suorsa, M., Regel, R. E., Paakkarinen, V., Battchikova, N., Herrmann, R. G., & Aro, E. M., (2004). Protein assembly of photosystem II and accumulation of subcomplexes in the absence of low molecular mass subunits PsbL and PsbJ. *Eur. J. Biochem. 271*, 96–107.
10. Rokka, A., Suorsa, M., Saleem, A., Battchikova, N., & Aro, E. M. (2005). Synthesis and assembly of thylakoid protein complexes: multiple assembly steps of photosystem II. *Biochem J. 388*, 159–168.
11. Heinemeyer, J., Eubel, H., Wehmhöner, D., Jänsch, L., Braun, H. P., (2004). Proteomic approach to characterize the supramolecular organization of photosystems in higher plants. *Phytochemistry. 65*, 1683–1692.
12. Ciambella, C., Roepstorff, P., Aro, E. M., & Zolla, L. (2005). A proteomic approach for investigation of photosynthetic apparatus in plants. *Proteomics. 5*, 746–757.
13. Krause F. (2006). Detection and analysis of protein-protein interactions in organellar and prokaryotic proteomes by native gel electrophoresis: (Membrane) protein complexes and supercomplexes. *Electrophoresis. 27*, 2759–2781.

14. Dekker, J. P., & Boekema, E. J. (2004). Supramolecular organization of thylakoid membrane proteins in green plants. *Biochim Biophys Acta. 706*, 12–39.
15. Eubel, H., Heinemeyer, J., Sunderhaus, S., & Braun, H. P. (2004a). Respiratory chain supercomplexes in plant mitochondria. *Plant Phys Biochem. 42*, 937–942.
16. Wittig, I., Carrozzo, R., Santorelli, F. M., & Schagger, H., (2006). Supercomplexes and subcomplexes of mitochondrial oxidative phosphorylation. *Biochim Biophys Acta. 1757*, 1066–1072.
17. Boekema, E. J., & Braun, H. P., (2007). Supramolecular structure of the mitochondrial oxidative phosphorylation system. *J Biol Chem. 282*, 1–4.
18. Dudkina, N. V., Heinemeyer, J., Sunderhaus, S., Boekema, E. J., & Braun, H. P. (2006). Respiratory chain supercomplexes in the plant mitochondrial membrane. *Trends Plant Sci. 11*, 232–240.
19. Sabar, M., Balk, J., & Leaver, C. J. (2005). Histochemical staining and quantification of plant mitochondrial respiratory chain complexes using blue-native polyacrylamide gel electrophoresis. *Plant J. 44*, 893–901.
20. Krause, F., Reifschneider, N. H., Vocke, D., Seelert, H., Rexroth, S., & Dencher, N. A. (2004a). "Respirasome"-like supercomplexes in green leaf mitochondria of spinach. *J Biol Chem. 279*, 48369–48375.
21. Krause, F., Scheckhuber, C. Q., Werner, A., Rexroth, S., Reifschneider, N. H., Dencher, N. A., Osiewacz, H. D., (2004b). Supramolecular organization of cytochrome c oxidase- and alternative oxidase-dependent respiratory chains in the filamentous fungus Podospora anserine. *J Biol Chem. 279*, 26453–26461.
22. Eubel, H., Heinemeyer, J., & Braun, H. P. (2004b). Identification and characterization of respirasomes. *Plant Physiol. 134*, 1450–1459.
23. Schägger, H., & von Jagow, G. (1991). Blue native electrophoresis for isolation of membrane-protein complexes in enzymatically active form. *Anal Biochem. 199*, 223–231.
24. Schägger, H., Cramer, W. A., & von Jagow, G., (1994). Analysis of molecular masses and oligomeric states of protein complexes by blue native electrophoresis and isolation of membrane protein complexes by two-dimensional native electrophoresis. *Anal Biochem. 217*, 220–230.
25. Jänsch, L., Kruft, V., Schmitz, U. K., & Braun, H. P. (1996). New insights into the composition, molecular mass and stoichiometry of the protein complexes of plant mitochondria. *Plant J. 9*, 357–368.
26. Boekema, E. J., Hifney, A., Yakushevska, A. E., Piotrowski, M., Keegstra, W., Berry, S., Michel, K. P., Pistorius, E. K. & Kruip, J. (2001). A giant chlorophyll protein complex induced by iron deficiency in cyanobacteria. *Nature. 412*, 745–748.
27. Takahashi, H., Iwai, M., Takahashi, Y., & Minagawa, J., (2006). Identification of the mobile light-harvesting complex II polypeptides for state transitions in Chlamydomonas reinhardtii. *Proc Natl Acad Sci USA. 103*, 477–482.
28. Desveaux, D., Allard, J., Brisson, N., & Sygusch, J. (2002). A new family of plant transcription factors displays a novel ssDNA-binding surface. *Nat Struct Biol. 9*, 512–517.
29. Desveaux, D., Subramaniam, R., Despres, C., Mess, J. N., Levesque, C., Fobert, P. R., Dangl, J. L., & Brisson, N. (2004). A "whirly" transcription factor is required for salicylic acid-dependent disease resistance in Arabidopsis. *Dev Cell. 6*, 229–240.

30. Desveaux, D., Maréchal, A., & Brisson, N., (2005). Whirly transcription factors: defense gene regulation and beyond. *Trends in Plant Sci. 10,* 95–102.

31. Krause, K., Kilbienski, I., Mulisch, M., Rodiger, A., Schafer, A., & Krupinska, K. (2005). DNA-binding proteins of the Whirly family are targeted to the organelles. *FEBS. 579,* 3707–3712.

32. Hellens, R. P., Edwards, E. A., Leyland, N. R., Bean, S., Mullineaux, P. M. (2000). pGreen: a versatile and flexible binary Ti vector for Agrobacterium-mediated plant transformation. *Plant Mol Biol. 42,* 819–832.

33. Clough, S. J., Bent, A. F. (1998). Floral dip: a simplified method for Agrobacterium-mediated transformation of Arabidopsis thaliana. *Plant J. 16,* 735–743.

34. Varotto, C., Pesaresi, P., Jahns, P., Leßnick, A., Tizzano, M., Schiavon, F., Salamini, F., & Leister, D., (2002). Single and double knockouts of the genes for photosystem I subunits G, K and H of Arabidopsis: Effects on photosystem I composition, photosynthetic electron flow and state transitions. *Plant Physiol. 129,* 616–624.

35. R Rajagopal, S., Buknov, N. G., & Carpentier, R. (2002). Changes in the structure of chlorophyll-protein complexes and excitation energy transfer during photoinhibitory treatment of isolated photosystem I submembrane particles. *J Photochem Photobiol Biol. 67,* 194–200.

36. Rajagopal, S., Buknov, N. G., Tajmir-Riahi, H. A., & Carpentier, R. (2003). Control of energy dissipation and photochemical activity in photosystem I by NADP-dependent reversible conformational changes. *Biochemistry. 42,* 11839–11845.

37. Danielsson, R., Suorsa, M., Paakkarinen, V., Albertsson, P. A., Styring, S., Aro, A. M., & Mamedov, F. (2006). Dimeric and monomeric organization of photosystem II: distribution of five distinct complexes in the different domains of the thylakoid membrane. *J Biol Chem. 281,* 4241–14249.

38. Wollman, F. A., (2001). State transitions reveal the dynamics and flexibility of the photosynthetic apparatus. *EMBO J. 20,* 3623–3630.

CHAPTER 12

METHODS TO IMPROVE GERMINATION IN SUGARCANE

S. P. SINGH, PRIYANKA SINGH, and B. L. SHARMA

UP Council of Sugarcane Research, Shahjahanpur – 242001, India

CONTENTS

12.1   INTRODUCTION

Germination has a special significance in vegetative propagated crop like sugarcane, where poor germination leads to gap resulting in low tonnage. Workers of all over the world have stressed the importance of germination. Out of different factors affecting important sugarcane yield, germination accounts only three percent, but its impact in cane culture is significant as on the basis of the crop from the physiological points of view, it is equally important as any other phases of the crop [11].

Clements [5] regards germination denoting as activation of sprouting of undamaged buds together with initiation of roots from root primordial on the piece of stalk used as planting material. Van Dillewijin [49] considers bud as a miniature stem with its growing points and the primordia of leaves and root, which develop in to new shoots and set roots. The latter function as such till the young shoots produces their own root system. Moreover, the success of bud germination depends on a number of factors *viz.*, temperature, light and water, a certain minimum values of these factors are essential for proper germination [51].

The transition of the dormant buds into active stage also constitutes a complex phenomenon character, which includes hormonal balance and food constituents of seed cane. The morphological and bio-chemical changes, which ultimately result in the germination of buds are mainly controlled by two factors, i.e., internal and external. The former comprise of the health, age of cane, water content in sett, food reserve, auxin, enzymes, etc., where as the later includes the soil conditions and environment, namely aeration, compactness, nutrients, water content, temperature, humidity, wind velocity, etc., when these factors are favorable the germination is maximum.

A good germination lays the foundation of the subsequent ratoon crop as in north India more than 50% of the total cane is ratoon [35]. Unfortunately germination is low in north India [24]. It is 30–45% of the planted buds as compared to 70% or more under central and southern India and still lower in the fields of cultivars [35]. Considering the value of good

12.2 VARIOUS ASPECTS OF GERMINATION IN SUGARCANE

Large number of workers across the country has paid attention to various aspect of the germination of sugarcane.

12.2.1 APICAL DOMINANCE

Under normal conditions, a bud does not develop until it forms the part of the living stalk. As long as the growing apex is in the functioning state, the growing points of the stalk exerts an inhibiting effect on the lateral buds which are thus kept in the state of dormancy. It is observed that the upper bud exerts inhibitory effects on lower buds, delaying their germination till it is complete in upper one. This is known as top dormancy or apical dormancy. This phenomenon is closely related with the moisture content of setts. Punching the internodes and pouring water in to it reduce the effect of top dominance. Treatments bringing about lessening of auxin or its disintegration have enhanced the extent of germination further, factors aiding the reduction of natural hormones either by riding the action of natural IAA oxidase system by providing a time lag or by some pre planting treatments like soaking in hot water or by antagonistic action of synthetic hormones or by lessening or nullifying the effect of apical dominance by planting one bud setts or by such cultural practices a topping have also been stressed [44].Germination was maximum in single bud setts where the apical dominance does not exist where as the minimum germination was in case of full canes, where the apical dominance was at a high degree [34]. Apical dominance was related to water content of the setts. In dry setts it was more distinct.

12.2.2 ROOT INITIATION

The germinating bud is initially dependent upon the sett roots originated from root primordia, for nutrient and water. In some varieties they develop prior to shoot in others the situation is reverse. In a comparative study of one, two and three budded setts, It was observed that rooting and sprouting in one bud sett was more in the beginning but at the end the and three budded setts proved superior [32]. Developing shoots develop their own root system after about three weeks under favorable condition. The mortality of root was attributed to the competition of shoot roots. It was pointed out that roots but not the sett roots are associated with growth and yield.

12.3 FACTORS AFFECTING GERMINATION

12.3.1 VARIETY

Wide differences in rate extant of germination of different varieties have been observed [35] and consequently grouped as early and late sprouting ones. Variation in germination of varieties may be due to varying proportions of dry scale on buds, high fiber content, hardness and thickness, waxiness of rind and even certain genetic metabolic and enzymatic characters [53]. However, varietal differences, morphological characteristics as well as dynamics of metabolic and enzymatic activities and consequently the genetic base of varieties [44] CoS 767, CoS 96261, CoS 8279, CoSe 01434, have shown good germination.

12.3.2   SETT SIZE

The most suitable length or size of sett (i.e., one, two, three and four budded) for germination and has been studied by the workers from early days (Figure 12.1). Since planting is done with cane setts having one or more buds, the intrinsic stalk and soil characteristics the time of planting are also of much significance [52]. A positive correlation between average internodal length and germination percentage of different varieties have been noted by Suba Rao et al. [45]. The thickness of setts is related to food reserves of the planting material. Thick cane known to give 5%

FIGURE 12.1   Different sizes of sett size affects germination.

more germination than the thin cane. Similarly the effect of cane setts ranging from one budded to eighteen bud has been extensively explored on germination. The one bud setts results in early start of sprouting, but after three to four weeks of planting the overall germination is poor owing to exhaust of food reserves. The two- and three-bud setts not only give more germination, as also they prove economical and give a good stand of the crop.

12.3.3   SEED QUALITY OR HEALTH OF CANE

Normally the seed cane quality is judged by the size, thickness, food reserve and health of the setts, besides the age and standard of crop that is proposed to be used as planting material. Initially the sprouts draw their food material from the setts. It is believed that the setts sown from the crop grown on high nitrogen and water level content did not only give over 20% more germination, but also improves the yield and sugar content of the cane [18]. Setts from well-manured and irrigated field gave 23% higher germination [10]. Thick canes gave 5% more germination than thin canes [46].

Sugar, minerals, amide, and water are some of the important constitutes which have been established beyond doubt to influence germination. Presence of adequate moisture in the setts is necessary for good germination, its effect is particularly marked under late plantings. In summers, when the setts are socked in water the germination is increased appreciably.

12.3.4   AGE OF SEED CANE

Variation in the age of seed crop also play an important role in germination. It is believed the portion of seed cane, which contains fairly good quantity of amids nitrogen, glucose and water gives better germination. It is because of this planting material of the top immature half cane is advocated for good germination. Seed crop of only 8 months age is normally recommended for planting purposes. Anakapalle Research Station has developed a concept of seed where a crop of 8 months duration is harvested and utilized for seed purposes.

12.3.5 SOURCE

In a study seed material taken from plants, first, second and third ratoon crops were studied and it was appeared that the seed setts from plant crop was favorable, there was decline in germination with adverse the age of the stubble crop. The water shoots were found to be favorable in germination, therefore, they can also be utilized as a planting material [42].

12.3.6 PART OF CANE

Top portion of the immature young canes of the top are better seed material [20, 22] as they germinate quicker than the lower ones [47]. Top setts are more useful for dry condition and autumn planting, bottom and mature part are inferior in germination (Figure 12.2).

Although top setts are widely recommended for seed material reports of yield increase by their use are less common. The advantages of top setts are for sprouting and apparent better growth of the crop.

FIGURE 12.2   Top portion of the immature young canes of the top are better seed material.

12.3.7 LIGHT

Experiments on the role of light were done at the IISR Lucknow. The early results indicated that the darkness is necessary for the germination of bud under certain condition. Exposure of light favored germination, darkness appeared to counteract apical dominance but was perhaps devious [30]. Out of the different factors affecting germination, a certain minimum value of light was necessary [51].

12.3.8 TEMPERATURE

The temperature of the soil perhaps is the most important factor for the germination of sugarcane [53]. High temperature favors catabolic reaction, which promote cell division and early growth process [6]. At Shahjahanpur, for proper germination a range from 17.4–22.9°C with minimum variation was considered while the optimum appeared to be 22.9°C. Low temperature had a depressing effect on germination and different varieties showed varying degree of susceptibility [7].

In Punjab, 23.9–29.4°C soil temperature was optimum for rapid germination [34]. Later on, from detailed study minimum temperature from 16.8–28.06°C was reported to be for good germination. Solomon et al. [41] found that 30–35°C to is more conductive for both sprouting of bud as well as growth different moisture levels.

12.3.9 SOIL MOISTURE

Soil Moisture influences germination to a greater extent by affecting aeration and soil temperature [22]. Moreover, soil moisture plays the role of keeping the buds moist and this helps in sprouting and also keeping the young sprouts well supplied with water. The percentage of the germination increases with an increase in the soil moisture. High moisture content in soil, particularly in late planting when humidity is low and temperature is high, was found to have beneficial effect on germination. So immediate irrigation was suggested when planting was done late in April–May [10, 48]. Earlier, it was reported that 15% soil moisture was most suitable for germination of sugarcane [19]. The optimum soil moisture for maximum germination varies for different localities depending on edaphic, biotic and climatic conditions. Soil aeration plays very important role in germination and growth of the plant (Figure 12.3).

12.3.10   SETT MOISTURE

Moisture content of setts is one of the important factors affecting germination. For the sprouting of bud and root primordia a critical sett water content is essential. In Bihar, the critical water content in the setts for optimum germination was 50.3%. A study carried out at Lucknow, reviled that critical water content of nodal tissues required for bud germination was 70% and the optimum level for rapid germination was 72–74%. Germination was extremely poor below critical level of sett moisture irrespective of adequate soil moisture [6]. Bud and root primordia appeared to compete for water in the setts. Similarly a study in one-bud setts indicated that sprouting and early shoot growth was influenced more by sett water than the sett water potential [33].

12.3.11   NUTRITIONAL STATUS OF THE SETT

The energy required for sprouting of the bud and for growth and development of the young sprout, until the seedlings are established is drawn from the nutrients stored in the sett. The nutrient status of cane therefore

FIGURE 12.3   Sett and soil moisture affects germination.

has marked influence in germination and growth of newly developed shoots. High nitrogen, starch and glucose contents are essential for good seed cane though their contents vary with different varieties [12]. Dutt and Narasimhan [8] have emphasized the importance of starch in germination. Canes from the highly fertilized plots have been reported to give high germination and early vigorous growth. Nitrogen content of setts has special significance as controlling factor in germination [27]. The top cuttings of a stalk, which contained more glucose showed early and better germination than stalk cutting which have more of sucrose.

12.3.12   SPACING AND DEPTH OF PLANTING

Spacing between two rows did not influenced germination, but depth of planting influences germination up to the marked extent (Figure 12.4). Shallow planting in poorly aerated and improperly drained soil give more germination. The depth may be increased from winter to summer, wet to dry areas and from heavy to light soil. In ideal soil a planting depth of 5–7 cm is recommended. In heavy soil the shallow planting may give 6–13% more germination. Ten-eye bud is recommended for 1-meter length and 90 cm row to row distance for maximum germination (fine seed cane of two budded setts).

FIGURE 12.4   Spacing and depth of planting.

12.3.13   ENZYMATIC ACTIVITY

Besides physical and chemical factors germination is mainly controlled by endogenous biochemical factors, which are not fully understood. Since photosynthesis reserve sugar is mainly sucrase enzymes of carbohydrate metabolism, i.e., acid invertase and amylases play an important role in sprouting of setts. Activation in amylase activity and changes in protein, amino acid and carbohydrate contents in the germinating bud were also noticed. The metabolic role of reducing sugars acid invertase and amylase in the sprouting of sugarcane cuttings has been clearly delineated. Solomon and Singh [40] reported that a dramatic increase in the invertase activity was noticed in setts as well as in the buds 10–30 day after planting, it was also revealed that high acid invertase activity was correlated with quicker and higher germination. Amylase activity in the germinating organs probably remobilized starch formed as a result of sugarcane reserve interconversion process. The starch formation in the storage tissue was probably responsible for root development during germination of setts (Solomon et al., 1988b). Besides acid invertase and amylase, other enzymes like starch phospharylase, per oxidase and IAA oxidase play an important role in the sprouting of seed cane [41]. In sugarcane plant growth regulators (PGRS) have been extensively used to promote sprouting and early growth of aerial buds. The activity of acid invertase, amylase and starch phosphorylase showed a marked increase in the treated setts, whereas peroxidase and IAA oxidase recorded a marginal change in their specific activity.

12.3.14   HEAT THERAPY

Sugarcane setts treated with hot water gave higher germination at Coimbatore. Improvement in some germination of varieties (Co 313, Co 859, CoS 443) was observed at Shahjahanpur also by treating the sugarcane setts in hot water at 45°C. Inhibitory effect was marked when the temperature and duration were increased further. The hot therapy might induce such metabolic processes, which lead to better germination of buds. Setts treated through aerated steam system [9] had higher percentage of germination. The delay in planting of the treated seed cane upto 72 hours did not impair the germination significantly if the seed cane is stored in shade and setts are dipped in fungicide solution before planting.

12.3.15 SETT SOAKING

The effects of germination sett soaking in nutrient water and hormonal solution have extensively been explored. IAA, NAA, 2,4-D, IBA, GA, CCCA, Ethrel, and Dormex, at varying concentration and socking durations are known to improve germination in sugarcane. This is carried out by increased intake of water, leaching out of germination inhibitors and greater retention of water by the setts, [10, 19, 23, 36, 38, 39]. Effect of water soaking of setts on germination indicated that the increasing level of sett moisture gave better germination under adverse condition.

12.3.16 SETT DIPPING

Dipping of setts in organo-mercurial compounds such as areton, aglal, tafan Ganna beej Ghol, Bavistin results 10–15% higher germination. They stimulate the sett root growth, water uptake by these roots and probably mitigate the effect of germination inhibitors present in the buds. This technique to improve germination is extensively being used by farmers (Figure 12.5).

FIGURE 12.5 Soaking of setts in bavistin (0.1%) before planting.

12.4   CONCLUDING REMARKS

Germination in sugarcane is one of the most important factors contributing towards the stand and yield of the crop in Northern India. To improve germination it is suggested that:

1. Healthy and thick canes from well mannured and irrigated field should be selected for planting.
2. The two or three budded setts preferably from the top portion of canes should be planted soon after cutting, when soil moisture and temperature are 8–15% and 16–30°, respectively.
3. In heavy soils with poor aeration, shallow drainage and during autumn shallow planting is beneficial whereas in light soil in spring season deep planting is useful. Depending on the soil and climatic condition the depth should be regulated from 5–7 cm.
4. The setts should be treated with organo-mercurial compounds and planted setts should be sprayed with insecticides.
5- Under adverse condition (inadequate moisture, high temperature late plantings the staled cane) it is advisable that the setts must be soaked for a period not exceeding 24 hrs., and field must be irrigated soon after planting. These practices are immensely important and may improve the germination by 30–40%.

KEYWORDS

- **germination**
- **seed**
- **sett**
- **soil moisture**
- **sugarcane**
- **temperature**
- **variety**

REFERENCES

1. Annual Report (1961–62). U.P. Council of Sugarcane Research, Shahjahanpur.
2. Annual Report (1983). U.P. Council of Sugarcane Research, Shahjahanpur.
3. Annual Report Sugarcane Research Scheme Annakapalli 1934–35.
4. Chaudhari, R. S., & Bhatnagar, V. B. (1953). *Journal of Science and Research (BHU)* 4, 149–156.
5. Clemants, H. F. (1940). *Hawaiian Planters Rec. 44*, 117–147.
6. Divedi, R. S., & Sinha, O. K. (1993). A Scenario of Research on Physiology and Bio-chemistry of Sugarcane in Subtropical India. G. B. Singh, O. K. Sinha (eds.). IISR Sci. Club Lucknow, 143–190.
7. Dutta Bunnal & Vijay Saradhym (1947). *Proceedings of Indian Congress. 34.*
8. Dutta, N. L., & Narsimahan, R. (1949). *Current Science. 18*, 246–347.
9. Edison, S. (1977). *Indian Sugar crops Journal. 4*(2), 32–34.
10. Gahlot, K. N. S. (1956). *Indian Sugar. 6*(9), 652–671.
11. Jasdeep, A. Bendigiri, A. V., & Hapase, D. G. (1988). *Proceedings of 38ᵗʰ Annual Convention of DSTA*, India, pp. A364–A380.
12. Kakde, J. R. (1985). *Germination in Sugarcane.* Sugarcane Production Metro-politan Book Co. Pvt. New Delhi, pp. 169–175.
13. Khanna, K. L. (1933a). *Proceedings of Indian Science Congress. 20*, 56.
14. Khanna, K. L. (1933b). Department of Agriculture Bihar and Orissa, 5.
15. Kirtikar & Annag Nath (1966). *Proceedings of 54ᵗʰ Annual Convention of Sugar Technologists Association, India. 7–18.*
16. Krishna Iyenger, C. V. (1951). *Nature*, London. 168–252.
17. Krishnamurthy Rao (1937). *Proceedings of Indian Science Congress. 24.*
18. Mathur, R. H., & Haidar, I. U. (1940). *Proceedings of 9ᵗʰ Annual Convention of Sugar Technologists of India.*
19. Mathur, R. N. (1940). *Proceedings of 9ᵗʰ Annual convention of Sugar Technolo-gists Association of India.*
20. Mishra, G. N. (1957). *Proceedings of 3ʳᵈ Biennial conference of Sugar Research and Development.* I: 136–142.
21. Mishra, G. N. (1965b). *Indian Sugar. 15*(9), 599–612.
22. Mishra, G. N. (1965). *Indian Sugar. 15*(6), 419–424.
23. Narsimha Rao, N. (1974). Sixty years of Agriculture Research at Annakapllae.
24. Panje, R. R., & Gill, P. S. (1962). Annual Reeport, IISR, Lucknow. 20–21.
25. Panje, R. R., Mathur, P. S., & Motibule. (1966). *Proceedings of 13ᵗʰ ISSCT Tai-wan.*
26. Punje, R. R., Mathur, P. S., & Motibule. (1969). *Indian Journal of Agriculture Sciences. 39*(12), 1142–1149.
27. Rage, R. D., & Wagle, P. V. (1939). *Indian Journal of Agriculture Sciences. 9*, 423–448.
28. Rao, J. T., Natrajan, B. V., & Bhagya Luxmi, K. B. (1983). Sugarcane, ICAR New Delhi.
29. Singh, H., Adlakha, P. A., & Gill, H. S. (1960). *Indian Journal of Sugar Research and Development. 5* (1), 1–8.

30. Singh, K. (1977). IISR Lucknow Silver Jubilee 25 years' Results. pp. 1–2.
31. Singh, N., & Singh, H. (1956). *Proceedings of ISSCT Congress. 1*(1), 283–301.
32. Singh, R. G., & Ali, S. A. (1983). *Indian Sugar Crops. 9*(1), 1–4.
33. Singh, S. (1972). *Indian Sugar. 22*(5), 425–428.
34. Singh, S., & Gill, H. S. (1951). Proceedings *of 1ˢᵗ Bienu Congress of Sugar Research and Development,* Coimbatore part II. *4,* 85–89.
35. Singh, S., Gill, H. S., & Sidhi, B. S. (1964). *Indian Sugar Journal. 9*(1), 56–59.
36. Singh, S., & Srivastava, K. K. (1969). *Experintia. 25,* 1262–1263.
37. Singh, U. S. (1967a). *Proceedings of National Academics of Sciences,* India. 13 *37*(2), 189–191.
38. Singh, S. P. M., Shahi Lal, V. K., & Sharma, B. L. (2016). *Agrica. 5,* 66–68.
39. Singh, S. P. M., Johari, D., & B. L. Sharma. (2015). *Agrica. 4,* 38–40.
40. Solomon, S., & Singh, K. (1987). *Indian Journal of Plant Physiology and Biological Research. 6,* 7–25.
41. Solomon, S., Singh, K., Srivastava, K. K., & Madan, U. K. (1988b). *Indian Journal of Biological Research and Development. 6,* 7–25.
42. Srikhande, J. D., & Gahlot, K. N. S. (1960). *Proceedings of 4ᵗʰ All India Conference sugarcane Research and Development Workers* Part 1. p. 10.
43. Srivastava, R. S. (1963). *Indian Journal of Sugarcane Research and Development. 7*(4), 219–225.
44. Subbha Rao, M. S., & Prasad, R. B. (1960). *Indian Journal of Plant Physiology. 3*(2), 181–194.
45. Subbha Rao, M. S. Prasad, R. B., & Khanna, K. L. (1959). *Indian Journal of Sugarcane Research and Development. 4*(1), 22–26.
46. Tandon, R. K., & Mishra, G. N. (1956). *Indian Sugar. 6* (6), 379–387.
47. Tondon, R. K. (1955). Serving Uttar Pradesh through in Sugarcane Bureau of Agricultural Information, Lucknow (UP).
48. Vaghalkar. (1934). Annual Report Sugarcane Research Station, Padegaon.
49. Van Devellewijn, C. (1952). *Botany of Sugarcane.* The Chron, Bota Co Waltham Mars, USA.
50. Venkatraman, T. S. (1926). *Agriculture Journal of India. 21*(2), 101–106.
51. Whitema, P. C., Bull, T. A., & Glasziou, K. T. (1963). *Australian Biological Sciences. 16*(2), 416–428.
52. Yadav, R. L. (1981). *Indian Journal of Agronomy. 26*(2), 130–136.
53. Yadav, R. L. (1991). Sugarcane Production technology constraints and potentialities. Oxford and IBH Publishing Co. Pvt Ltd. New Delhi, pp. 44–52.

CHAPTER 13

SECONDARY METABOLITES: THE PLANT PHYSIOLOGICAL TALE OF STRESS RESPONSE

R. K. UPADHYAY

Department of Botany, Haflong Government College
(Assam University Affiliation), Haflong – 788891, Assam, India,
E-mail: rku.univ@yahoo.com

CONTENTS

ABSTRACT

Secondary metabolites include a wide variety of organic compounds which are not directly involved in plant physiological and biochemical processes of normal growth and development. These metabolites are also known as natural products, accumulated often in plants acclimatized to stresses, since they help in overcoming stress conditions. Secondary metabolites appear to function primarily in defense against predators and pathogens and in providing reproductive advantage—attractants of pollinators and seed dispersers. Plants are capable of synthesizing an overwhelming variety of small organic molecules, called secondary metabolites, usually with very complex and unique carbon skeleton structures, with many of them having high interests to the pharmaceutical and chemical industries. These compounds are an extremely diverse group of natural products synthesized by plants as well as fungi, bacteria, algae, and animals. They may act to create competitive advantage as poisons of rival species. Secondary metabolites are used in the pharmaceutical industry as flavorants, dyes, and perfumery. Further, the secondary metabolism is an integral part of the developmental program of plants, and in accumulation of which it often marks the onset of developmental stages. This mini review article summarizes the role of different stress factors, including abiotic ones, particularly on secondary metabolites and pharmaceuticals in plants.

13.1   INTRODUCTION

Secondary metabolites are classified into several groups based on their chemical structure and biosynthesis, of which three main groups are: terpenes (such as plant volatiles, cardiac glycosides, carotenoids, and sterols), phenolics (such as phenolic acids, coumarins, lignans, stilbenes, flavonoids, tannins, and lignin) and nitrogen containing compounds (such as alkaloids and glucosinolates) (Figure 13.1). Terpenoids are composed of five-carbon units synthesized by way of the acetate or mevalonate pathway or the glyceraldehydes 3-phosphate or pyuvate pathway. These are the largest and most diverse families of natural products, ranging in structure from linear to polycyclic molecules and in size from the five-carbon hemiterpenes to natural rubber, comprising thousands of isoprene units. Many plant terpenoids are toxins and feed deterrents to herbivores or are attractants of various sorts. Secondary metabolites are not essential for the growth and development of a plant, but are rather required for the interaction of plants with their environment [1]. It is reported that a large family of N-containing secondary metabolites is found in approximately 20% of the species of vascular plants [2], most frequently in the herbaceous dicot and relatively a few

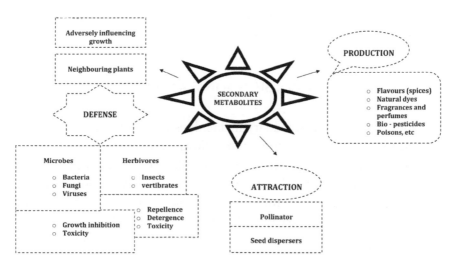

FIGURE 13.1   Secondary metabolites and its adaptation in plants.

in monocots and gymnosperms. Most of them, including, of course, the pyrrolizidine alkaloids (PAs), are generally toxic to some degree and appear to serve primarily in defense against microbial infection and herbivoral attack [3]. They are usually synthesized from one of the few common amino acids, in particular, aspartic acid, lysine, tyrosine and tryptophan [4]. Alkaloids are synthesized principally from amino acids. These nitrogen containing compounds protect plants from a variety of herbivorous animals and many of them possess pharmacologically important activity. Because of their bioactive properties, alkaloids are often sequestered in the vacuoles of plant cells to avoid toxic effects. Many studies have already been carried out to characterize the mechanism of vacuolar accumulation of alkaloids using cell cultures, protoplasts, and isolated native vacuoles. Phenolic compounds, which are primarily synthesized from products of shikimic acid pathway are widely distributed in nature and have carried out several important roles in plants. Tennins, lignans, flavonoids, and some simple phenolic compounds serve as defenses against herbivores and pathogens. In addition, lignins strength cell walls mechanically, and many flavonoid pigments are important attractions for pollinators and seed dispersers. Some phenolic compounds have allelopathic activity and may adversely influence the growth of neighboring plant.

13.2   STRESS RESPONSE, SIGNALING AND PRODUCTION

The formation of plant secondary metabolites is complex, and its process is dynamic, involving multiple sub-cellular compartments such as the cytosol, endoplasmic reticulum, and vacuole. The metabolites are also used in the pharmaceutical industry as flavorants, dyes and perfumery, besides acting as attractants for pollinating insects, for seed dispersing animals and for root nodule bacteria. Plant secondary metabolites are a large resource of natural medicines and exhibit various pharmacological and biological activities.

The seasonal variation of climate also affects composition and production of secondary metabolites in plants. Saponins and melatonin play an important role in this regard [5, 6], protecting plant responses to different environmental changes [7, 8], in addition to acting as a tolerant to environmental pollutions to increasing the survival of plants and serotonin, improving the physiological functions in plants, protecting from environmental constraints and pathogenic infection besides ensuring protection against ROS in plants [9]. As conventional growing of medicinal plants is relatively expensive, production of medicinal and other compounds can be done in vitro plant. Hypoxoside produced by Hypoxis species is used as anticancer drug. Red-pigmented anthocyanins, used as a food colourant, have been produced from the callus of oxalis species. Genes involved in ubiquinone biosynthesis in the *Escherichia coli* (ubiC and ubiA) and another plant gene such as HMG COA reductase, were used for the transformation of *Nicotiana tabacum* and *Lithospermum erythrorhizon*. In L. erthrorhizon, the expression of the enzymes led to an upregulation of certain biosynthetic reaction steps involved in the formation of the pharmaceutical (and dye) shikonin. Secodary metabolites of *R. Rosea* are produced in the root. These compounds, designated as secondary metabolites, differ from primary metabolites, such as carbohydrates, amino acids, and nucleotides that have basic functions to maintain the life cycle, cell division and acquiring energy, in their restricted distribution in the plant kingdom, i.e., occurring

In some limited plant families or even in specific species. Many secondary metabolites have diverse biological activities, and function in defense responses against pathogens and herbivores and in protecting plants from damage by UV light. Some secondary metabolites have also pharmacological importance, such as producing natural medicines for humans. Intensive studies have been carried out to characterize the nature of secondary metabolites from the viewpoints of their biological activities, chemical structures, biosynthesis, and their functions in plants as well as their clinical usage [10]. Secondary metabolites are often found to be accumulated in particular tissues at a high concentration. This accumulation in such appropriate compartments should be regulated in a highly sophisticated manner, in view of the outbreak toxicity plants due to mislocalization.

Most of the secondary metabolites are translocated from source cells to sink organs via long distance transport or transporters (both primary and secondary transporters) which are involved in such translocation process, and many transporter genes, especially genes belonging to the multidrug and toxin extrusion type transporter family, also known as MATE transporter in case of Arabidopsis particularly, are all responsible for the membrane transport of secondary metabolites, performing different physiological functions in different plant species (Table 13.1). Some of the secondary metabolites are transported in intercellular fashion [11–13].

TABLE 13.1 Physiological Function of Transporters in Different Plants During Secondary Metabolite Translocation [41]

Plant species	Locations	Physiological functions
Arabidopsis thaliana	Plasma membranes, vacuoles	Efflux of endogenous, metabolites, xenobiotics, and citrate
Hordeum vulgare	Plasma membranes, root and shoots	Efflux of citrate and tolerance to aluminum
Sorghum bicolor	Plasma membranes and roots	Efflux of citrate and tolerance to aluminum
Lupinus albus	Plasma membranes and roots	Not yet determined for MATE transporters

(Reprinted with permission from Yazaki, K., Sugiyama, A., Morita, M., & Shitan, N. (2008). Secondary transport as an efficient membrane transport mechanism for plant secondary metabolites. Phytochemistry Review, 7, 513–524. © 2007 Springer Nature.)

Climate change, cold, temperature, light, plant growth hormones also influence the production of secondary metabolites. During cold or low temperature, synthesis of sugar alcohol like sorbital and soluable sugars, Anthocyanin, suberin, etc., are also induced in plants [14–16]. An extensive collection of abiotic stimuli like salinity, drought, heavy metal, etc., are capable to trigger changes in the plant metabolism, resulting in enhanced production of plant secondary products. Different studies on comparative analysis state that the amount of secondary plant products is maximum in plants, and severely affect or grow under abiotic stress, unlike other plants cultivated under optimal conditions of growth and productivity. Abiotic stresses basically lead to the production of the following natural secondary metabolites in plants (Table 13.2). As a result, the recognized increase in concentration of secondary plant products could not be barred as compared to stressed plants due to the fact that the total amount of secondary metabolites per plant is more or less the same in both traits, whereas the biomass is significantly lower in the stressed plants.

Plant stress, like excess light, wounds, pathogen attack, nutrient deficiencies, etc., usually increased the accumulation of secondary metabolites like phenylpropanoids, phenylamides, polyamines, etc., [17, 18]. Salt and drought stress are the two of the most serious factors limiting the productivity of different crops and especially quantity and quality of

TABLE 13.2   Secondary Metabolites Produced Under Different Types of Stress in Plants

Plant species	Secondary metabolite present	References
Taxus candensis	Paclitaxel	[42]
Cakile maritima	Polyphenol	[43]
Tulipa gesneriana	Anthocyanin	[44]
Oryza sativa	Polyamines	[45]
Polygonum hydropiper	Flavanol	[46]
Hyoscyamus muticus	Sesquiterpenes	[47]
Lycopersicon esculentum	Sorbitol	[48]
Glycine max	Isoflavonoid, Trigonelline	[49, 50]
Coffea Arabica	Alkaloids	[51]
Digitalis thapsi	Cardenolide	[52]
Cenchrus pennisetiformis	Starch and Sucrose	[53]

their metabolic (secondary plant products) products to a greater extent. Generally, plants produce secondary plant products in nature as a defense mechanism against attack by pathogens and insects, while others promote defensive action against herbivores and pathogens [19]. Secondary metabolites have various functions, including protection against pathogens and UV light in plants, and have been used as natural medicines for cure of diseases as well as humans towards utilizing their diverse biological activities. Many of these natural compounds are accumulated in a particular compartment such as vacuoles, and some are even translocated from source cells to sink organs via long distance transport. The plant secondary metabolites of plants have been a fertile area of chemical investigation for many years, driving the development of analytical chemistry and of new synthetic reactions and methodologies. Identifying key elements involved in these processes will allow generating novel tools for metabolic engineering and molecular mechanisms of plant cells. However, technological advances in analytical chemistry, in particular in the development of high-field nuclear magnetic resonance spectroscopy and Fourier transform-ion cyclotron mass spectrometry, have facilitated the elucidation of structures of secondary metabolites that are conspicuously present even at low levels within a plant. Anthocyanins are reported to increase in response to salt and cold stress [20, 21]. However, drought often causes oxidative damages and overproduction of reactive oxygen species (ROS), besides reportedly increasing the amount of flavanoids and phenolic acids in plants like *Willow* leaves, as well as decreasing saponin contents in *Chenopodium quinoa* [22, 23]. Flavanoids are also reported to provide protective mechanisms to plant growing under drought and aluminum stress [24].

Some metals, like cadmium, copper, silver, etc., are also induced by secondary metabolites like shikonin and production of digitalin [25–27]. Copper also stimulated the production of betalains and betacyanins in plants like *Beta vulgaris* and *Amaranthus caudatus* [28, 29]. Melatonin in seeded plants like *Cucumis sativa* L. also improved responses in germination during cold stress [30]. Likely temperature stress also influenced physiological activity in plant like premature leaf senescence [5]. Light, too initiates the production of secondary metabolites such as, gingerol and zingiberene [31]. Radiations like UV-B have been reported to increase flavanoids, vinblastrin and vincristine production in plants [32–36].

However, growth hormones like IAA, cytokinin, kinetin, calcium, etc., were also found to be enhancing the production of secondary metabolites in different plant species [37–39]. About 25% of today's pharmaceuticals contain at least one active ingredient of plant origin. Even so, only a minute fraction of the enormous biosynthetic potential of plant cells is being exploited. Secondary metabolites, being in valuable to humans are very often difficult to isolate in large quantities. Plants like *Arabidopsis*, have been identified to be responsible for the membrane transport of secondary metabolites. Better understanding of membrane transporters like ATP binding cassette (ABC) transporters, as well as of the biosynthetic genes of secondary metabolites will be important for metabolic engineering, aiming to increase the production of the commercially valuable secondary metabolites in plant cells. In plant cells, vacuoles, which occupy as much as 90% of the cells volume, play a central role in the accumulation of secondary metabolites such as alkaloids and flavonoids, and a large number of transporters, channels and pumps reside in the vacuolar membrane (tonoplast) [13, 40]. Increasing secondary metabolite yield by cloning related genes for rate limiting enzymes, large scale production of commercially useful proteins and industrial compounds, tissue specific expression and chloroplast transformation for high-level expression of desired compounds, are the major research areas in the present. Better understanding of secondary transporters with their regulatory mechanisms by biosynthetic pathways will provide us with better strategies for engineering towards commercial applications.

13.3   CONCLUSION

Though secondary metabolites have no specific role for life processes in plants, yet they specifically interact with the adaptation and defense processes in plants against herbivores and pathogens. Of course their influence might depend on prevailing ecosystem will certainly provide a long way for their pathways of production, application and regulation. The biochemical and molecular approaches towards understanding their pathways of production, application and regulation will certainly provide a long way for flooding the market of production of secondary metabolites vigorously as well as for introduction of better plant defense system.

KEYWORDS

- climate change
- physiological changes
- plant stress
- secondary metabolites
- transporters

REFERENCES

1. Kutchan, T., & Dixon, R. A. (2005). Secondary metabolism: nature's chemical reservoir under deconvolution. *Current Opinion in Plant Biology, 8,* 227–229.
2. Hegnauer, R. (1988). Biochemistry, distribution and taxonomic relevance of higher plant alkaloids. *Phytochemistry, 27,* 2423–2427.
3. Schafer, H., & Wink, M. M. (2009). Medicinally important secondary metabolites in recombinant microorganisms or plants: progress in alkaloid biosynthesis. *Biotechnology Journal, 4*(12), 1684–1703.
4. Pearce, G., Strydom, D., Johnson, S., & Ryan, C. A. (1991). A polypeptide from tomato leaves induces wound inducible protienase inhibitor proteins. *Science, 253,* 895–898.
5. Morison, J. I. L, & Lawlor, D. W. (1999). Interactions between increasing CO_2 concentration and temperature on plant growth. *Plant Cell Environment, 22,* 659–682.
6. Szakiel, A., Paczkowski, C., & Henry, M. (2010). Influence of environmental abiotic factors on the content of saponins in plants. *Phytochemistry Review, 2,* 25.
7. Arnao, M. B., & Hernandez-Ruiz, J. (2006). The physiological function of melatonin in plants. *Plant Signaling and Behavior, 3,* 89–95.
8. Lin, J. T., Chen, S. L., Liu, Sc, & Yang, D. J. (2009). Effect of harvest time on saponins in Yam (*Dioscorea pseudojaponica* Yamamoto). *Journal of Food and Drug Analysis, 17,* 116–22.
9. Tan, D. X, Manchester, L. C., Helton, P., & Reiter, R. J. (2007). Phytoremediative capacity of plants enriched with melatonin. *Plant Signaling and Behavior, 2,* 514–516.
10. Croteau, R., Kutchan, T. M., & Lewis, N. G. (2000). Natural products (secondary metabolites). In: Buchanan B., Gruissem W., Jones R., (Eds.). *Biochemistry and Molecular Biology of Plants.* American Society of Plant Physiologists, Rockville, MD, pp. 1250–1268.
11. Sakai, K., Shitan, N., Sato, F., Ueda, K., & Yazaki, K. (2002). Characterization of berberine transport into *Coptis japonica* cells and the involvement of ABC protein. *Journal of Experimental Botany, 53,* 1879–1886.

12. Terasaka, K., Sakai, K., Sato, F., Yamamoto, H., & Yazaki, K. (2003). Thalictrum minus cell cultures and ABC-like transporter. *Phytochemistry, 62*, 483–489.

13. Rea, P. A. (2007). Plant ATP-binding cassette transporters. *Annual Review in Plant Biology, 58*, 347–375.

14. Cristie, P. J., Alfenito, M. R., & Walbot, V. (1994). Impact of low temperature stress on general phenylpropanoid and Anthocyanin pathways: enhancement of transcript abundance and Anthocyanin pigmentation in maize seedlings. *Planta, 194*, 541–549.

15. Griffith, M., & Yaish, M. W. F. (2004). Antifreeze proteins in overwintering plants: a tale of two activities. *Trends in Plant Science, 9*, 399–405.

16. Janska, A., Marsik, P., Zelenkova, S., & Ovesna, J. (2010). Cold stress and acclimation–what is important for metabolic adjustment? *Plant Biology, 12*, 395–405.

17. Dixon, R. A., & Paiva, N. (1995). Stressed induced phenyl propanoid metabolism. *Plant Cell, 7*, 1085–1097.

18. Edreva, A. M., Velikova, V., & Tsonev, T. (2000). Phenylamides in plants. *Russian Journal of Plant Physiology, 54*, 287–301.

19. Harborne, J. B., & Williams, C. A. (2000). Advances in flavonoid research since 1992. *Phytochemistry, 55*, 481–504.

20. Chalker-Scott, L. (1999). Environmental significance of anthocyanins in plant stress responses. *Photochemistry and Photobiology, 70*, 1–9.

21. Parida, A. K., & Das, A. B. (2005). Salt tolerance and salinity effects on plants: a review. *Ecotoxicology Environmental Safety, 60*, 324–349.

22. Larson, R. A. (1988). The antioxidants of higher plants. *Phytochemistry, 27*, 969–978.

23. Soliz-Guerrero, J. B., de Rodriguez, D. J., Rodriguez–Garcia, R., Angulo-Sanchez, J. L., & Mendez-Padilla, G. (2002). Quinoasaponins: concentration and composition analysis. In: Janick, J., Whipkey, A., (Eds.). Trends in New Crops and New Uses. Alexandria: ASHS Press, p. 110.

24. Winkel-Shirley, B. (2001). Flavonoid biosynthesis, A colourful model for genetics, biochemistry, cell biology and biotechnology. *Plant Physiology, 26*, 485–493.

25. Mizukami, H, Konoshima, M., & Tabata, M. (1977). Effect of nutritional factors on shikonin derivative formation in *Lithospermum* callus cultures. *Phytochemistry, 16*, 1183–1186.

26. Ohlsson, A. B., & Berglund, T. (1989). Effect of high $MnSO_4$ levels on cardenolide accumulation by *Digitalis lanata* tissue cultures in light and darkness. *Journal of Plant Physiology, 135*, 505–507.

27. Marschner, H. (1995). *Mineral Nutrition of Higher Plants*, Academic Press, London, p. 889.

28. Mizukami, H., Konoshima, M., & Tabata, M. (1977). Effect of nutritional factors on shikonin derivative formation in *Lithospermum* callus cultures. *Phytochemistry, 16*, 1183–1186.

29. Obrenovic, S. (1990). Effect of Cu (11) D-penicillanine on phytochrome mediated betacyanin formation in *Amaranthus caudatus* seedlings. *Plant Physiol Biochemistry, 28*, 639–646.

30. Posmyk, M. M., Balabusta, M., Wieczorek, M., Sliwinska, E., & Janas, K. M. (2009). Melatonin applied to cucumber (*Cucumis sativus* L.) seeds improves germination during chilling stress. *Journal of Plant Research, 46*, 214–223.

31. Anasori, P., & Asghari, G. (2008). Effects of light and differentiation on gingerol and zingiberene production in callus culture of *Zingiber officinale* Rosc. *Research in Pharmaceutical Science, 3*, 59–63.

32. Kramer, G. F., Norman, H. A., Krizek, D. T., & Mirecki, R. M. (1991). Influence of UV–B radiation on polyamines, lipid peroxidation and membrane lipids in cucumber. Phytochemistry, *30*, 2101–2108.

33. Liu, L., Dennis, C., Gitz, III., Jerry, W., & McClure. (1995). Effects of UV-B on flavonoids, ferulic acid, growth and photosynthesis in barley primary leaves. *Physiologiae Plantarum, 93*, 734–738.

34. Fischbach, R. J., Kossmann, B., Panten, H., Steinbrecher, R., Heller, W., Seidlitz, H. K., et al. (1999). Seasonal accumulation of ultraviolet-B screening pigments in needles of Norway spruce (*Picea abies* (L.) Karst). *Plant Cell and Environment, 22*, 27–37.

35. Shiozaki, N., Hattori, I., Gojo, R., & Tezuka, T. (1999). Activation of growth and nodulation in symbiotic system between pea plants and leguminous bacteria by near UV radiation. *Journal of Photochemistry and Photobiology B. Biology, 50*, 33–37.

36. Bernard, Y. K., Binder, Christie A. M., Peebles Jacqueline, V., & Shanks, Ka-Yiu San. (2009). The effects of UV-B stress on the production of terpenoid indole alkaloids in *Catharanthus roseus* hairy roots. *Biotechnol Programme, 25*, 8615.

37. Narayan, M. S., Thimmaraju, R., & Bhagyalakshmi, N. (2005). Interplay of growth regulators during solid-state and liquid-state batch cultivation of anthocyanin producing cell line of *Daucus carota*. *Process Biochem. 40*, 351–358.

38. Tuteja, N., & Mahajan, S. (2007). Calcium signaling network in plants: an overview. *Plant Signaling and Behaviour, 2*, 79–85.

39. Chen, Q., Qi W, Reiter, R. J., Wei, W., & Wang, B. (2009). Exogenously applied melatonin stimulates root growth and raises endogenous indoleacetic acid in roots of etiolated seedlings of *Brassica juncea*. *Journal of Plant Physiology, 166*, 324–328.

40. Martinoia, E., Maeshima, M., & Neuhaus, H. E. (2007). Vacuolar transporters and their essential role in plant metabolism. *Journal of Experimental Botany, 58*, 83–102.

41. Yazaki, K., Sugiyama, A., Morita, M., & Shitan, N. (2008). Secondary transport as an efficient membrane transport mechanism for plant secondary metabolites. *Phytochemistry Review, 7*, 513–524.

42. Ketchum, R. E., Tandon, M., Gibson, D. M., Begley, T., & Shuler, M. L. (1999). Isolation of labeled 9-dihydroxybaccatin III and related taxoids from cell cultures of *Taxus Canadensis* elicited with methyl jasmonate. *J. Nat. Prod. 62*, 1395–1398.

43. Ksouri, R., Megdiche, W., Debez, A., Falleh, H., Grignon, C., & Abdelly, C. (2007). Salinity effects on polyphenol content and antioxidant activities in leaves of the halophyte *Cakile maritima*. *Plant Physiology and Biochemistry, 45*, 244–249.

44. Saniewski, M., Miszczak, A., Kawa-Miszczak, L., Wegrzynowicz-Lesiak, E., Miyamoto, K., & Ueda, J. (1998). Effects of methyl jasmonate on anthocyanin accumulation, ethylene production, and CO_2 evolution in uncooled and cooled tulip bulbs. *Journal of Plant Growth Regulation, 17*, 33–37.

45. Krishnamurthy, R., & Bhagwat, K. A. M. (1989). Polyamines as modulators of salt tolerance in rice cultivars. *Plant Physiology, 91*, 500–504.

46. Nakao, M., Ono, K., & Takio, S. (1999). The effects of calcium on flavanol production in cell suspension cultures of *Polygonum hydropiper*. *Plant Cell Report, 18*, 759–763.

47. Curtis, W. R., Wang, P., & Humphrey, A. (1995). Role of calcium and differentiation in enhanced sesquiterpene elicitation from calcium alginate immobilized plant tissue. *Enz. Microb. Technol. 17*, 554–557.

48. Tari, I., Kiss, G., Deer, A. K., Csiszar, J., Erdei, L., & Galle, A., et al. (2010). Salicylic acid increased aldose reductase activity and sorbitol accumulation in tomato plants under salt stress. *Biologiae Plantarum, 54*, 677–683.

49. Stab, M. R., & Ebel, J. (1987). Effects of Ca^{++} on phytoalexin induction by fungal elicitor in soybean cells. *Arch. Biochem. Biophys., 257*, 416–423.

50. Cho, Y., Lightfoot, D. A., & Wood, A. J. (1999). Trigonelline concentrations in salt stressed leaves of cultivated *Glycine max*. *Phytochemistry, 52*, 1235–1238.

51. Joye, L. B., & David, J. G. (1991). Calcium and phosphate effect on growth and alkaloid production in *Coffea arabica*: experimental results and mathematical model. *Biotechnology and Bioengineering, 37*, 859–868.

52. Margarita, C., Margarita, M., Jorge, F. T., & Purificacion, C. (1995). Calcium restriction induces cardenolide accumulation in cell suspension cultures of *Digitalis thapsi* L. *Plant Cell Report, 14*, 786–789.

53. Ashraf, M. (1997). Changes in soluble carbohydrates and soluble proteins in three arid-zone grass species under salt stress. *Tropical Agriculture, 74*, 234–237.

CHAPTER 14

EFFECT OF WASTEWATER IRRIGATION ON CROP HEALTH IN THE INDIAN AGRICULTURAL SCENARIO

ABHIJIT SARKAR,[1] SUJIT DAS,[1] VAIBHAV SRIVASTAVA,[2] POOJA SINGH,[2] and RAJEEV PRATAP SINGH[2]

[1] Laboratory of Applied Stress Biology, Department of Botany, University of Gour Banga, Malda – 732103, West Bengal, India

[2] Institute of Environment and Sustainable Development, Banaras Hindu University, Varanasi – 221 005, India, Tel.: +91-993-591-2997, E-mail: rajeevprataps@gmail.com

CONTENTS

ABSTRACT

Following the unplanned urbanization, industrialization, and global population burst, natural water storage is day-by-day experiencing critical crisis worldwide. The generation of contaminated water and its proper treatments has become a major environmental issue. In India, a major agriculture region utilizes wastewater or contaminated water for the irrigation of agricultural lands due to shortage of available fresh water. Present chapter mainly concentrated on the availability and utilization dynamics of wastewater/contaminated water in Indian agricultural sectors. Furthermore, the impacts of wastewater/contaminated water on agriculture crops were catalogued. As the wastewater is a rich source of plant's requisite organic matter, in many cases wastewater irrigated crops showed higher vigor and yield; but with anomalous physiological and biochemical response. We believe the information catalogued in present chapter should be useful for working in this specific sphere.

14.1   INTRODUCTION

Water is the most essential natural resource for maintaining an adequate food supply, and a productive environment for the existing life on Earth. We know that the Earth is called 'Blue Planet' because 70% of its surface is water. Out of the total global water storage 97.5% Saline Ocean and other 2.5% fresh water; and within that fresh water storage, 0.3% forms surface water, 30% ground water, and the rest in icebergs, ice sheets, and glaciers, etc. Among the total surface water, 87% is available from lakes, 11% from swamps and 2% from rivers (Figure 14.1).

Due to rapid increase in population growth and economic development, water resources in many parts of the world are pushed to their natural limits. In the last 50 years, the global population has more than doubled, from 3 billion in 1959 to 6.7 billion in 2009. It is predicted that the human population will reach 8.7–11.3 billion by the year 2050 [4]. This rapid growth in human population definitely increase the water demand from various users, namely: domestic, municipal, agricultural, horticultural, power, and industrial sectors, and it also put tremendous pressure on the global water

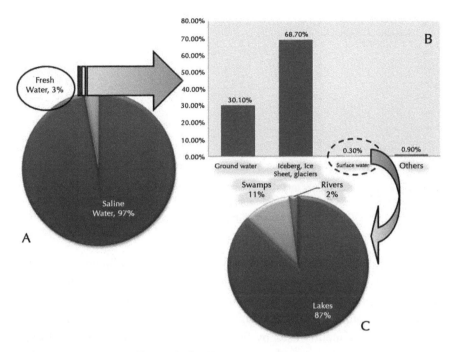

FIGURE 14.1 Availability and distribution of global water storage. A: nature and distribution of global water storage; B: distribution of fresh water storage; C: distribution of available surface water.

resources; hence, increase the volume of wastewater generation. There-fore, there is an urgent need to conserve the fresh water and to use the contaminated water for irrigation of agricultural crops to overcome this deficit condition. India is the seventh largest geographical area and second largest populous country in the world. The total population in India was 1,210 million people based on 2011 census report, Ministry of Statistics and Programme Implementation, Govt. of India., but after looking back, in the year 1950 it has at about 359 million people. This huge increase in population also increases the demand of fresh water and wastewater generation simultaneously. In India only 30% of the wastewater is treated before it's discharged. Thus, untreated water finds its way into water sys-tem such as rivers, lakes, groundwater, and causing water pollution. But due to economic and technical constrains the wastewater treatments vary between different cities in India (Figure 14.2).

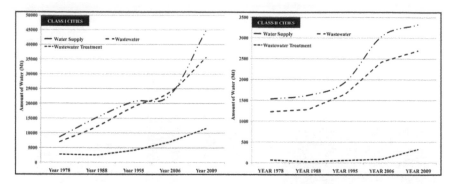

FIGURE 14.2   Water supply and usage, wastewater generation and wastewater treatment dynamics in class I and II type cities of India (Data source: MOSPI, GoI, 2011, http://www.mospi.gov.in/sites/default/files/publication_reports/climateChangeStat2015.pdf).

Fresh water resources are becoming scarce and are allocated for urban water supply. On the other hand large amounts of water are needed for agricultural production. If the contaminated water can be used as a substitute water source for irrigation both the issues can be solved. Domestic wastewater is frequently used by farmers to cultivate vegetables and salad crops like carrot, lettuce, cabbage and others for nearly urban markets [20]. But there is always a concern about the contamination and concentration of potentially toxic elements (Cd, Pb, As, F, etc.) from both domestic and industrial sources by the vegetables [21].

14.2   AVAILABILITY OF FRESH WATER SOURCES IN THE INDIAN SCENARIO

The availability of the usable fresh water resources is the most pressing of many other environmental challenges in Indian scenario. Geometric increase in population coupled with rapid industrialization, urbanization, agricultural and economic development has resulted in severe impact on the quality and quantity of water resources in India.

The total geographical area of India is about 3,287,240 km^2 (2.4% of the Earth's land mass) and its population is 1,210 million people based on 2011 census report, which is about 17.5% of that of the global population. Due to this increasing population growth rate, the amount of contaminated water generation increases considerably and while the per capita water availability in the country reduced remarkably. The average annual per capita availability of water taking into consideration the population in various censuses has come down 70% from 1951 to 2011 in a span of 60 years. According to census 1951 the per capita availability of water was 5,177 cubic meters. This is down to 1,545 cubic meters as per 2011 census. According to food and agricultural organization the value of renewable internal fresh water resources per capita in India was 1,116 cubic meters as of 2014.

In India, the national per capita annual availability of water is 2,208 cubic meters, while the average availability of in Barak and Brahmaputra Basin is as high as 16,589 cubic meters and in contrast Sabarmati Basin have as low as 360 cubic meters, but the rest of the country is about 1,583 cubic meters [38].

14.3 WATER RESOURCES AND THEIR QUALITY IN INDIA

With the increasing population, the amount of contaminated water generation increases continuously, which is often, discharged into the main water bodies and most often trough the unlined draining to join the rivers, lakes etc. This causes a crucial problem of contamination to rivers, lakes as well as underground waters. The most critical situation has been found in urban rivers which receives treated as well as untreated was water discharge because of urbanization [39]. During the last century the urban population increased over 26 million to 236 million.

As per CPCB report about 38,000 million liters a day of wastewater is generated by urban population and it is also estimated that 75–80% of water contamination by volume is from domestic sewages. According to CPCB, the critical condition of water body is due to the contamination of organic and bacterial discharge of domestic, municipal wastewater mostly in unprocessed type from the municipal centers of the country [38].

The water quality management in India is accomplished under the provision of Water (Prevention and Control of Pollution) Act, 1974. The crucial aim of this Act is to maintain and restore the wholesomeness of national aquatic resources by prevention and control of pollution. It was considered ambitious to maintain or restore all natural water body at pristine level. Planning pollution control activities to attain such a goal is bound to be deterrent to developmental activities and cost prohibitive. Since the natural water bodies have got to be used for various competing as well as conflicting demands, the objective is aimed at restoring and/or maintaining natural water bodies or their parts to such a quality as needed for their best uses (Table 14.1).

TABLE 14.1   Classification of Utilizable Water Source

Types of Water	Class	Quality criteria
Drinking water (*source*: without conventional treatment but after disinfection)	A	• Total coliforms organisms MNP/100 mL should be 50 or less • pH between 6.5 and 8.5 • Dissolved oxygen – 6 mg L^{-1} or more • BOD (5 days 20ºC) – 2 mg L^{-1} or less
Outdoor bathing (organized)	B	• Total coliforms organisms MNP/100 mL should be 500 or less • pH between 6.5 and 8.5 • Dissolved oxygen – 5 mg L^{-1} or more • BOD (5 days 20ºC) – 3 mg L^{-1} or less
Drinking water (*source*: after conventional treatment and disinfection)	C	• Total coliforms organisms MNP/100 mL should be 50 or less • pH between 6 to 9 • Dissolved oxygen – 4 mg L^{-1} or more • BOD (5 days 20ºC) – 3 mg L^{-1} or less
Propagation of wild life and fisheries	D	• pH between 6.5 and 8.5 • Dissolved oxygen – 4 mg L^{-1} or more • Free ammonia (as N) – 1.2 mg L^{-1} or less
Irrigation, Industrial cooling, controlled waste disposal	E	• pH between 6.0 and 8.5 • Electrical conductivity at 25ºC < 2250 µ mhos cm^{-1} • Sodium absorption ratio < 26 • Boron < 2 mg L^{-1}

(Adapted from Central Pollution Control Board Ministry of Environment and Forests, Government of India, "Status of Water Quality in India 2011, http://cpcb.nic.in/upload/NewItems/NewItem_198_Status_of_WQ_in_India_2011.pdf)."

14.4 CHARACTERISTICS FEATURES OF CONTAMINATED WATER

Contaminated water is usually that part of the water supply to the community, industry or other sources, which have been utilized for different purposes and mixed with solids waste, either suspended or dissolved. In general wastewater contains about 99.9% water and 0.1% solids. Wastewater is generally divided in three categories depending on the availability of several characteristics factors.

14.4.1 PHYSICAL CHARACTERISTICS

Fresh wastewater discharged from different sources may have the color in light brownish to grey, and with time it may change to dark grey or black. Generally, the odor of wastewater is produced by gas production due to the decomposition of organic matter or by substances added to the wastewater. Wastewater also contains different types of solids particles such as total solids (TD), settle able solids, suspended solids (SS), and filterable solids (FS). In case of total solids all the matter that remains as reside upon evaporation at 103°C to 105°C. On the other hand, settle able solids can be removed by primary sedimentation. The normal temperature of wastewater is commonly higher than that of fresh water and it is depended upon the contaminants present on the water.

14.4.2 CHEMICAL CHARACTERISTICS

In wastewater, organic matter is derived from animals, plants and man-made activities. It includes proteins (40–60%); carbohydrates (25–50%) and fats, oils and grease (10%), and various natural and synthetic organic chemicals from the process industries. The inorganic matters present on contaminated water are chlorides, nitrogen, phosphorous, sulphur, toxic inorganic compounds (copper, lead, silver, chromium, arsenic, and boron). Wastewater also contains several gases like N_2, O_2, CO_2, H_2S, NH_3, CH_4, etc. The hydrogen ion concentration is an important parameter in both natural and wastewaters. It is a very important factor in the wastewater treatment. Wastewater can be classified as neutral (pH 7), alkaline (pH > 7) and acidic (pH < 7).

14.4.3   BIOLOGICAL CHARACTERISTICS

Biological characteristics are an important factor in wastewater. The main concerns of the microorganism group in wastewater are bacteria, fungi, algae, protozoa, viruses, and pathogenic microorganisms. Bacteria present in wastewater may be spheroid, rod, curved, spiral, and filamentous. Some important bacteria are – *Pseudomonas sp.*, which reduces NO_3 to NO_2; *Zoogloea sp.*, which helps through its slime production in the formation of flocks in the aeration tanks; *Sphaerotilus natuns*, which causes sludge bulking in the aeration tanks; *Bdellovibrio sp.*, which mainly destroy pathogen; *Actinetobacter sp.*, which stores large amounts of phosphates under aerobic conditions and release it under an anaerobic condition. Wastewater also contains several fungal species, which mainly helps in decomposing the complex organic matter to its simple forms. The microbial population of wastewater also includes different types of algal species, which executes eutrophication phenomenon and oxidation of ponds; protozoa, which mainly feed on bacteria and help in the purification of treated wastewater; and viruses.

14.5   PLANT RESPONSES UNDER CONTAMINATED WATER IRRIGATION

14.5.1   MORPHOLOGICAL EFFECT OF CONTAMINATED WATER ON PLANTS

Several previous reports demonstrated that contaminated water irrigation significantly affects the morphological traits of diverse plants worldwide. Sunflower (*Helianthus annuus* L.) is an important cash crop, which produces a major percentage of edible oil throughout the world. The study done by Heidari [7] in Iran with cloth detergent contaminated water on Sunflower plants stated that 2 and 20 gm L^{-1} of detergent contamination severely reduced the germination and early growth traits like plant height, leaf number per plant, total biomass and stem weight of oil seed. The plants grown under 20 gm L^{-1} detergent contaminated water showed a decrease of 84, 60, and 63% in seed germination, shoot length and root length respectively as compared the control [7]. Several reports from India, also demonstrated similar responses in different plants under contaminated water irrigation [15, 18]. Though, the magnitude of changes in diverse morphological traits due to the contaminated water irrigation may vary among crops. After transplantation at IIT, Delhi campus, India, Thapliyal et al. [18] observed the survival rate of the ladyfinger (*Abelmoschus esculentus* L.) increased up to 87% in contaminated water irrigation followed by 67% rainwater. According to Thapliyal et al. [18], this increased survival rate under contaminated water possibly due to the more availability of free nutrients to the plants. On the other hand root length was tremendously increased at rainwater irrigation site and that can be attributed to the absence of nutrients in rain water leading to plants deeper into soil for nutrients. Singh and Agrawal [15] in their study on ladyfinger at Varanasi, India also reported significant increase in shoot length and decrease in root length under contaminated water.

14.5.2 PHYSIOLOGICAL AND BIOCHEMICAL EFFECTS OF CONTAMINATED WATER ON PLANTS

The use of contaminated water for irrigation in developing country including in India is an important source to overcome the scarcity of ground water needed for agricultural crop production. Continuous use of wastewater leads to the enrichment of soil with essential macro- and micronutrients [22]. Micronutrients are beneficial for the growth and development of the plants at lower concentrations, but become toxic at excess than the requirement [28]. Contaminated water may carry some toxic metals like Lead (Pb), Cadmium (Cd), Copper(Cu), Arsenic (As), Fluoride (F), Chromium (Cr), etc., and its long term application on agricultural production is known to produce a significant physiological and biochemical responses in plants, these includes membrane damage, structural disorganization of organelles, impairment in the metabolic function and ultimately growth retardation [27, 29, 39].

The main physiological effect of contaminated water on plants is its oxidative stress. Singh and Agrawal [15] in their study on *Beta vulgaris* at Dinapur district of Varanasi, with untreated wastewater reported an increase of 15%, 7%, 47%, 38%, and 20% in proline, protein, thiol, ascorbic acid and phenol content respectively. Heavy metal increase the generation of reactive oxygen species (ROS) and free radicals, which can cause the peroxidation of lipid membrane leading to increase permeability and oxidative stress of plants by transferring electron involving metal cations or by inhibiting the metabolic reactions controlled by metals (Nada et al., 2007). Ascorbic acid is a natural antioxidant that scavenges free radicals generated by heavy metals [26]. Peroxidase is an antioxidative enzyme also showed increment in its activity in the plants grown at wastewater as compared to ground water [15]. Peroxidase plays an important role in plant defense against oxidative stress. So it serves as an indicator of metal toxicity [36]. Plants grown at wastewater irrigation site showed higher levels of antioxidants [15] material to nullify the negative effect of heavy metals. Increased levels of heavy metals present on contaminated water affect permeability of membrane potentiality, which may lead to a water stress like condition that induces proline production remarkably [23]. Some other possible roles of proline are the stabilization of proteins [25], scavenging of hydroxyl radicals [34], and regulation of NAD/NADH ratio of plants [24]. Another important factor is thiol, which do not represent a single compound, they are sulphur containing polypeptides, also known as phytochelatins. Phytochelatins are mainly involved in the detoxification of heavy metals [30] and facilitating their further transport to the vacuolar portion [32]. Total chlorophyll and carotenoid are higher in plants of wastewater irrigation sites. Carotenoid is a photosynthetic pigment, also function as non enzymatic antioxidant protecting plants from oxidative stress by changing the physical properties of photosynthetic membranes with involvement of xanthophyll cycle [33]. An increase in carotenoid content is suggested a defence strategy of the plants to combat metal stress [35].

Photosynthetic rate and stomatal conductance are higher in plants at wastewater irrigation sites as compared to ground water irrigated ones. The higher level of antioxidants in plants at wastewater-irrigated site may have reduced the negative effects of ROS on photosynthesis. The significant increase in photosynthetic as well as growth rate of plants grown at wastewater sites compared to ground water led to higher uptake and translocation of heavy metals in plants. Higher bioavailability of heavy metal in wastewater may have reduced the nutrient availability to plants that may be cause for not showing significant increment in biomass of the plants at wastewater irrigated. The favorable physiological and growth responses are not translated into increment in the biomass accumulation and yield of plants, as the photosynthates are utilized in the formation of secondary metabolites to ameliorate the negative influence of heavy metals of contaminated water.

Dissolved salts present in contaminated water increase the osmotic potential of soil water and an increase in osmotic pressure of the soil solution increases the amount of energy which plants must expend to take up water from the soil. As a result, respiration is increased and the growth and yield of most plant decline progressively as osmotic pressure increases. Many of the ions present in contaminated water are harmless or even beneficial at relatively low concentration may become toxic to plants at higher concentration, either through direct interference with metabolic processes or through indirect effects on other nutrients, which might be rendered inaccessible. The salinity of soil water is related to, and often determined by the salinity of the irrigation water. Accordingly, plant growth, crop yield and quality of produce are affected by the total dissolved salts in the irrigation. Irrigation water could be a source of excess sodium in the soil solution and it interfere with germination and seedling emergence. 100–500 ppm fluoride contaminated irrigation water decreases the following physiological characteristics: growth, leaf expansion, photosynthetic CO_2 assimilation, stomatal conductance, chlorophyll fluorescence yield, plant biomass, and harvest index. A 500-ppm level induced marked interveinal chlorosis, leaf margin necrosis, leaf-curl on younger leaves. Cloth detergent powder in high concentration reduced seed germination (20 and 2 g L^{-1}), seedling vigor, plant height, leaf number per plant, total biomass and stem weight by means of high osmotic potential, oxidative stress, salinity stress and heavy metal stress. Germination stage is more sensitive to detergent than the early growth stage [7]. On the other hand contaminated water released from domestic households contains essential plant nutrients such as N, P, K, and micronutrients, which are beneficial for plant growth [15].

14.6 CONCLUSION

Till the onset of 21st century, natural water resources are facing a severe crisis worldwide. Due to superfluous population hike, the crisis of availability of fresh water is getting to be the most important issue in coming future. So, utilization of wastewater has become an integral part of modern day sustainable development. Even, in countries like India, Bangladesh, Sri Lanka, and most parts of Africa, where technological advancements were not at their best, wastewater can be a major source of agricultural irrigation. In this chapter, we mainly reviewed the Indian scenario of wastewater utilization in agriculture, and their impact on plants health and efficacy. Due to availability of higher concentrations of free nutrition, plants prefer to grow faster under wastewater irrigation, but the available heavy metals prevails health risks. So, it is important to treat the wastewater at least to remove toxic metals, and then they should be used for agricultural irrigation purpose.

ACKNOWLEDGMENT

The authors would like to thank the HoD, Department of Botany, University of Gour Banga, and the Director, IESD, Banaras Hindu University for providing necessary facilities.

KEYWORDS

- agricultural production
- Indian scenario
- irrigation
- wastewater

REFERENCES

1. Abdelkader, A. F. A. (2013). Physiological Response of *Brassica rapa* plants to irrigation using underground well water and non reclaimed wastewater of ABU-Rawash Drainagr, Egypt. *The Egyptian Society of Experimental Biology 9*(1), 1–17.

2. Abedin, Md. J., Howells, C., & Meharg, A. A. (2002). Arsenic uptake and accumulation on rice (Oryza sativa L.) irrigated with contaminated water. *Plant and Soil 204*, 311–319.

3. Agrawal, V., Bhagat, R., & Thikare, N. (2014). Impact of Domestic Sewage for Irrigation on properties of soil. *International Journal of Research Studies in Science, Engineering and Technology 1*(5), 60–64.

4. Bengtsson, M., Shen, Y., & Oki, T. (2006). A SRES-based global population dataset for 1990–2100. *Population and Environment. 28*, 113–131.

5. Darvishi, H. H., Manshouri, M., & Farahan, H. A. (2010). The effect of irrigation by domestic wastewater on soil properties. *Journal of Soil Science and Environmental Management 1*(2), 030–033.

6. Hassanein, R. A., Hashem, H. A., El-Deep, M. H., & Shouman, A. (2013). Soil contamination with heavy metals and its effect on growth, yield and physiological responses of vegetable crop of plants (Turnip and Lettuce). *Journal of Stress Physiology & Biochemistry 9*(4), 145–162.

7. Heidari, H. (2013). Effect of irrigation with contaminated water by cloth detergent on seed germination, traits and early growth of sunflower (*Helianthus annuus* L.). *Notulae Scientia Biologicae 5*(1), 86–89.

8. Khai, N. M., Tuan, P. T., Vinh, N. C., & Oborn, I. (2008). Effects of using wastewater as nutrient source on soil chemical properties in peri-urban agricultural system. *VNU Journal of Science 24*, 87–95.

9. Khurana, M. P. S., & Singh, P. (2012). Wastewater use in crop production: A Review. *Resources and Environment 2*(4), 116–131.

10. Ladwani, K. D., Ladwani, K. D., Manik, V. S., & Ramtake, D. S. (2012). Impact of domestic wastewater irrigation on soil properties and crop yield. *International Journal of Scientific and Research Publications 2*(10), 1–7.

11. Mahmoud, E. K., & Ghoneim, A. M. (2016). Effect of polluted water on soil and plant contamination by heavy metals in El-Mahla El-Kobra, Egypt. *Solid Earth 7*, 703–711.

12. Nayek, S., Gupta, S., & Saha, R. N. (2010). Metal accumulation and its effects in relation to biochemical response of vegetables irrigated with metal contaminated water and wastewater. *Journal of Hazardous Materials 178*, 588–595.

13. Pimentel, D., Berger, B., Filiberto, D., Newton, M., Wolfe, B., Karabinakis, E., Clark, S., Poon, E., Abbett, E., & Nandagopal, S. (2004). Water Resources: Agriculture and Environmental Issues. *BioScience 54*(10), 909–918.

14. Rusan, M. J. M., Hinnawi, S., & Rousan, L. (2007). Long term effect of wastewater irrigation of forage crops on soil and plant quality parameters. *Desalination 215*, 143–152.

15. Singh, A., & Agrawal, M. (2010). Effect of municipal wastewater irrigation on availability of heavy metals and morphological characteristics of *Beta vulgaris* L. *Journal of Environmental Biology 31*(5), 727–736.

16. Singh, P. K., Deshbhratar, P. B., & Ramteke, D. S. (2012). Effects of sewage wastewater irrigation on soil properties, crop yield and environment. *Agricultural Water Management 103*, 100–104.

17. Singh, M., & Verma, K. K. (2013). Influence of Fluoride contaminated irrigation water on physiological responses of poplar seedlings (*Populus deltoides* L. Clone–S$_7$ C$_{15}$). *Fluoride 46*(2), 83–89.

18. Thapliyal, A., Vasudevan, P., Dastidar, M. G., Tandon, M., & Mishra, S. (2011). Irrigation with domestic wastewater: Responses on growth and yield of Ladyfinger *Abelmoschus esculentus* and on soil nutrients. *Journal of Environmental Biology 32*, 645–651.

19. Xu, J., Wu, L., Chang, A. C., & Zhang, Y. (2010). Impact of long-term reclaimed wastewater irrigation on agricultural soils: A preliminary assessment. *Journal of Hazardous Materials 183*, 780–786.

20. Kiziloglu, F. M., Turan, M., Sahin, U., Angin, I., Anapali, O., & Okuroglu, M. (2007). Effects of waste water irrigation on soil and cabbage-plant (*Brassica olerecea* var. capitate cv. yalova-1) chemical properties. *Journal of Plant Nutrition and Soil Science 170* (1), 166–172.

21. Kiziloglu, F. M., Turan, M., Sahin, U., Kuslu, Y., & Dursur, A. (2008). Effects of untreated and treated wastewater irrigation on some chemical properties of cauliflower *(Brassica* L. var. rubra) grown on calcareous soil in Turkey. *Agricultural water Management 95*, 716–724.

22. Kanan, V., Ramesh, R., & Sasikumar, C. (2005). Study on ground water characteristics and the effects of discharged effluents from textile units at Karur District. *Journal of Environmental Biology 26* (2), 269–272.

23. Basak, M., Sharma, M., & Chakraborty, U. (2001). Biochemical responses of *Camellia sinensis* (L.) O. Kuntze to heavy metal stress. *Journal of Environmental Biology 22*, 37–41.

24. Alia, & Pardha Saradhi, P. (1993). Suppression in mitochondrial electron transport is the prime cause behind stress induced proline accumulation. *Biochemical and Biophysical Research Communications 193*, 54–58.

25. Anjum, F., Rishi, V., & Ahmed, F. (2000). Compatibility of osmolyte with Gibbs energy of stabilization of proteins. *Biochimica et Biophysica Acta 1476*, 75–84.

26. Halliwell, B. & Gutteridge, J. M. C. (1993). Free radicals in biology and medicine clarendon. *Press, Oxford. London* pp. 96–98.

27. Kimbrough, D. E., Cohen, Y., Winer, A. M., Creelman, L., & Mabuni, C. (1999). A critical assessment of chromium in the environment. *Critical Reviews in Environmental Science and Technology 29* (1), 1–46.

28. Kocak, S., Tokusoglu, O., & Aycan, S. (2005). Some heavy metal and trace essential element detection in canned vegetable foodstuffs by differential pulse polarography (DPP). *Electronic Journal of Environmental, Agriculture and Food Chemistry 4*, 871–878.

29. Long, X. X., Yang, X. E., Ni, W. Z., Ye, Z. Q., He, Z. L., Calvert, D. V., & Stoffella, J. P. (2003). Assessing zinc thresholds for phytotoxicity and potential dietary toxicity in selected vegetable crops. *Communications in Soil Science and Plant Analysis 34*, 1421–1434.

30. Kneer, R., & Zenk, M. H. (1992). Phytochelatins protect plant enzymes from heavy metal poisoning. *Phytochemistry 31*, 2663–2667.

31. Nada, E., Ben, A. F., A, Rhouma., Ben, R. B., Mezghani, I., & Boukhris, M. (2007). Cadmium-induced growth inhibition and alteration of biochemical parameters in almond seedlings grown in solution culture. *Acta Physiologiae Plantarum 29* (1), 57–62.

32. Ortiz, D. F., Ruscitti, T., McCue, K. F., & Ow, D. W. (1995). Transport of metal – binding peptides by HMT1, a fusion yeast ABC type vacuolar membrane protein. *Journal of Biological Chemistry 27*, 4721–4728.

33. Gruszecki, W. I., & Strzatka, K. (1991). Does the xanthophylls cycle take part in the regulation of fluidity of the thylakoid membrane? *Biochimica et Biophysica Acta 1060*, 310–314.

34. Smirnoff, N., & Cumbes, Q. J. (1989). Hydroxyl radical scavenging activity of compatible solutes. *Phytochemistry 28*, 1057–1060.

35. Sinha, S., Gupta, A. K., & Bhatt, K. (2007). Uptake and translocation of metals in fenugreek grown on soil amended with tannery sludge: involvement of antioxidants. *Ecotoxicology and Environmental Safety 67*, 267–277.

36. Radotic, K., Ducic, T., & Mutavdzic, D. (2000). Changes in peroxidase activity and isozymes in spruce needles after exposure to different concentrations of cadmium. *Environmental and Experimental Botany 44*, 105–113.

37. Ekanayake, I. J., Toole, J. C. O., Garrity, D. P., & Jan, T. M. M. (1983). Inheritance of root characters and their relations to drought resistance in rice. *Crop Science Society of America 25*, 927–933.

38. CPCB (2011). http://www.cpcb.nic.in, Accessed on 11th January 2011.

39. Chen, C. R., Xu, Z. H., Mathers, N. J., (2004). Soil carbon pools in adjacent natural and plantation forests of subtropical Australia. *Soil Science Society of American Journal* 68, 282e291.

INDEX

9 781774 630761